Human and Environmental Security in the Era of Global Risks

Mohamed Behnassi • Himangana Gupta
Olaf Pollmann

Editors

Human and Environmental Security in the Era of Global Risks

Perspectives from Africa, Asia and the Pacific Islands

 Springer

Editors
Mohamed Behnassi
Faculty of Law, Economics and Social
Sciences, Center for Research on
Environment, Human Security and
Governance (CERES)
Ibn Zohr University
Agadir, Morocco

Himangana Gupta
Ministry of Environment, Forest
and Climate Change
New Delhi, Delhi, India

Olaf Pollmann
Environmental Solutions
SCENSO GbR - Scientific
Sankt Augustin, Nordrhein-Westfalen
Germany

ISBN 978-3-030-06527-0   ISBN 978-3-319-92828-9   (eBook)
https://doi.org/10.1007/978-3-319-92828-9

This Springer imprint is published by the registered company Springer Nature Switzerland AG
The registered company address is: Gewerbestrasse 11, 6330 Cham, Switzerland

About the Publishing Institution

The Center for Research on Environment, Human Security and Governance (CERES)

The CERES, previously the North-South Center for Social Sciences (NRCS), 2008–2015, is an independent and not-for-profit research institute founded by a group of researchers and experts from Morocco and other countries. The CERES aims to develop research and expertise relevant to environment and human security and their governance from a multidimensional and interdisciplinary perspective. As a think tank, the CERES aspires to serve as a reference point, both locally and globally through rigorous research and active engagement with policy-making processes. Through its research program, the CERES aims to investigate the links between environmental/climate change, their implications for human security, and the needed shifts to be undertaken in both research and policy. The CERES, lead by Dr. Mohamed Behnassi and mobilizing a large international network of researchers and experts, aims to undertake original research, provide expertise, and contribute to effective science and policy interactions through its publications, seminars, and capacity building.

Preface

Lives and livelihoods of millions of people worldwide, especially in vulnerable regions, are being severely threatened by numerous environmental, climatic, economic, geopolitical, societal, and technological risks. The impacts of such complex, interconnected, and multidimensional risks continue to reveal our shared vulnerabilities and the declining resilience of our social-ecological systems. Similarly, such impacts are fostering a growing perception that threats to security are widespread, crosscutting, and increasingly associated with intractable and interrelated crises. More precisely, it is being gradually perceived that these impacts will be felt not just in the immediate regions and by affected generations, but also across the international community and by future generations. In the Anthropocene era, the viability of human systems, as well as their reproductive capacity, are consequently questioned due to the significant and unprecedented impact that humans have on the Earth's climate and ecosystems. This viability, depending on an ecological interconnection, is increasingly challenged by the environmental abuses and pressures generated by anthropogenic actions for the sole goal of reaching and controlling resources.

Therefore, the complex and uncertain evolution of current risks should alter our understanding and perception of security with the potential to manage current dynamics effectively. Many scientists and agencies are indeed starting to perceive the concept of "security" more broadly, moving away from the inadequate state-centered concept of national security toward the concept of soft security. This shift is justified by the fact that emerging global risks are raising new and unavoidable questions with regard to human and environmental security. However, much remains to be done in this respect because the key scientific and political agendas are still substantially directed by conventional paradigms and the pace of change is still slow with marginal impacts in many areas. Undeniably, perceiving the current risks – such as climate risks – from a human-environmental nexus connects these risks to a myriad of issues and areas such as human rights, poverty, vulnerability, equity, redistribution, accountability, geopolitical stability, resilience, adaptation, etc.

It is true that during the last decades the links between emerging risks and the security of humans and nature have been the object of considerable research and deliberations, but it is only recently becoming an important focus of policy-making and advocacy. Therefore, we presume in this contributed volume that our ability – or lack thereof – to make innovative conceptual frameworks, institutional and policy arrangements, and technological advances for managing the current emerging risks will foster or undermine the environmental security and consequently determine the future human security. Moreover, taking into account the links between environmental/climate security, human security and sustainability will help frame a new research agenda and potentially develop a broad range of responses to many delicate questions.

Within such a perspective, this contributed volume has been conceived as an opportunity to deepen the debate on the linkages between global risks and human and environmental security. Compared to relevant existing publications, this volume's approach consists of questioning the ability of existing concepts, regulatory frameworks, technologies, and decision-making mechanisms to accurately deal with emerging risks to human and environmental security and help act in the direction of effectively managing their impacts and fostering the resilience of concerned systems and resources. Thirty-eight authors from various disciplines and contexts have been provided with the opportunity to share relevant research, insights, and successful practices, and most of them have explored innovative options to guide future processes of change. In addition, the volume's content and approach are multidisciplinary (for a fruitful interaction between numerous scientific fields) and relevant to policy-making processes (enabling interactions among experts and decision makers from different levels and spheres).

More precisely, the core research subject has been approached in three parts comprising 19 chapters. The *first part* focuses on the management of environmental and climate change induced risks from a human security perspective and the need for a paradigm and governance shift. Cases from the Pacific Island Countries and Malaysia have been presented. The *second part* comprises a variety of case studies from India, Ghana, Sudan, Morocco, Thailand, which provide empirical examples of how environmental and climate change generates many risks with the potential to compromise human security and generate conflicts. The *third part* comprises a variety of case studies from Africa exploring the linkages between environment and development.

Part I*: The selected contributions examine the management of environmental and climate change induced risks from a human security perspective with a focus on the need for a paradigm and governance shift.*

In Chap. 1, Mohamed Behnassi analyses the relevance and usefulness of the mainstreaming of a rights-based approach in the global climate regime. According to this research, not only are the human implications of climate change serious, but also the global climate regime is not sufficiently shaped to reduce them and, in some instances, existing responses may even exacerbate these implications. Therefore, the significant challenge currently being faced is how to ensure that human rights are widely recognized and genuinely mainstreamed in the global climate regime.

Based on this assumption, the analysis dissects the potential overlap, convergence, and synergies between the international human rights framework and the global climate regime. It further assesses the advantages of mainstreaming a rights-based approach into this regime and what should be done on the ground to effectively achieve this objective.

In Chap. 2, Hartmut Sommer analyzes the risks to human and environmental security, well-being, and welfare with a focus on the "right" indicators, how they are measured and why they are only rarely used to guide policies. After classifying different forms of individual and systemic risks and their basic measurement, the author analyses the existing perceptions and attitudes toward these risks and claims that the actual problems of non-sustainable development are partially due to ill-conceptualized and misleading indicators for risks. Alternatively, the author presents a set of existing. sustainability-oriented indicators and discusses the reasons why these indicators are only rarely used in official politics. The author also suggests the measures needed to overcome the underlined deficiencies claiming for risk management systems, robust decision-making, and ethical principles guiding individual and collective behavior.

In Chap. 3, Amine Amar focuses on the socio-ecological coviability as an efficient way to combat poverty, reduce inequality, and address insecurity risks. For the author, addressing such challenges efficiently requires the perception and empirical interpretation of the linkages between social and ecological problems from a coviability perspective. Taking poverty eradication as example, the author shows how traditional approaches often disregard environmental degradation and biodiversity loss as externalities, whereas the social and ecological approach considers not only the social but also the environmental causes of poverty. Such a perspective has both research and policy implications when it comes to poverty alleviation. Based on this, the author addresses these issues while recommending some tools adapted from economic analyses and development economics and presenting some success stories and empirical cases.

In Chap. 4, Laurent Weyers analyzes the states' discretion to shape their climate actions and claims that this ability is not without limit since it can arguably be reconsidered, especially if we make a reference to the reasoning followed by the International Court of Justice (ICJ) in the *Whaling in the Antarctic case*. The author attempts to analyze the analogy that can be drawn between the kind of legal questions raised in both cases: whether and how a state's discretionary power can be subject to judicial review; whether and how disputed scientific facts can be arbitrated upon; whether states, when determining the appropriate course of their actions, may be expected to give regard to certain elements of fact and law? From the *Whaling case*, the author asserts that a new framework of analysis can be drawn, allowing to better articulate the politics of climate change with the law and to transcend some of the main shortcomings of the international climate regime.

In Chap. 5, Aline Treillard analyzes the biodiversity, human security, and climate change nexus from a legal perspective. This research provides the opportunity to identify the theme of security in climate change and biodiversity related multilateral agreements. The author argues that despite a rather anthropocentric construction of international environmental law, human security is not explicitly taken into account in different legal regimes, thus revealing an unusual feature: the fragmentation of the law. After highlighting the shortcomings, the author provides some suggestions to improve the international system based on the idea of interdependence by focussing, for example, on intergenerational justice, on the management of institutional vulnerabilities, on the adoption of a convention on environmental displaced persons, etc.

In Chap. 6, Arunesh Asis Chand refers to the case of Pacific Island Countries in order to explore the relevant governance and risk management options, especially in a context where adaptation and mitigation are becoming a major concern for these countries to deal with global environmental change and human security issues. According to the author, these response mechanisms need to be analyzed and addressed on different scales through a multilevel and multi-actor governance perspective. The analysis elaborates on how an appropriate governance system can help in effective designing, implementing, and in particular managing adaptation and mitigation policies. By doing so, this research further incites important discussion on the availability of a wider range of options to effectively design and implement climate policies than those embodied in the international environmental regime.

In Chap. 7, Maizatun et al. take us to Malaysia, assessing the progression of policies and laws toward addressing climate change and sustainability issues. As a developing country, Malaysia is not subjected to any mitigation commitment; nevertheless, from a domestic compliance perspective, its contribution is evident through various climate actions even if the country remains aware that any response to climate change must be balanced with its continuing need to grow, to increase its per-capita income, and to raise its living standards, in accordance with the sustainability principles. The authors focus on the changes that have taken place within the Malaysia's policy and legal frameworks in dealing with the compelling climate issue, while taking into consideration its various needs. The analysis questions the potential of these frameworks to reinforce the country's resilience to climate change implications while pursuing its priority for continued development.

Part II: *The selected contributions comprise a variety of case studies from India, Ghana, Sudan, Morocco, and Thailand, which provide empirical examples of how environmental and climate change generates many risks with the potential to compromise human security and generate conflicts.*

In Chap. 8, Himangana Gupta and Raj Kumar Gupta demonstrate how grassroots environment movements in India are facing full force of state machinery to protect corporate interests. For the authors, the problem of environmental and human security in the Indian context can be linked to the deficit of democracy in recent times. With the climate change agenda moving into the hands of corporate interests, what was earlier done in the name of development is now done in the name of environment. Indian citizens marginally contribute to environmental policy-making although the country is among the most vulnerable to climate change and loss of biodiversity, in addition to the strong tradition of living in harmony with nature. In previous decades, many successful grassroots resistance movements were active and powerful, but such spontaneous movements are failing now in the absence of political support and wide media coverage, which is often controlled by the corporate sector. From this perspective, the analysis traces the history of popular and successful environment movements and discusses the reasons behind their failure in present times.

In Chap. 9, titled "Real or hyped? Linkages between environmental/climate change and conflicts," Bukari et al. analyze the violent conflicts between farmers and Fulani pastoralists in Ghana, commonly interpreted as linked to environmental and climate change, thus reiterating the environmental scarcity/security and neo-Malthusian postulations. This empirical work examines the perceptions of farmers and Fulani pastoralists in Ghana regarding the key causes of conflicts, and assesses the extent to which environmental/climate change has contributed to these conflicts. Based on the analysis of interview outcomes, the authors note that farmer communities and pastoralists perceive environmental and climate factors as indirectly influencing conflicts between them, especially through increased pastoralists' migrations and competition for pasture lands. By reference to the data sets (of rainfall), however, it was revealed that despite climate variability, there were basically no major changes in rainfall figures. The authors also found that the abundance of resources and increases in the value of land were major drivers of conflicts between farmers and pastoralists in the study area.

In a similar vein, Deafalla et al. assess in Chap. 10 the environmental change impacts and human security in semi-arid region by investigating the case of Nuba mountains of Sudan. According to the authors, the links between climate change, environmental degradation, resource scarcity, and conflict are highly complex and poorly understood as yet. Hence, the aim of their study is to raise understanding of the links between patterns of local-level economic and demographic changes of Nuba Mountains region, specifically of those systems in poor condition under environmental change. The analysis shows that there are increasingly rapid changes in land cover, social structure, and institutional and livelihood transformation across broad areas of the state. Therefore, more information exchange is needed to inform actors and decision makers regarding specific experiences, capacity gaps, and knowledge to address the environmental change.

In Chap. 11, El-Abbas et al. undertake a spatial assessment of environmental change in Blue Nile region of Sudan. Taking into account the fact that innovative technologies are becoming progressively interlinked with the issue of environmental change, the authors believe that these technologies provide a systematized and objective strategy to document, understand, and simulate the change process and its associated drivers. Based on this, the authors try to develop spatial methodologies that can assess the environmental change dynamics and their associated drivers. The study concludes with a brief assessment of an "oriented" framework, focused on the alarming areas where serious dynamics are located and where urgent plans and interventions are most critical, guided with potential solutions based on the identified driving forces.

In Chap. 12, El Morabet et al. assess the vulnerability to climate change and adaptive capacity of social-ecological systems in Kenitra and Talmest, which are located in north and central Morocco. Focusing on flooding as a climatic extreme event that is causing loss of life and property in both Moroccan rural and urban areas, the authors demonstrate through the vulnerability assessment that this problem is also exacerbated by unsustainable development practices and poor land-use planning in the country. More specifically, the increased vulnerability to climatic changes recently observed in the study area can be explained both by its high exposure to the perturbation and by its socioeconomic situation that intensifies sensitivity and hinders adaptive capacity.

In Chap. 13, Seksak Chouichom and Lawrence M. Liao assess the participation of female farmer groups in *Kai* algal processing and production in northern Thailand. This is because the role of female farmers has increased in recent years to supplement family income in many rural households. The authors used an interview schedule with open-ended and closed questions and descriptive statistics. The responses were scored on a five point Likert's scale in order to assess the level of participation among respondents. Results of this study revealed that female farmers were mostly old, with a low educational attainment, receive low incomes, belong to a family of three to four members, and most of them were members of the locally organized *kai* algal processing group of the village. The participation of female farmers in group activities got established strongly within five areas, namely planning, production and implementation, group activities evaluation, selling activities, and benefit distribution. The respondents indicated that the raw material was insufficient for the whole year production due to the highly seasonal nature of the algae. Moreover, the group needed more logistic and financial support for marketing and promotion both inside and outside the province.

Part III: *The selected contributions comprise a variety of case studies from Africa exploring the interlinkages between environment and development.*

In Chap. 14, Pollmann et al. analyze the risks and opportunities of sustainable biomass and biogas production for the African market. Based on the international trends to support developing countries by establishing a market for biomass production, with the aim of advanced energy production, resource efficiency, and emission reduction, the authors argue that this has to be balanced with the internationally supervised and important sector of food security, especially in African context. Since biomass residues – such as cow manure and other organic substances – are available as a cheap or free by-product in Africa, the authors reveal that turning these natural resources into a valuable commodity (mainly biogas usable for cooking or electricity and fertilizer usable for private gardening) requires a simple conversion by fermentation. Current knowledge has proved how effective and simple biomass and biogas production can be. The reduction of greenhouse gases as a side benefit of biomass production provides an indication of how positive this strategy could be for both the agricultural sector as well as general public.

In the same perspective, Lucia Beran and Harald Dyckhoff attempt in Chap. 15 to analyze the linkages between global biomass supply and sustainable development. The authors observe that biomass is not accounted for in conventional energy statistics since it constitutes a neglected energy carrier although it has ever since provided the basis for human life and activity. In this work, the authors assess the current global draw on the Earth's biomass resources by examining the indicators "Ecological Footprint " and "Human Appropriation of Net Primary Production, " quantifying humankind's biomass demand and the Earth's biomass supply. They reveal that humankind appropriates about 20–30% of the ecosystem's supplying capacity even if other definitions partly suggest lower and higher values. When using the energetic metabolism accounting concept to acquire data on biomass supply for the past centuries to complement conventional energy statistics, the authors find that the actual energy supply to humankind is about twice as high as conventional energy statistics essentially suggest. Against these results, ideas like substituting fossil resources with biomass in the future for the provision of energy services to mitigate the current energy and climate crisis might be controversial to the achievement of sustainability.

In Chap. 16, Mark Matsa attempts to think from the periphery when analyzing the Tonga community development from a climate change perspective. This minority community is suffering from many problems, including climate impacts, which threaten peoples' livelihoods. In spite of this, and with little outside assistance, this subsistence, semi-pastoral community seems determined to prevail. In this empirical work, Matsa assesses the relationship between climate/environmental change and rural community development. Results show that although the Tonga community is getting some assistance from NGOs and the central government, the assistance is not sustainable partly because it does not incorporate Tonga traditional knowledge systems which have been the bedrock of Tonga community resilience for generations. This study posits that for meaningful climate-compatible development to take place in Binga, a community derived development "basket of priorities" be used as a basis for sustainable community development. An identify-define-initiate-lead (IDIL) and a protect-empower-capacitate-facilitate (PECF) model is thus suggested to help the Tonga community cope with climate change impacts more sustainably.

In Chap. 17, Ahrabous et al. note that natural resources have historically provided location-specific advantages for communities at various stages of their development and that agriculture and rural communities are regarded as producers of commodities and environmental and social services. Based on this, they tried to analyze the role of environmental amenities in rural economic development by investigating the case of Todgha oasis in Morocco. Amenities can be broadly defined as qualities of a region that make it an attractive place to live and work. For the most part, amenities represent goods and assets that are not effectively regulated by markets. Their supply cannot be easily increased, while the demand grows significantly with development. In many cases, amenities are public goods, and it is difficult to make users pay to benefit from them. Many of the beneficiaries from the promotion of amenities may live in urban areas, while most of the costs associated with this development are borne by residents in rural areas. The authors support the argument that in the search for forward-looking policy strategies, building on environmental amenities should emerge as an important area of policy action, thus supplementing traditional, agriculture-oriented rural policies and placing rural policy in the broader field of regional development.

In Chap. 18, Aïcha EL Alaoui and Hassane Nekrache attempt to empirically investigate the links between economic growth and environment in four countries from the Middle East and North Africa (Morocco, Algeria, Tunisia, and Egypt), since the main objective of these countries in the coming years is to improve the economic growth, which is perceived as necessary to meet the increasing demand of their populations, to improve their well-being, and to help manage existing environmental challenges. The results of applied models in the analysis show that the linkages between economic growth and environment are still uncertain, complex, and ambiguous. Therefore, the authors recommend that these countries, through policy-making and the involvement of private actors, should apply preventive and precautionary measures to reduce environmental damage. Such measures must be adapted to specific economic and environmental conditions benefiting from the experiences and good practices developed in other regions and avoiding others' past mistakes related to pollution, regional development, and natural resource management.

Finally, in Chap. 19, Peter Karácsony takes us to Hungary to investigate the role of corporate social responsibility (CSR) in environmental sustainability. Given the evolution of corporate ethics, many private companies have accepted the responsibility to do no harm to the environment. CSR provides many principles and tools which permit companies to limit and even mitigate their negative externalities. Within such a context, the author attempts to present, to some extent, current practices and approaches to environmental aspects of CSR in the case of some Hungarian corporations.

This volume contributes to the scientific and policy debate about the implications of global risks for both human and environmental security with a focus on the necessary shifts to be made to effectively manage these implications. The analysis focuses on some countries from Africa, Asia, and the Pacific Islands given their specific situations in terms of climate vulnerability, environmental degradation, resource decline, and security implications of existing and future trends. Through the insights shared in this publication, the editors aim to contribute to the growing academic literature pertained to human and environmental security while enhancing political discussions and policy agendas on how to effectively address current and future relevant challenges. Interested scholars, students, practitioners, and decision makers from concerned regions and worldwide will find the publication a useful and instructive reading.

Agadir, Morocco   Mohamed Behnassi
New Delhi, India   Himangana Gupta
Potchefstroom, South Africa   Olaf Pollmann

Acknowledgments

This contributed volume is based on the best papers presented during the International Conference on *"Human and Environmental Security in the Era of Global Risks (HES2015),"* organized on November 25–27, 2015, in Agadir by the Environment and Human Security Program (EHSP), North-South Center for Social Sciences, Morocco (now the Center for Research on Environment, Human Security and Governance (CERES)), the French Institute of Research for Development (IRD), UMR ESPACE-DEV, France, and the Project 4C/IKI-Morocco of the Deutsche Gesellschaft für Internationale Zusammenarbeit (GIZ) GmbH, Germany, in collaboration with the Interdisciplinary Research Centre for Environmental, Urban Planning and Development Law (CRIDEAU), University of Limoges, France, the Universidade Federal do Pará (UFPA) and the Universidade Federal da Bahia (UFBA), Brazil, and the CLIMED Project (ANR: 2012-2015/CIRAD), France.

I have been honored to chair the HES2015 and to share the editorship of this volume with my colleagues Dr. Himangana Gupta (Programme Officer, National Communication Cell, Ministry of Environment, Forest and Climate Change, New Delhi, India) and Dr. Olaf Pollmann (North-West University, School of Environmental Science and Development, South Africa). Their professionalism, expertise, and intellectual capacity made the editing process an exciting and instructive experience and definitely contributed to the quality of this publication.

The chapters in this volume are also the result of the invaluable contributions made by our peer-reviewers, who generously gave their time to provide insight and expertise to the selection and editing process. On behalf of my co-editors Himangana Gupta and Olaf Pollmann, who actively participated in the peer-review process, I would specifically like to acknowledge, among others, with sincere and deepest thanks to the following colleagues: Raj Kumar Gupta (Independent Journalist and Policy Analyst, New Delhi, India), Pooja Pal (Panjab University, Chandigarh), and Szilárd Podruzsik (Corvinus University of Budapest, Hungary).

I would also like to seize this opportunity to pay tribute to all institutions that made this book project an achievable objective. In particular, we thank the funding institutions of the HES2015 Conference which, in addition to NRCS (now CERES), include the French Institute of Research for Development (IRD), UMR

ESPACE-DEV, France, and the Project 4C/IKI-Morocco of the Deutsche Gesellschaft für Internationale Zusammenarbeit (GIZ) GmbH, Germany. Special thanks are due to all chapters' authors and co-authors without whom this valuable and original publication could not have been produced.

Agadir, Morocco   Mohamed Behnassi

Contents

List of Figures

List of Tables

List of Abbreviations and Acronyms

ABS	Antiblockiersystem
AF	Adaptation Fund
AFD	Agence Française de Développement (*French Development Agency*)
AFDB	African Development Bank
ALBA	Bolivarian Alliance for the Americas
ALNAP	Active Learning Network on Accountability and Performance in Humanitarian Action
BAT	Best Available Techniques Economically Achievable
BJP	Bharatiya Janata Party
CAS	Complex Adaptive Systems
CDB	Convention on Biological Diversity
CDM	Clean Development Mechanism
CERES	Center for Research on Environment, Human Security and Governance
CIEL	Center for International Environmental Law
COP	Conference of Parties
CRIDEAU	Center for Interdisciplinary Research in Environmental law, Land planning and Urban law
CRPF	Central Reserve Police Force
CSR	Corporate Social Responsibility
DIW	German Institute for Economic Research (Deutsches Institut für Wirtschaftsforschung)
DRM	Disaster Risk Management
DRR	Disaster Risk Reduction
EC	Environmental Change
ECOSOC	Economic and Social Council
EIT	Economies In Transition
EOS	Earth Observing System
FAO	Food and Agriculture Organization
FAO	Food and Agriculture Organization

FCPF	Forest Carbon Partnership Facility
FPIC	Free, Prior and Informed Consent
GBRMPA	Great Barrier Reef Marine Park Authority
GCF	Green Climate Fund
GCMs	General Circulation Models
GDA	Gushiegu District Assembly
GDP	Gross Domestic Product
GEF	Global Environment Facility
GEF	Global Environment Facility
GHG	Greenhouse Gases
GMOs	Genetically-Modified Organisms
GNH	Gross National Happiness indicator
GNI	Gross National Income
HDI	Human Development Index
HPI	Happy Planet Index
HRC	Human Rights Council
IACHR	Inter-American Commission on Human Rights
ICC	Inuit Circumpolar Conference
ICCPR	International Covenants on Civil and Political Rights
ICESCR	International Covenants on Economic, Social, and Cultural Rights
ICHRP	International Council on Human Rights Policy
ICJ	International Court of Justice
ICRW	International Convention for the Regulation of Whaling
IDNDR	International Decade for Natural Disaster Reduction
IIASA	International Institute for Applied Systems Analysis
ILO	International Labor Organization
IMF	International Monetary Fund
IPCC	Intergovernmental Panel on Climate Change
IWC	International Whaling Commission
JARPA	Whale Research Program under Special Permit in the Antarctic
JI	Joint Implementation
KP	Kyoto Protocol
LAGOS	Laboratory for Territorial Governance, Human Security and Sustainability
LMOs	Living Modified Organisms
LPI	Legatum Prosperity Index
MATE	Morocco, Algeria, Tunisia, and Egypt
NAMA	Nationally Appropriate Mitigation Action
NCCAS	National Climate Change Adaptation Strategy
NDA	National Democratic Alliance
NDC	Nationally Determined Contributions
NOAA	National Oceanic and Atmospheric Administration
NTPC	National Thermal Power Corporation
NWP	Numerical Weather Prediction
OAS	Organization of American States

OECD	Organization for Economic Cooperation and Development
OHCHR	Office of the UN High Commissioner for Human Rights
PA	Paris Agreement
PACC	Pacific Adaptation to Climate Change
PCA	Post Classification Analysis
PCD	Post Change Detection
PE	Political Ecology
PICCAP	Pacific Islands Climate Change Assistance Programme
PIL	Public Interest Litigation
PNG	Papua New Guinea
REDD	Reducing Emissions from Deforestation and Forest Degradation
SCCF	Special Climate Change Fund
SDG	Sustainable Development Goals
SDM	Sustainable Development Mechanism
SEDA	Sustainable Energy Development Authority
SES	Socio-Ecological Systems
SMEs	Small and Medium-Sized Enterprises
SPM	Summary for Policy Makers
SPREP	South Pacific Regional Environment Programme
SUV	Sport Utility Vehicle
UNCED	United Nations Conference on Environment and Development
UNCTAD	United Nations Conference on Trade and Development
UNCTs	Guidance Note for UN Country Teams
UNDG	United Nations Development Group
UNDG-HRM	UN Development Group's Human Rights Mainstreaming Mechanism
UNDHR	UN Declaration on Human Rights
UNDP	United Nations Development Program
UNDRIP	UN Declaration on the Rights of Indigenous Peoples
UNEP	United Nations Environment Program
UNESCAP	Economic and Social Commission for Asia and the Pacific
UNFCCC	United Nations Framework Convention on Climate Change
UNGA	United Nations General Assembly
UNHRC	United Nations Human Rights Council
UNPFII	UN Permanent Forum on Indigenous Issues
UPA	United Progressive Alliance
USAID	United States Agency for International Development
WANEP	West African Network for Peacebuilding
WHO	World Health Organization
WMO	World Meteorological Organization

About the Editors and Contributors

Editors

Mohamed Behnassi, Ph.D. is specialist in Environment and Human Security Law and Politics. After the obtention of his Ph.D. in 2003 from the Faculty of Law, Economics and Social Sciences, Hassan II University of Casablanca for a thesis titled: *Multilateral Environmental Negotiations: Towards a Global Governance for Environment*, he accessed to the Faculty of Law, Economics and Social Sciences, Ibn Zohr University of Agadir, Morocco as Assistant Professor (2014). In 2011, he obtained the status of Associate Professor and in 2017 the status of Full Professor. He served as the Head of Public Law Department (2014–2015) and the Director of the Research Laboratory for Territorial Governance, Human Security and Sustainability (LAGOS) (2015–at present). In addition, Dr. Behnassi is the Founder and Director of the Center for Environment, Human Security and Governance (CERES) (former North-South Center for Social Sciences (NRCS), 2008–2015). Dr. Behnassi is also Associate Researcher at the UMR ESPACE-DEV, Research Institute for Development (IRD), France. In 2011, he completed a US State Department-sponsored Civic Education and Leadership Fellowship (CELF) at the Maxwell School of Citizenship and Public Affairs, Syracuse University, USA and in 2014 he obtained a Diploma in Diplomacy and International Environmental Law from the University of Eastern Finland and the United Nations Environment Programme (UNEP), Finland. Dr. Behnassi has pursued several post-doctoral trainings since the completion of his Ph.D.

His core teaching and expertise areas cover: environmental change, human security, sustainability, climate change politics and governance, human rights, CSR, etc. He has published numerous books with international publishers such as: *Environmental Change and Human Security in Africa and the Middle East* (Springer 2017); *Vulnerability of Agriculture, Water and Fisheries to Climate Change* (Springer 2014); *Science, Policy and Politics of Modern Agricultural System* (Springer 2014); *Sustainable Food Security in the Era of Local and Global Environmental Change* (Springer 2013) ; *Global Food Insecurity* (Springer, 2011); *Sustainable Agricultural Development* (Springer, 2011); *Health, Environment and Development* (European University Editions, 2011) ; and *Climate Change, Energy Crisis and Food Security* (Ottawa University Press, 2011). He has also published numerous research papers and made presentations on these at international conferences. In addition, Dr. Behnassi has organized many international conferences covering the above research areas in collaboration with national and international organizations and managed many research and expertise projects on behalf of various national and international institutions. Behnassi is regularly requested to contribute to review and evaluation processes and to provide scientific expertise nationally and internationally. Other professional activities include Social Compliance Auditing and consultancy by monitoring human rights at work and the sustainability of the global supply chain.

Himangana Gupta, Ph.D. is an expert in climate change and biodiversity policy and diplomacy. She is part of the National Communication Cell of the Indian Ministry of Environment, Forest and Climate Change. Her current position, as a Programme Officer, involves coordination with scientists, climate experts, and ministries to compile India's National Communications to the UNFCCC. She is doctorate in environment science with specialization in climate change and biodiversity policy and was a University Gold medalist in Masters. She was an expert reviewer of IPCC AR5 WGII report. She has written research papers in reputed international and national journals on current state of climate negotiations, forestry, industrial efficiency, rural liveli-hoods, and women in climate change mitigation and adaptation. She has presented her research at various national and international forums, particularly on climate policies and adaptive capacity of local people, energy efficiency, biodiversity conservation, and women empowerment.

Olaf Pollmann, Ph.D. is Deputy Head of the section "African Service Centers" in West (WASCAL) and Southern Africa (SASSCAL) on behalf of the Federal Ministry of Education and Research (BMBF). He is responsible for the coordination of the entire process-management of all Africa-related activities. Especially the coordination, arrangements, and result dissemination of the consultation results for the final preparation and implementation of the perpetuation of the African Centers by the African partners are essential duties. But also coordinated diplomatic negotiations with WASCAL- and SASSCAL-stakeholders and further actors of the African network are covered by the work of Dr. Pollmann. Beside these responsibilities Dr. Pollmann is also CEO of the company SCENSO – Scientific Environmental Solutions in Germany.

From 2007 till 2010, Dr. Pollmann worked as a post-doctoral research fellow in the field of sustainable resource management, waste reduction, and water purification at the Department for Environmental Sciences and Development at the North-West University (NWU), Potchefstroom Campus in South Africa.

Before that research stay abroad Dr. Pollmann studied civil engineering majoring in water-supply, particularly environmental engineering and technical waste-management as well as additional environmental management techniques at the Gottfried Wilhelm Leibniz University of Hannover in Germany. Subsequent to that position he worked as a research scientist at the Leuphana University of Lueneburg, Germany.

In 2006, he finished his Doctorate (Dr.-Ing./Ph.D.) in the field of environmental-informatics at the Institute of Waste Management in collaboration with the Institute of Numerical Methods and Informatics in

Civil Engineering at the Technical University of Darmstadt, Germany.

In 2011, he finished an additional dissertation in the field of sustainable resource management with the title "Reduction of Anthropogenic Environmental Influences by Advanced and Optimized Technologies" at the North-West University (NWU), South Africa, and achieved another Ph.D. in Environmental Sciences. From 2012 till 2015 he was employed at the NWU as a guest scientist and extraordinary senior lecturer in the field of Environmental Sciences.

Dr. Pollmann has also published numerous papers in national and international journals and gave several oral presentations at international conferences. He is continuously requested to be partner in various review and evaluation processes, e.g., international journals and conferences (Scientific Advisor) as well as EU- and African calls (Scientific Expert).

Contributors

Mhamed Ahrabous Hassan II Institute of Agronomy and Veterinary Medicine, Rabat, Morocco

Amine Amar National Institute of Statistics and Applied Economics (INSEA), Mohamed V University of Rabat, Rabat, Morocco
Researcher at the Center for Research on Environment, Human Security and Governance (CERES), Rabat, Morocco

Mohamed Aneflouss Department of Geography, Research Laboratory on Dynamics of Space and the Society (LADES), Faculty of Arts and Human Sciences, Hassan II University of Mohammedia, Rabat, Morocco

Fatima Arib Faculty of Law, Economics and Social Sciences, Cadi Ayyad University of Marrakech / Researcher at the Center for Research on Environment, Human Security and Governance (CERES), Marrakesh, Morocco

Mohamed Behnassi Faculty of Law, Economics and Social Sciences, Center for Research on Environment, Human Security and Governance (CERES), Ibn Zohr University, Agadir, Morocco

Lucia Beran Lehrstuhl für Unternehmenstheorie, Nachhaltige Produktion und Industrielles Controlling, RWTH Aachen, Aachen, Germany

Kaderi Noagah Bukari Center for Development Research (ZEF), University of Bonn, Bonn, Germany

Arunesh Asis Chand Faculty of Science Technology and Environment, School of Geography, Earth Science and Environment, University of the South Pacific, Suva, Fiji

Seksak Chouichom Kasetsart University, 50 Ngamwongwan Rd., Ladyao, Chatuchack, Bangkok, Thailand

Elmar Csaplovics Researcher Professor, Faculty of Environmental Sciences, TU Dresden, Dresden, Germany

Taisser H. H. Deafalla Institute of Photogrammetry and Remote Sensing, University of Dresden, Dresden, Germany

Harald Dyckhoff Lehrstuhl für Unternehmenstheorie, Nachhaltige Produktion und Industrielles Controlling, RWTH Aachen, Aachen, Germany

Mustafa M. El-Abbas Researcher Professor, Faculty of Environmental Sciences, TU Dresden, Dresden, Germany

Faculty of Forestry, University of Khartoum, Khartoum, Sudan

Aïcha EL Alaoui Sultan My Slimane University, Chief of Research group of "the social economy and social justice", Beni Mellal, Morocco

Rachida El Morabet Department of Geography, Research Laboratory on Dynamics of Space and the Society (LADES), Faculty of Arts and Human Sciences, Hassan II University of Mohammedia; Researcher at Center for Research on Environment, Human Security and Governance (CERES), Rabat, Morocco

Zhar Essaid Department of Geography, Research Laboratory on Dynamics of Space and the Society (LADES), Faculty of Arts and Human Sciences, Hassan II University of Mohammedia, Rabat, Morocco

Aziz Fadlaoui Environmental Economist, National Institute of Agronomic Research (INRA), Meknes, Morocco

Himangana Gupta National Communication Cell, Ministry of Environment, Forest and Climate Change, New Delhi, India

Raj Kumar Gupta Journalist and Independent Analyst on Environmental and Social Policy, New Delhi, India

Sharifah Zubaidah Syed Abdul Kader Civil Law Department, Ahmad Ibrahim Kulliyyah of Laws, International Islamic University Malaysia, Kuala Lumpur, Malaysia

Peter Karácsony Department of Economics, Faculty of Economics, J. Selye University, Komárno, Slovakia

Lawrence M. Liao Graduate School of Biosphere Science, Hiroshima University, Higashi-Hiroshima, Japan

Mark Matsa Department of Geography and Environmental Studies, Faculty of Social Sciences, Midlands State University, Gweru, Zimbabwe

Said Mouak Department of Geography, Research Laboratory on Dynamics of Space and the Society (LADES), Faculty of Arts and Human Sciences, Hassan II University of Mohammedia, Rabat, Morocco

Maizatun Mustafa Legal Practice Department, Ahmad Ibrahim Kulliyyah of Laws, International Islamic University Malaysia, Kuala Lumpur, Malaysia

Hassane Nekrache Statistician and Demographer Engineer, The High Commission for Planning, Rabat, Morocco

Mostafa Ouadrim Department of Geography, Research Laboratory on Dynamics of Space and the Society (LADES), Faculty of Arts and Human Sciences, Hassan II University of Mohammedia, Rabat, Morocco

Szilárd Podruzsik Corvinus University of Budapest, Department of Agricultural Economics and Rural Development, Budapest, Hungary

Olaf Pollmann North-West University, School of Environmental Science and Development, Potchefstroom, South Africa

Jürgen Scheffran Researcher Professor, Head of the Research Group Climate Change and Security (CLISEC), Institute of Geography, University of Hamburg, Hamburg, Germany

Hartmut Sommer Department of Life Sciences and Engineering, University of Applied Sciences Bingen, Bingen, Germany

Papa Sow Center for Development Research (ZEF), University of Bonn, Bonn, Germany

Azlinor Sufian Legal Practice Department, Ahmad Ibrahim Kulliyyah of Laws, International Islamic University Malaysia, Kuala Lumpur, Malaysia

Aline Treillard Center for Interdisciplinary Research in Environmental Law, Land Planning and Urban law (CRIDEAU-OMIJ), Faculty of Law and Economics, University of Limoges, Limoges, France

Leon van Rensburg North-West University, School of Environmental Science and Development, Potchefstroom, South Africa

Laurent Weyers Université Libre de Bruxelles (ULB), Bruxelles, Belgium

Part I
Managing Environmental and Climate Change Induced Risks: The Need for a Paradigmatic and Governance Shift

Chapter 1
Mainstreaming a Rights-Based Approach in the Global Climate Regime

Mohamed Behnassi

Abstract The scientific and policy debate on environment and human rights linkages increasingly perceives climate change as a risk multiplier and a key crosscutting issue. Recent research has shown that climate change is putting both human security and several fundamental rights at risk. Not only are the human implications of climate change serious, but also the global climate regime is not sufficiently shaped to reduce them and a large part of response mechanisms, including on domestic and regional levels, do not systematically refer to justice, equity, and human rights frameworks and, in some instances, may even exacerbate environmental damage and human rights violations. Therefore, the significant challenge currently being faced is how to ensure that human rights are widely recognized and genuinely mainstreamed in the global climate regime. A key issue is how to bring the discourses of human rights and climate change together into the climate multilateral negotiation process without importing additional burdens, setbacks or unnecessary complications. Bringing human rights into the process is also about power, ambition and resilience; further, it is about endeavoring to change the power dynamics so that the movement may be progressively propelled by vulnerable countries. This chapter intends to dissect the potential overlap, convergence, and synergies between the international human rights framework and the global climate regime. The analysis assesses the advantages of mainstreaming a rights-based approach into this regime and what should be done on the ground to effectively achieve this objective. To this end, the chapter provides many research and policy-oriented recommendations.

Keywords Global climate regime · Rights-based approach · Human rights instruments · Paris agreement · UNFCCC · COPs

M. Behnassi (✉)
Faculty of Law, Economics and Social Sciences, Center for Research on Environment, Human Security and Governance (CERES), Ibn Zohr University, Agadir, Morocco

© Springer International Publishing AG, part of Springer Nature 2019
M. Behnassi et al. (eds.), *Human and Environmental Security in the Era of Global Risks*, https://doi.org/10.1007/978-3-319-92828-9_1

3

1 Introduction

People in the world's affluent countries are still either misperceiving or knowing a little about the massive and severe poverty that still persists in many regions such as South Asia, Africa, and Latin America. According to Jodoin and Lofts (2013:1), "most of them regard such poverty as a mild reason for charitable giving, but they typically do not recognize it as an injustice – let alone an injustice in which they are personally involved. They believe that questions of justice do not arise in this context because they do not see themselves as acting in ways that foreseeably harm the global poor or as involved with the poor in joint productive activities whose fruits are then divided among the collaborators".

Jodoin and Lofts (2013:1) presume that this common belief among the world's affluent is mistaken and the most relevant way of appreciating this fact can be showed by a basic understanding of anthropogenic climate change. By engaging in a consumption lifestyle that emits considerable greenhouse gases and pollutants, affluent societies directly and massively contribute to global warming and environmental degradation which are associated with substantial harms to poor and vulnerable populations through their multiple impacts and associated risks (i.e. increasing frequent extreme-weather events, spread of tropical diseases, desertification, rising sea levels, biodiversity loss, etc.). Poor people are much more vulnerable to these impacts and risks because they tend to live in the most exposed areas, typically cannot protect themselves, and also lack the means to cope once some threat has materialized. In addition, these people are not to blame for environmental degradation and carbon emissions since their contribution is universally sustainable in the sense that, if all human beings contributed to pollution at their level, we would have no climate change problem[1]. Assuming that the global distribution of income and wealth will remain as dramatically uneven as it is today, present excess emissions will inevitably cause vastly greater harms to poor populations in the future[2] than they are causing today.

Within this context, the scientific and policy debate on environment and human rights linkages increasingly perceives climate change as a risk multiplier and a key cross-cutting issue. Many recent studies and assessments have demonstrated that climate change is putting both human security and several fundamental rights at risk. Indeed, the Intergovernmental Panel on Climate Change (IPCC) (2007; 2013), the body of the world's leading climate scientists, has concluded in its previous and recent assessments that climate change is both occurring and is attributable to increasing atmospheric concentrations of greenhouse gases (GHGs) resulting from anthropogenic activity. The United Nations Human Rights Council (UNHRC 2008) has recently acknowledged that the environmental changes brought about by global

---

[1] The consumption of the world's affluent produces vastly more pollution and emission: i.e. a US resident produces over 10 times as much CO_2 compared to a resident of India, and over 50 times compared to a resident of Kenya (Jodoin and Lofts 2013:1).

[2] The greenhouse gases rapidly accumulating in the atmosphere will remain there for centuries and even millennia – continuing to warm the planet even if all emissions ceased today.

warming can interfere with the realization and enjoyment of fundamental, internationally recognized human rights – including those protected by International Covenants on Civil and Political Rights (ICCPR) and Economic, Social, and Cultural Rights (ICESCR).

In fact, climate change is not just a mere environmental issue, but it swiftly becomes the greatest challenge to the promotion of fundamental rights since its human implications are already serious and alarming. Undeniably, climate change is rapidly becoming an issue of justice and inequality for millions of people around the world and for the future generations, which will suffer severe loss and damage as well. The international community's past failure – and insufficient current actions – to mitigate and adapt to climate change according to available scientific evidence is further threatening human rights (CIEL and CARE International 2015). More specifically, the rights of vulnerable peoples, who are already experiencing the adverse effects of climate change, are more threatened than ever (CIEL 2013).

Besides, not only human implications of climate change are serious and already violating the rights of vulnerable individuals and groups in many parts of the world, but also the global climate regime and the internationally prescribed actions to address climate change are not sufficiently shaped to reduce these new violations; during implementation, many response mechanisms may even worsen the situation. On other scales, most climate policies being elaborated and implemented in many countries are based on different considerations and objectives and do not thoroughly refer to the frameworks of justice, equity, and human rights. The response measures designed to address climate impacts are likely to have serious repercussions on the lives and livelihood of people around the globe, especially the poorest and most vulnerable communities: i.e. Reducing Emissions from Deforestation and Forest Degradation (REDD+) activities may negatively affect the rights of local communities and indigenous peoples who live in and manage forests; and large-scale alternative energy projects may result in communities being forced to leave their homes and ancestral lands without adequate consultation or consent (CIEL and CARE International 2015). At the extreme, climate impacts and response measures threaten to destroy many cultures of peoples around the world, render their lands uninhabitable, and deprive them of basic means of subsistence.

Therefore, the significant challenge currently being faced when designing climate policies and governance systems involves simultaneously minimizing and halting climate change impacts and ensuring that all response measures consider human rights in systematic way. This implies, among other measures, the recognition and the substantial mainstreaming of a rights-based approach in the evolving global climate regime since potential synergies exist between the current human rights instruments and climate agreements adopted so far. A key issue, therefore, is how to bring the discourses of human rights and climate change together into the multilateral negotiation process without importing additional burdens, setbacks or unnecessary complications. In a post-2015 regime, the need to achieve climate justice, including urgent action on mitigation and real safeguards to prevent serious violations of human rights, is increasingly a social demand. The narrative of human rights and climate justice is becoming powerful and legitimate; therefore there is a real opportunity to use this narrative to engage relevant actors. Additionally, bringing human rights into the process is also about power, ambition and resilience; it is about trying to change the power dynamics so that the movement may be progressively propelled by vulnerable countries.

Currently, the linkages between climate change and human rights are being documented; and therefore becoming less controversial and seem beyond dispute. The challenge now lies in introducing a rights-based approach to the development and implementation of effective and equitable solutions to climate change. Within such perspective, this chapter intends to identify the potential overlap, convergence, and synergies between the international human rights framework and the global climate regime[3]. Firstly, an overview of the human implications of climate change is undertaken (Sect. 2). Then, the analysis assesses the extent to which human rights have been considered in the evolving climate regime since its genesis (Sect. 3). Thirdly, relevant approaches for considering human rights while developing solutions to climate change are presented and evaluated (Sect. 4). Finally, the analysis assesses the advantages of mainstreaming a rights-based approach into the global climate regime and what should be done on the ground to effectively reach this objective (Sect. 5). Research and policy-oriented recommendations will be provided throughout the analysis as well.

2 Overview of Climate–Related Human Rights Concerns

2.1 Human Implications of Climate Change

Climate change is currently perceived as a massive threat to human security and development. There is also a growing certainty that, as the climate system warms, poorer nations and their poorest populations, will be the worst affected. Climate impacts documented by IPCC (such as the increase of the severity of droughts, land degradation, desertification, the intensity of floods and tropical cyclones, the incidence of malaria and heat-related mortality, and the decrease of biodiversity, crop yield and food security) are likely to undermine the realization of a range of civil, political, economic, social, and cultural protected human rights. The rights to life and to health provide useful examples in this regard:

* The Human Rights Committee (HRC), which is a body of independent experts that monitors implementation of the ICCPR, noted that the 'inherent right to life' cannot be interpreted in a restrictive manner, and that the protection of this right requires states to take positive measures. The Committee also noted that states have the supreme duty to prevent wars, acts of genocide, and other acts of mass violence causing arbitrary loss of life (HRC 1982). According to Rajamani

---

[3] The global climate regime is defined here as the set of international, national and sub-national norms, institutions, and actors involved in addressing climate change.

(2010), this duty could arguably be extended to cover specific human-induced climatic incidents that cause arbitrary loss of life. The IPCC fourth assessment report (i.e. IPCC 2007) had early confirmed that extreme weather events have the potential to increase the risk of mortality (especially for the elderly, chronically sick, very young, and the socially isolated), injuries, and infections.

- The 'right to the highest attainable standard of health' is considered indispensable for the enjoyment of other human rights, and it is widely protected in international and regional instruments, and under national constitutions as well. The right to health has been broadly defined as an 'inclusive right' including timely and appropriate health care, access to safe and potable water, adequate sanitation, an adequate supply of safe food, nutrition and housing, healthy occupational and environmental conditions, and access to health-related education and information (HRC 2000). The World Health Organization (WHO), in various reports (i.e. WHO 2013), has confirmed that climate change is a serious challenge that undermines and places the fundamental determinants of good health (such as clean air, fresh water, food security, and freedom from disease) at risk. Addressing climate change is, therefore, perceived as a huge health opportunity, including reducing the over 6.5 million annual deaths from air pollution (most victims are poor, elderly, women and children including rural households and people living in burgeoning low-income cities). This is the reason behind citing the 'right to health' in Paris Agreement, and recognizing the opportunities to obtain health co-benefits from mitigation – and adaptation – actions (UNFCCC 2015).

Direct climate impacts, such as extreme weather events and rising sea levels, also threaten millions of people in coastal and low-lying areas, while melting snow and ice threaten the welfare and security of many indigenous peoples of the Arctic. Loss of freshwater resources and glacial melt put communities at risk in the Andes and the Himalayas. At the same time, seawater intrusion[4] contaminates groundwater in coastal communities, negatively affecting agricultural production and potable water availability. Ocean acidification and changes in weather patterns alter ecosystems and their capacity to provide services to human communities. Increasing weather extremes constrain food security and access to nutritious forms of food while changing the prices of commodities in global markets, making food more expensive and harder to access for the world's poorest people (Table 1.1).

According to CIEL (2013), numerous climate change impacts have direct impacts on human populations and their livelihood:

- Sea-level rise and storms cause flooding, population displacement, salinization of fresh-water resources, and the diminishment of habitable or cultivable land.
- Rising surface temperatures lead to greater occurrence of diseases such as scrub typhus, diarrheal diseases, and other mosquito-borne diseases.

---

[4] Seawater intrusion is the movement of seawater into fresh water aquifers due to natural processes or human activities. It is caused by decreases in groundwater levels or by rises in seawater levels.

Table 1.1 Climate change impacts on selected human rights

Climate impact	Human impact	Rights implicated
Sea-level rise	Loss of land	Self-determination (ICCPR[a]; ICESCR[b], 1)
Flooding	Drowning, injury	Life (ICCPR,6)
Sea surges	Lack of clean water, disease	Health (ICESCR,12)
Erosion	Damage to coastal infrastructure, homes, and property	Water (CEDAW,14; ICRC,24)
Salination of land and water	Loss of agricultural lands	Means of subsistence (ICESCR,1)
	Threat to tourism, lost of beaches	Adequate standard of living (ICESCR,12)
		Adequate housing (ICESCR,12)
		Culture (ICCPR,27)
		Property (UDHR[c],17)
Temperature increase	Spread of disease	Life (ICCPR,6)
Change in disease vectors	Changes in traditional fishing livelihoods and commercial fishing	Health (ICESCR,12)
Coral bleaching	Threat to tourism, lost coral and fish diversity	Means of subsistence (ICESCR,1)
Impacts on fisheries		Adequate standard of living (ICESCR,12)
Extreme weather events	Dislocation of populations	Life (ICCPR,6)
High intensity	Contamination of water supply	Health (ICESCR,12)
Storms	Damage to infrastructure: Delays in medical treatment, food crisis	Water (CEDAW[d], 14; ICRC[e], 24)
Sea surges	Psychological distress	Means of subsistence (ICESCR,1)
	Increased transmission of disease	Adequate standard of living (ICESCR,12)
	Damage to agricultural lands	Adequate and secure housing (ICESCR,12)
	Disruption in educational services	Education (ICESCR,14)
	Damage to tourism sector	Property (UDHR,17)
	Massive property damage	
Changes in precipitation	Outbreak of disease	Life (ICCPR,6)
Change in disease vectors	Depletion of agricultural soils	Health (ICESCR,12)
Erosion		Means of subsistence (ICESCR,1)

Source: Submission by the Maldives to the OHCHR in September 2008, as part of OHCHR's consultative study on the relationship between climate change and human rights, Reproduced in CIEL (2013)

[a]ICCPR: International Covenant on Civil and Political Rights
[b]ICESCR: International Covenant on Economic, Social and Cultural Rights
[c]UDHR: Universal Declaration of Human Rights
[d]CEDAW: Convention on the Elimination of all Forms of Discrimination Against Women
[e]ICRC: International Committee of the Red Cross

- The increasing number and intensity of weather events endanger life, health, and housing. Changing coastlines and melting permafrost cause damage to land, houses, and other infrastructure.
- Changing precipitation patterns and the melting of glaciers affect access to water, which in turn affects the ability to irrigate lands and secure access to food.

In addition, the continuous delay in mitigating climate change means the costs of adapting to its adverse impacts are increasing, and the risk of experiencing severe and irreversible loss and damage will be much higher in the future. A recent United Nations Environmental Program (UNEP) report (2014) estimates that adaptation costs in developing countries are two to three times higher than projected in previous studies. Climate change has even further aggravated poverty for vulnerable people around the world – and will continue to do so – even though they are the least responsible for causing it (CIEL and CARE International 2015).

While different populations may face similar risks of exposure to the negative effects of climate change, their actual vulnerability is specific and socially constructed; it is dependent on their socio-economic conditions and the available resources, infrastructure, and knowledge. Some categories such as women, children, the elderly, indigenous communities, and persons with disabilities may excessively suffer due to their high vulnerability to climate impacts.

Indeed, in the poorest regions of the world, women often bear the primary responsibility for gathering the essential food, water, and fuel supplies for their families. Recurrent droughts caused by climate change make their work extremely hard as wells run dry, biodiversity and crop production decline, and fuel wood has to be collected from remote locations with potential risks to their safety and health. Just as they are more severely affected because of their social role, women will bear the burden of adapting despite their own insignificant contribution to greenhouse gas emissions (Gupta 2015). Similarly, food and water shortages will increase malnutrition among children, therefore affecting their physical development and diminishing their chances to receive school education. Children are also more vulnerable to natural disasters as they lack physical strength and know-how, and are often orphaned or separated from their families as a post-disaster consequence. In addition, people with disabilities, who represent 20% of the poorest people worldwide (WHO 2011), are disproportionately affected by climate adverse impacts, which are even expected to cause increasing hardship for this vulnerable group (Lewis and Ballard n.d.). Indigenous communities are also particularly vulnerable to climate change, since their way of life is often inextricably tied to ecosystems and natural resources (Watt-Cloutier n.d.). Consequently, environmental changes often impact their ability to access water, food, and shelter. For many indigenous communities, lands and natural resources are not a mere commodity, but a central element of their spiritual and cultural identity; thus serious environmental changes resulting from climate change can affect both their physical and cultural survival.

2.2 To What Extent Response Mechanisms to Address Climate Change May Affect Human Rights?

Actions to address human rights-related climate impacts – such as adaptation, mitigation, finance, and science and technology transfer – may, in some instances, further affect, perhaps profoundly, the livelihood, lifestyles, living conditions and cultures of vulnerable communities (CIEL 2013):

2.2.1 Adaptation

Adaptation measures, as one of the key measures to deal with climate change impacts, are in principle supposed to help increase the resilience and capacity of social and ecological systems to cope with potential risks. In this regard, the governments' duty in international human rights instruments to protect people from harm, which is universally recognized, inevitably implies the mainstreaming of human rights into adaptation policy and governance. Climate adaptation policies have the potential to infringe on human rights in case their conception and implementation are not adequate or disconnected from some concerns[5] such as, *inter alia*, justice, equity, poverty alleviation, social inclusion, and redistribution. Indeed, a variety of adaptation policies may have human rights implications, such as those pertained to food, water, forest, and the availability of other resources to support the adaptation needs of vulnerable populations.

Moreover, due to the nature of climate system and the long-term effects of increased carbon concentrations, climate change will continue over several decades, and therefore adaptation strategies should be maintained by governments as a key to continuously protecting fragile ecosystems and vulnerable populations from severe impacts. Any attempt to reorient financial resources to other areas may exacerbate the vulnerability of concerned ecosystems and communities. Similarly, disaster risk management should be dedicated to address the particular situation of the most vulnerable and marginalized.

2.2.2 Mitigation

Regarding mitigation, it is well known that the global climate regime (mainly the United Nations Framework Convention on Climate Change (UNFCCC), the Kyoto Protocol (KP), the Bali Action Plan, and the Paris Agreement) is shaped in such a

---

[5] Adaptation actions, such as relocation in response to sea-level rise or other environmental factors, may impact the right to culture, particularly for indigenous peoples, local communities and other vulnerable groups. For example, relocation can have a particular impact on the right to culture of indigenous peoples whose cultural and spiritual practices are tied to the land, or for local communities who might lose access to significant sites such as ancestral burial grounds. For more information, refer to Jodoin and Lofts (2013).

way so as to develop the mitigation potential of different countries using a myriad of strategies and mechanisms. Within the framework of KP, and according to the so-called 'Clean Development Mechanism (CDM)' (as well as 'Joint Implementation' as a form of CDM between Northern countries), industrialized countries were able to buy their way out of their obligation to reduce GHG emissions with 'climate projects' in developing countries, without actually reducing the volume of carbon dioxide released.

Other mitigation strategies, such as REDD+ programs, could provide funds to developing countries, indigenous peoples, and forest-dependent communities involved in forest conservation. The Bali Action Plan clearly contemplates additional emission reduction commitments for developed countries, and introduces the concept of Nationally Appropriate Mitigation Actions (NAMAs)[6] to be taken by developing countries, which are to be supported by technology, financing, and capacity building. Recently, prior to the adoption of Paris Agreement, the Nationally Determined Contributions (NDCs)[7] have been introduced as a flexible mechanism (referred to in the Paris Agreement as the 'ambition or ratchet mechanism') to promote worldwide mitigation potential.

Although the contours of the mitigation regime are still unclear, the mitigation measures necessary to address climate change have the potential to impact human rights (CIEL 2008a, b). The KP's market mechanisms (mainly CDM which enables the trading in carbon offsets) do support projects in developing countries that result in emission reductions, generating carbon credits that developed countries can use to offset their own emissions. In addition to using the CDM to even promote carbon-intensive industries (including coal-fired power plants), some investments projects tended not to benefit the countries that need them most urgently (those in sub-Saharan Africa, for example) but emerging economies such as China and India. Also, many CDM projects were often marked by negative environmental and human implications (such displacement of communities). Denial of free and prior informed consent by indigenous peoples and other local communities could further aggravate these impacts, with respect to their rights, lands, and territories. Therefore, considering these human implications while designing and implementing mitigation policies and projects is vital.

Current CDM modalities and procedures contain some tools that help promote a rights-based approach, such as disclosure of environmental assessments and channels for public participation. However, the CDM has yet to fully adopt a rights-based approach (CIEL 2013) to ensure that its operations contribute to the advancement of sustainability, mainly its human dimension. The Sustainable

---

[6] The NAMAs are mitigation measures proposed by developing country governments to reduce emissions below 2020 business-as-usual levels and to contribute to domestic sustainable development. It can take the form of regulations, standards, programs, policies or financial incentives. The term was coined in the Bali Action Plan of 2007, and later, in 2009 in Copenhagen, developing countries submitted NAMAs to the UNFCCC.

[7] According to Art. 4/2 of the Paris Agreement, each Party shall prepare, communicate and maintain successive NDC that it intends to achieve. Parties shall pursue domestic mitigation measures, with the aim of achieving the objectives of such contributions.

Development Mechanism (SDM), recently established in the Paris Agreement, is meant to reflect the learning from these mistakes even though the design of this mechanism is left to future climate negotiations. This market mechanism is now supposed to ensure that the projects actually reduce carbon emissions beyond what the climate plans of the respective countries already promise. Transparency, accountability, respect for human rights, additionality,[8] and public participation are all elements that must be made mandatory.

In the same vein, policies and measures adopted by governments in connection with energy, forests, and land use can significantly impact indigenous and local communities.[9] Land tenure, traditional use of resources, and benefit-sharing considerations may all be implicated by mitigation efforts. For example, governments establishing protected areas over forests occupied or otherwise used by indigenous peoples and other forest-dependent communities could potentially displace traditional occupants, and implicate relocation and traditional resource-use rights. Therefore, because land and livelihoods may be impacted by REDD+ activities, it is important to ensure respect and protection of the rights of forest-dependent communities. This protection is essential to ensure the success and permanence of measures taken on the ground. This implies the right of forest-dependent communities to participate in and to share the benefits of REDD+ programs and measures. Participation is more than a mere consultation, as it involves the right to free and prior informed consent of concerned communities, in accordance with their customs and traditions, including their traditional governance structures.

Notably, in 2010, Parties to the UNFCCC agreed to promote and support the following safeguards for REDD+ activities: consistency with international obligations; respect for the rights of indigenous peoples and local communities; full and effective participation of stakeholders; good governance systems; and avoided damage to biodiversity and ecosystems. Parties also agreed to develop a system of information sharing on how safeguards are being implemented. In the end, a rights-based approach to forests and land use can ensure that relevant policies and projects do not interfere with the rights of those who have preserved the forest since time immemorial (Meridian Institute 2009).

---

[8] The principle of 'Additionality' is the requirement that the greenhouse gas emissions after implementation of a CDM project activity are lower than those that would have occurred in the most plausible alternative scenario to the implementation of the CDM project activity (http://www.cdm-rulebook.org/84.html)

[9] For example, REDD-related projects or large hydroelectric projects frequently result in the mass displacement of communities. The dispersal of such communities may negatively impact their enjoyment of the right to culture, particularly in the case of Indigenous peoples, who often have a special cultural and spiritual relationship to their land. See Jodoin and Lofts (2013).

2.2.3 Climate Finance

In terms of finance, lower-income and climate vulnerable countries are not generally in a financial position to efficiently deal with climate change and fully protect their populations from its adverse impacts. Their limited public budgets are usually dedicated to cover vital sectors, such as infrastructure, health, nutrition, and education. Any attempt to allocate the available resources to fund adaptation policies may negatively impact these sectors; therefore it is increasingly critical to assist these countries in accomplishing this duty by funding the necessary climate strategies. In addition, the development and application of finance safeguards are also necessary to prevent social and environmental harm and maximize participation, transparency, accountability, equity, and rights protection. In 2010, Parties to the UNFCCC established the Green Climate Fund (GCF) (currently the largest multilateral climate fund), and agreed, among other things, to develop mechanisms to ensure that social and environmental safeguards apply to the fund. To do so, institutions involved in funding climate-related activities should provide transparent processes, maintain policies and procedures that respect internationally recognized rights, and allow meaningful opportunities for public participation.

However, on the ground, the outlook is not very promising. In mid-October 2016, the industrialized countries presented a roadmap, outlining how they intend to provide the pledged $100 billion for climate protection in developing countries by 2020. The initiative is to be welcomed. However, it has one major weakness: while the roadmap promises to double the funding available for adaptation, this is from a very low base, as currently fully four fifths of the $100 billion are earmarked for mitigating emissions and only one fifth for adaptation (i.e. protecting crops, prevention and response to natural disasters, and safeguarding water supplies). The Paris Agreement had called for a 'balance' – ideally a 50–50 split (which has not been established in terms of agreed policy, however).

The developed countries will benefit if developing countries go further in curbing their own carbon emissions – they will then have to do less themselves. However, developed countries must not shirk their responsibilities toward countries that are most heavily and already affected by climate change but have contributed the least to it. This also applies to the objective of developed countries to generate a third of the pledged $100 billion by leveraging resources from the private sector. Therefore, the Green Climate Fund (GCF) should be fed with additional pledges even if new pledges are not expected during the coming negotiations.

Often, however, the problem is less a lack of money but that many countries simply do not have a way of accessing it directly. That has become apparent over the recent period during which the GCF has been approving project and program proposals. Hence, the GCF must offer developing countries direct access, a greater voice, and more country ownership rather than distributing funds primarily through existing multilateral implementers (such as the Adaptation Fund, the Global Environment Facility (GEF), and the World Bank) (Fuhr et al. 2016). Other multilateral regimes and institutions – such as the UN Environment Program (UNEP), the UN Development Program (UNDP), the Food and Agriculture Organization (FAO), and the International Labor Organization (ILO) – should also be associated and involved in the process. The principal efforts and leadership, however, should stem from the climate change and human rights regimes, which are most directly involved (CIEL 2013). In addition, many countries require administrative assistance from abroad[10] in order to fully benefit from available resources (Fuhr et al. 2016).

2.2.4 Scientific and Technological Transfer

In addition to finance, many developing countries are still lacking the scientific and technological capacities to deal appropriately and efficiently with climate change. Thus, science and technology transfer is increasingly considered critical to supporting sustainability and avoiding the shifting of polluting industries from developed countries to the developing world (CIEL 2008a, b). Establishing a workable institutional mechanism for science and technology transfer is crucial to reaching and successfully implementing any future climate framework. A rights-based approach can help ensure that the science and technology agendas are oriented towards promoting human rights worldwide and that latest valuable innovations are fairly shared. In terms of effective implementation of adaptation strategies, a rights-based approach to science and technology transfer can help ensure that scientific inputs required by the most vulnerable peoples and communities are systematically considered a priority.

3 Global Climate Regime and Human Rights: A Late Convergence

Despite the magnitude of the climate challenge, the advancement of the scientific evidence about the anthropogenic source of the problem, and a long, extensive, and intensive multilateral negotiation process, an effective and equitable framework to address it is still under development. The existing framework before the adoption of Paris Agreement in December 2015 (UNFCCC, Kyoto Protocol and all COPs' outcomes such as the Bali Action Plan and the Copenhagen Accord) has been considered both insufficient and inadequately implemented, with limited potential in terms of climate governance.[11] There are significant barriers to effective collective action on climate change and vast differences exist between countries in terms of contributions to the stock of carbon in the atmosphere, industrial advancement and wealth redistribution, nature of emissions use, and climate vulnerabilities. In addition,

---

[10] Germany currently has recently announced a new initiative on how to help developing countries adapt to climate change and promote renewable energy and energy efficiency.

[11] Climate governance is the diplomacy, mechanisms and response measures aimed at steering social systems towards preventing, mitigating or adapting to the risks posed by climate change (Jagers and Stripple 2003).

poverty is increasing in some parts of the world, along with persistent inequalities within and among countries. And even if climate impacts continue to prejudicially affect the poor, disempowered, culturally distinct, and the geographically disadvantaged across the globe, there is still a marked reluctance in many polities and societies to substantially modify existing lifestyles and development pathways, and opinion is divided on the promise of technological and market solutions to complex climate change problems (McInerney et al. 2011).

Operating within these constraints, states have over the past two decades and a half established a global regime, albeit an evolving one, to address climate change and its impacts. The existing instruments have attracted near-universal adherence. Their basic purpose is to set in place an international legal framework for common but differentiated responsibilities for the reduction of GHG emissions, and support for national adaptation efforts, with a particular concern for developing and vulnerable countries' special needs. However, unlike the international human rights regime, the UNFCCC and the Kyoto Protocol did not include express provisions for remedial measures for individuals or communities in light of a particular environmental harm. However, subsequent agreements have called for consideration of the social and economic consequences of response measures as well as enhanced international cooperation (McInerney et al. 2011).

3.1 Overview of the Progress Prior to the Adoption of Paris Agreement

In 2008, UN Secretary-General Ban Ki-moon affirmed: "there is virtually no aspect of our work that does not have a human rights dimension. Whether we are talking about peace and security, development, humanitarian action, the struggle against terrorism, climate change, none of these challenges can be addressed in isolation from human rights" (UN News Centre 2008). The United Nations in 2003 adopted a Statement of Common Understanding on Human Rights-Based Approaches to Development Cooperation and programming (UN 2003), and in 2009 created the UN Development Group's Human Rights Mainstreaming Mechanism (UNDG-HRM) (UNDG 2009).

Despite this progress in the UN system, and even if the linkages between climate change and human rights were progressively recognized by multilateral climate negotiations, UN human rights bodies, the scientific community, and civil society actors, the deliberations of past negotiations did not confirm the relevance of a rights-based approach. Indeed, it is only in the recent past that this approach has been explicitly brought to bear on the climate change problem. In the early days of climate negotiations, developed countries sought to introduce and privilege the 'right to development'. More recently, some countries and interest groups have sought to widen the range of human rights of relevance in the climate negotiations. Scholars and human rights bodies have begun to advocate a rights-centered approach to climate change; an approach which would place the individual at the center of inquiry, and draw attention to the impact that climate change could have on the enjoyment and protection of human rights.

Moreover, many developing countries have advanced an equity perspective on climate change – a perspective which is underpinned by human rights concerns. The crux of their argument rests on an appreciation of differences between countries in terms of contributions to the carbon stock and flow in the atmosphere (*historical versus current and future*), nature of emissions (*survival versus luxury*), economic status (*poverty versus wealth*), and physical impacts, including their ability to cope with them (*severe versus adaptable*). In their view, these differences suggest that developing countries should only be expected to contribute to solving the problem, to the extent that they are empowered to do so. It is in this context that these countries advanced their right to development, which remains deeply disputed. However, and whilst the UNFCCC does not endorse an explicit 'right' to development, disputed as it is, it does recognize the central role that development plays in the global climate regime[12] (Rajamani 2010).

The UNFCCC also recognizes in its Preamble that in the pursuit of social and economic development, emissions and energy consumption in developing countries will grow. This, read in conjunction with the global stabilization goal in Article 2, arguably suggests that the global climate regime envisions a redistribution of the ecological space, with developed countries reducing their emissions to make room for developing countries to grow. It could be further argued that such redistribution stems from a recognition that developing countries by extension of their populations, are entitled to an appropriate proportion of the ecological space so as to achieve and sustain a certain quality of life.

The discussions in the climate negotiations, and on the sidelines of it, on the right to an 'equitable sharing of atmospheric space', survival and luxury emissions, the contraction and convergence proposal (based on per capita CO_2 emission entitlements), and the Greenhouse Development Rights framework (based on the right of all people to reach a dignified level of sustainable human development), all draw on the equity-based right to development (Rajamani 2010). In the climate change context, the right to development takes within its fold the 'right to emit'. The desire to occupy a larger share of the ecological space, and an entitlement to it, is a legitimate one. Emissions not only produce the burden of marginalization, they also produce the benefit of power; and the right to use the atmosphere as a dumping ground represents a source of economic power. Disparity in access leads to disparity in economic opportunities; it partitions the world into winners and losers (Sachs 2007).

The exercise of a right to development, to the extent it is recognized and protected in the UNFCCC, is not unrestricted. Developing countries are required under the international climate regime to develop in a sustainable manner,[13] and while

---

[12] The Art. 2 (Objective) of the UNFCCC specifies that: stabilization of GHGs in the atmosphere must be achieved within a time frame sufficient to 'enable economic development to proceed in a sustainable manner'.

[13] Preamble, Arts. 3(4) and 4(1), UNFCCC.

doing so to address the adverse effects of climate change through adaptation. This is, however, a responsibility unique to developing countries in the sense that it requires them to take on board sustainable development at a period in the trajectory of their development; a comparable period in which the developed countries had no such restraints on their development. Although fettered, the exercise of such a right to develop will result in greater GHG emissions (Rajamani 2010).

During previous climate negotiations, the relevance of a rights-based approach has been established progressively. Much of the recent interest in the human rights dimensions of climate change has been sparked by the plight of the Inuit[14] and the Small Island States, at the frontlines, albeit different ones, of climate change. In their 2005 petition before the Inter-American Commission on Human Rights (IACHR), the Inuit claimed that climate impacts, which should be attributed to acts and omissions of the United States, have violated their fundamental human rights – in particular the rights to the benefits of culture, to property, to the preservation of health, to life, to physical integrity, to security, to a means of subsistence, to residence, to movement, and to the inviolability of the home (OAS and IACHR 2005). These rights, it was argued, were protected under several international human rights instruments, including the American Declaration of the Rights and Duties of Man (1948). The IACHR declined to review the merits of the petition, declaring that the information provided does not enable the Commission to determine whether the alleged facts would tend to characterize a violation of rights protected under the American Declaration (Jane 2006). Although the Inuit Petition did not fare well before the Commission, it drew attention to the links between climate change and human rights and led to a 'Hearing of a General Nature' on human rights and global warming.[15] The Hearing was held on 2007, and featured testimonies from the Chair of the Inuit Circumpolar Conference (ICC) and its lawyers, with the absence of US representatives. The Commission has taken no further action, but the ICC petition did generate considerable debate in the academic literature (McInerney et al. 2011).

During the climate negotiations process, indigenous groups have frequently delivered strong statements on the impacts of climate change on indigenous peoples' health, society, culture and well-being. Indigenous peoples' organizations have been admitted to the UNFCCC process as NGOs, with constituency status.[16] The Permanent Forum on Indigenous Issues under the Economic and Social Council (ECOSOC) at its second session (May 2003) recommended, without success, the establishment of an ad hoc open-ended working group on indigenous peoples and climate change. In its seventh session of April and May 2008, dedicated to climate

---

[14] The Inuit are a group of culturally similar indigenous peoples inhabiting the Arctic regions of Canada (Northwest Territories, Nunatsiavut, Nunavik, Nunavut, Nunatukavut), Denmark (Greenland), Russia (Siberia) and the United States (Alaska). Inuit means "the people" in the Inuktitut language.

[15] Press Release, Inter-American Commission on Human Rights, IACHR Announces Webcast of Public Hearings of the 127th Regular Period of Sessions, No 8/07 (Feb. 26, 2007), available at http:// www.cidh.org/Comunicados/English/2007/8.07eng.htm

[16] See Note, Secretariat of the UN Framework Convention on Climate Change, *Promoting Effective Participation in the Convention Process*, UNFCCC/SBI/2004/5 (Apr. 16, 2004) para. 39–47.

change, the Forum recommended that the Declaration on the Rights of Indigenous Peoples serves as a 'key and binding framework' in efforts to curb climate change, and that the rights-based approach guides the design and implementation of local, national, regional, and global climate policies and projects (UNPFII 2008).

Meanwhile, the Small Island States, and the Maldives in particular, launched a campaign to link climate change and human rights. Representatives of this group met in November 2007 to adopt the *Malé Declaration on the Human Dimension of Global Climate Change*, requesting, *inter alia*, that the UN Human Rights Council (UNHCR) convenes a debate on climate change and human rights.[17] The Council adopted a Resolution submitted by Maldives titled *Human Rights and Climate Change* in March 2008 that requested the Office of the UN High Commissioner for Human Rights (OHCHR) to conduct a detailed analytical study on the linkages between climate change and human rights[18]. The OHCHR published its study in 2009[19]. The study argued that climate change threatens the enjoyment of a wide spectrum of human rights, but fell short of finding that climate change necessarily or categorically *violates* human rights. However, the study concluded that states have obligations under human rights law to address climate change. After considering the report, the UNHCR decided by consensus in March 2009 to hear a panel discussion on the topic at its June 2009 session, and encouraged its Special Procedures mandate-holders to consider human rights implications of climate change within their mandates.

In addition to these developments, the Bali Action Plan, which in 2007 launched the post-2012 climate negotiations, identified five pillars on which a future climate regime should be built: shared vision, mitigation, adaptation, technology, and finance[20]. Chile suggested that a shared vision on climate change be based, *inter alia*, on a human rights perspective. Bolivia argued that the scientific basis for a climate regime must include a full assessment of social, economic, and environmental conditions (including the right to water, the protection of human rights, poverty eradication, etc.) in developing countries[21]. Argentina preferred to raise the human rights concern in the context of 'enhanced action on adaptation', arguing that further research on the impacts of climate change on human rights realization will be useful in ensuring that climate responses take place within a strong sustainability framework (UNFCCC AWG-LCA 2008).

---

[17] Malé Declaration on the Human Dimension of Global Climate Change (Nov. 14, 2007), *available at* http://www.meew.gov.mv/downloads/download.php?f=32

[18] United Nations Office of the High Commissioner for Human Rights, *Human Rights and Climate Change*, Resolution 7/23 (Mar. 28, 2008), *available at* http://ap.ohchr.org/documents/E/HRC/resolutions/A_HRC_RES_7_23.pdf

[19] United Nations Office of the High Commissioner for Human Rights, *Report on the Relationship Between Climate Change and Human Rights*, U.N. Doc. A/HRC/10/61 (Jan. 15, 2009).

[20] UNFCCC Conference of the Parties, 'Report of the Conference of the Parties on its thirteenth session, held in Bali from 3 to 15 December 2007, Addendum, Part Two: Action taken by the Conference of the Parties at its thirteenth session' (14 March 2008) FCCC/CP/2007/6/Add 1, 3.

[21] Ibid, 107.

In subsequent negotiating sessions in 2009, several Parties introduced the human rights discourse into the negotiating text. The least developed countries requested recognition in the 'shared vision' section of the text that the climate change adverse effects "have a range of direct and indirect implications for the full and effective enjoyment of human rights including the right to self-determination, statehood, life, food, health, and the right of a people not to be deprived of its own means of subsistence, particularly in developing countries" (UNFCCC AWG-LCA 2009). In the context of principles guiding adaptation action, Thailand proposed a reference to the UNDHR, the ICCPR and ICESCR, and Iceland to the UN Declaration on the Rights of Indigenous Peoples (UNDRIP) and the CEDAW (UNFCCC AWG-LCA 2009).

In the lead up to the COP15 in Copenhagen, December 2009, members of the Bolivarian Alliance for the Americas (ALBA)[22] proposed the inclusion of various rights in the negotiating text on 'shared vision', in particular the 'right to development' and the 'right to live well'. These were opposed by most developed countries who did not wish to refer in the climate texts to rights that were either not recognized in human rights treaties (such as the right to live well) or were disputed (such as the right to development). The Copenhagen Conference resulted in decisions to continue negotiations under the UNFCCC and Kyoto Protocol. States were tasked with continuing on the basis of the work that had been undertaken thus far[23] (Rajamani 2010).

In April 2010, Bolivia held a World People's Conference on Climate Change and the Rights of Mother Earth in Cochabamba, Bolivia, with an estimated participation of 35,000 people from social movements and organizations representing 140 countries. Bolivia submitted the outcome of this Conference to the UNFCCC process, and this submission was considered the most comprehensive exposition on rights done so far. In a submission liberally colored with rights discourse, Bolivia sought to introduce into the post-2012 negotiations the rights of developing countries (inter alia to 'equitable sharing of atmospheric space' and to 'development'), rights of all peoples including migrants (inter alia to life, food, housing, health, access to water, and to be protected from climate change adverse impacts), rights of indigenous peoples (inter alia to consultation, participation and prior, free and informed consent), and intriguingly to the rights of 'Mother Earth' (inter alia to live, to be respected, to regenerate its bio-capacity, and to integral health) (UNFCCC

---

[22] ALBA (Bolivia, Cuba, Ecuador, Nicaragua, and Venezuela), formally the Bolivarian Alliance for the Peoples of Our America (Spanish: Alianza Bolivariana para los Pueblos de Nuestra América), is an intergovernmental organization based on the idea of the social, political, and economic integration of the countries of Latin America and the Caribbean.

[23] Among the texts forwarded for further work is an overarching draft COP decision which contains several references to human rights. It 'note[s]' resolution 10/4 of the United Nations Human Rights Council is 'mindful' that "the adverse effects of climate change have a range of direct and indirect implications for the full enjoyment of human rights, including living well" and it "recognizes the right of all nations to survival". A forwarded draft decision on reducing emissions from deforestation in developing countries also contains a reference to the need to respect the knowledge and 'rights of indigenous peoples and members of local communities'.

AWG-LCA 2010). According to Rajamani (2010), the rights that the Bolivian submission referred to can be divided into three categories: rights recognized elsewhere in human rights instruments; rights recognized elsewhere but whose nature, content and extent are still disputed; and rights that are yet to be recognized. Most of the rights Bolivia highlighted in its submission fall into the first category.

In addition, building on the UNHRC's important recognition of climate change implications for the full enjoyment of human rights, the COP16 in Cancun, December 2010, took an initial step towards integrating human rights in the climate framework. For the first time, the UNFCCC process recognizes that rights obligations apply in the context of climate change, stating that, "Parties should, in all climate-related actions, fully respect human rights"[24]. Notably, the same decision also established the rights-based safeguards to be applied when financing and undertaking REDD+ activities[25]. However, and despite this progress, the UNFCCC has made minimal advancement in operationalizing rights protections from 2010 to 2015. The main exception is REDD+, where incremental improvements had been made in order to integrate further rights protections. These include a subsequent decision made by the UNFCCC stating that information on social, environmental and governance safeguard policies should be included in national communications, and that payments for results cannot occur without information showing that such safeguards are being addressed and respected.[26]

In general terms, for the most part, the negotiating Parties didn't take the necessary steps to ensure that human rights principles systematically guide the development and implementation of climate policies and response mechanisms established under the UNFCCC framework (Mary Robinson Foundation 2014). This seems to reflect differences of view between states (and regional and other groupings of states) on the so-called value-added of human rights in the climate change context, the comparative weight and focus to be given to human rights commitments within and beyond national borders, and perhaps also perceptions in various quarters that human rights might risk overloading an already fragile global climate agenda. Human rights have sometimes been characterized as a source of mistrust between the Global South and North, with certain developing countries expressing concern over human rights being used as a way of either preventing their development (should binding emissions reduction targets be applied to them) or operating as conditionalities on climate change adaptation funds (Limon 2009). Certain developed countries, correspondingly, have expressed concern at the possibility for an official recognition of human rights and climate change linkages to reinforce the case for extra-territorial human rights obligations or a collective and self-standing 'right to a safe and secure environment', or otherwise be used as a 'political or legal weapon against them'. These uncertain and evolving geopolitical dynamics form an important part of the context in which the international legal analysis and policy and operational implications fall to be considered.

---

[24] UNFCCC Decision 1/CP.16 (Dec. 2010), para 8.

[25] UNFCCC Decision 1/CP.16 (Dec. 2010), Appendix 1.

[26] UNFCCC Decision 9/CP.19 (Nov. 2013).

Therefore, much more work was needed to put human rights on both the political and negotiating agendas, as evidenced at the COP20 in Lima, December 2014. In the last days of negotiations, some Parties called for references to human rights, the rights of indigenous peoples, and gender equality to be included in the 2015 climate agreement. Despite these interventions and various submissions by UN Special Procedures and civil society organizations, calling for human rights to be fully integrated in the climate framework, the draft negotiating text included only one reference to rights in the preamble paragraphs (UNFCCC 2014).

To boost the process, the OHCHR's Key Messages on *Human Rights and Climate Change* (reflected in its submission, *Understanding Human Rights and Climate Change*, to the COP21 in Paris on December 2015) highlighted the essential obligations and responsibilities of states and other duty-bearers (including businesses) and their implications for climate change-related agreements, policies, and actions. In order to foster policy coherence and help ensure that climate change mitigation and adaptation efforts are adequate, sufficiently ambitious, non-discriminatory and otherwise compliant with human rights obligations, the OHCHR addressed various recommendations to be considered when developing the COP21's outcomes (OHCHR 2015).

Moving beyond the UNFCCC process, and even prior to the adoption of the Paris Agreement, there have been important advances in integrating human rights principles in climate funds and institutions. Some REDD+ initiatives have developed policies and guidance to operationalize rights and achieve consistency with the UNFCCC REDD+ safeguards. For example: the UN-REDD Programme adopted guidance for implementing the right to free, prior and informed consent (FPIC); and the UN-REDD Programme and the Forest Carbon Partnership Facility (FCPF) have adopted joint guidelines on stakeholder engagement.[27] In addition, and to varying degrees, at least two climate funds address human rights in their operational policies: first, the Adaptation Fund's (AF) Environmental and Social Policy provides that, "AF-supported programmes, projects and other activities shall respect and, where applicable, promote international human rights"[28]; second, the Green Climate Fund's (GCF) interim environmental and social safeguards – the Performance Standards of the International Finance Corporation – while controversial, contain some references to human rights, including FPIC in certain circumstances.[29] Similarly, the Guidance Note for UN Country Teams (UNCTs) has integrated a human rights-based approach, requiring that UNCTs consider in what ways, and to what extent, anticipated changes in climate will impede economic and social development at relevant levels, including consideration of poverty reduction,

---

[27] UN-REDD Programme, Guidelines on Free, Prior and Informed Consent (Jan. 2013); UN-REDD Programme and Forest Carbon Partnership Facility, Guidelines on Stakeholder Engagement in REDD+ Readiness with a Focus on Indigenous Peoples and Other Forest-Dependent Communities (Apr. 2012).

[28] Adaptation Fund, Environmental and Social Policy (Nov. 2013).

[29] International Finance Corporation, Environmental and Social Performance Standards (Jan. 2012).

strengthening human rights, and improving human health and well-being (UNDG 2010).

3.2   Paris Agreement: A Qualitative Move Forward

According to Bultheel et al. (2016), "the political process introduced by COP21 has enabled a new avenue for multilateral cooperation on climate action. This new process focuses largely on cooperation and inclusivity to encourage all actors, public and private, to commit and act for the climate. In contrast to simply sharing the burden of emissions reduction effort, this dynamic encourages actors to explore and capitalize on benefits and co-benefits of climate action. Overall, this new approach moves away from the constrained climate framework advocated by the Kyoto Protocol, and as a result has encouraged an unprecedented level of climate commitment from both states and non-state actors".

In this context, Höhne et al. (2016) claim that the adoption of the Paris Agreement moved the world a step closer to avoiding dangerous climate change after a long and slow negotiating process. Emerging economies agreed to assume significant commitments to climate action, while developed countries agreed to continue taking the leadership in the form of emission reductions and climate finance for developing countries. This compromise, in addition to other factors, has encouraged many developed countries to support a fast entry into force of the Paris Agreement.[30] It also illustrates how far the world has moved away from the binary North-South divide which gridlocked collective action to address many global challenges, including climate change. The current picture confirms the move away from global governance being led primarily by the United States in alliance with a group of advanced economies. It shows that, in climate governance, a new informal alliance among a group of developed countries and emerging economies is moving the process forward, despite the reticence of half of the members of the G20,[31] developed and emerging economies alike. In addition, the aggregated individual intended nationally determined contributions (INDCs) seemed insufficient to be consistent with the long-term goals of the Agreement of 'holding the increase in global average temperature to well below 2°C' and 'pursuing efforts' towards 1.5 °C. However, the

---

[30] On 5 October 2016, the threshold for entry into force of the Paris Agreement was achieved and officially declared on 4 November 2016. The first session of the Conference of the Parties serving as the Meeting of the Parties to the Paris Agreement (CMA 1) took place in Marrakech, Morocco from 15–18 November 2016. This marks one of the fastest entry into force processes in the history of international treaty law. This was generally possible due to the work of an informal coalition of countries encompassing many of those most vulnerable to climate change impacts, including Small Island States, and an important number of the fifteen largest GHG emitters, both developed as well as emerging countries. Currently, 120 Parties have ratified of 197 Parties to the Convention (http://unfccc.int/paris_agreement/items/9444.php)

[31] The G20 members are jointly responsible for over 74% of global GHG emissions and dominate both global financial resources and investments in technological research and development.

Paris Agreement gives hope that this inconsistency can be resolved. The preparation of the INDCs has also advanced national climate policy-making, notably in developing countries, and this process will be surely instructive and incremental in the long term (Höhne et al. 2016).

In addition to this progress, the Paris Agreement was an adequate opportunity to make explicit reference, inter alia, to non-state actors, climate justice, and human rights. This shift was due to numerous claims, the evolving receptiveness of negotiating parties to such claims, and the available evidence regarding the serious human implications of climate change.

3.2.1 Action by Non–State Actors

Climate change brings complexities to our world that require building strong partnerships between science, industry, community organizations, governments, and indigenous peoples. This is why action by non-state actors has increasingly been acknowledged as an important component of climate policy and governance (Savaresi 2016). Ahead of COP21, an unprecedented UNFCCC portal showcased voluntary emission reductions undertaken by actors such as companies, cities, and subnational governments. The preamble of the Paris Agreement builds upon this increased awareness, recognizing in a treaty for the first time the importance of engaging 'all levels of government' and 'various actors' in addressing climate change. In addition, while the UNFCCC already made generic reference to public participation in addressing climate change and developing adequate responses (UNFCCC, Art. 6), the Paris Agreement specifically emphasizes in its Article 6.8 enhanced public and private sector participation in the implementation of INDCs.

This renewed attention to the role of non-state actors is also evident in the acknowledgement that climate change adaptation "should be based and guided by", inter alia, "traditional knowledge, knowledge of indigenous peoples and local knowledge systems" (Paris Agreement, Art. 7.5). Even though the role of traditional knowledge had already been recognized in previous UNFCCC COP decisions, a reference in the Agreement potentially opens the way to greater inclusion of, and consideration for, traditional knowledge holders in international climate change governance. This potential is corroborated by the decision at the COP21 to establish a platform for the exchange of experiences and sharing of best practices on mitigation and adaptation of local communities and indigenous peoples[32]. Typically, under the climate regime these developments fall within the remit of the COP and subsidiary bodies assisting it with its work (Savaresi 2016).

---

[32] Decision -/CP.21, paras 134–136.

3.2.2 Climate Justice

The Paris Agreement's unprecedented preambular reference to 'climate justice' has been appreciated by the international community as well. During the COP21 negotiations, the climate justice concern attracted much attention. The term has been used to refer to distributive and corrective justice considerations associated both with climate change impacts and response measures. It is therefore inherently linked with equity in the climate regime, and ultimately revolves around how to share the burdens associated with a global transition towards low-carbon development models. In this connection, benefit-sharing may be conceptualized as well as a means to address climate justice.

Regarding the *corrective* justice, the Paris Agreement is the first climate multilateral instrument to make reference to the contentious matter of 'loss and damage' caused by climate change (Art. 8) without, however, shedding a great deal of light on how it will be addressed. During the negotiations, some developed countries successfully asked the COP to specify that the provision in the Agreement on loss and damage "does not involve or provide a basis for any liability or compensation" (Decision -/CP.21, at 58). Instead, the COP21 merely established a process to develop recommendations for approaches to avert, minimize and address displacement, and a clearinghouse for information on insurance and risk transfer (Decision -/CP.21, at 50).

Overall, as far as the *distributive* justice is concerned, the picture emerging from the Paris Agreement is even more opaque. The Agreement does not go very far in establishing new commitments around adaptation and means of implementation. Still, the accompanying COP decision does make reference to a collective quantified goal for support to developing countries from a floor of USD100 billion per year (Decision -/CP.21, at 54). The Agreement also subjects the implementation of developed Parties' obligations concerning the provision of finance (Art. 13.6) to a review process for the first time.

3.2.3 Human Rights

The Paris Agreement is also the first multilateral environmental agreement to make an explicit reference to human rights. It does so by specifying that Parties "should, when taking action to address climate change, respect, promote and consider their respective obligations on human rights, the right to health, the rights of indigenous peoples, local communities, migrants, children, persons with disabilities and people in vulnerable situations and the right to development, as well as gender equality, empowerment of women and intergenerational equity".

While the reference is confined to the preamble, and does not configure new and separate legal obligations for Parties, it is difficult to overlook its importance. While references to human rights had already appeared in previous COPs' decisions, inclusion in the preambular paragraphs of a binding treaty carries political and moral weight, and forging an explicit link between obligations under the climate regime and those under international human rights instruments Parties have already ratified, or may ratify in the future, as well as under domestic law. This means that Parties are expected to interpret and undertake their obligations under the Paris Agreement in light of their existing human rights obligations concerning matters such as public participation and the rights of women and indigenous peoples.

According to Savaresi (2016), the Paris Agreement seems to have opened a new season in international climate governance, fostering the emergence of a cooperative attitude that will potentially give greater consideration to the role of non-state actors in climate action, questions of justice, and the interplay between climate change and human rights instruments. How far Parties to the climate regime will use these new elements to address the challenging equity questions that have opposed them over many years, remains to be seen. Thus, it would be naïve to expect the Paris Agreement to be a miraculous remedy for all the imperfections of the global climate regime.

For Höhne et al. (2016), the next step for the global response to climate change is not only implementation but also strengthening of the Paris Agreement (especially with the current uncertain position of United States regarding their international climate commitments). To this end, national governments must formulate and implement policies to meet their INDC pledges, and at the same time consider how to raise their level of ambition. For many developing countries, implementation and tougher targets will require financial, technological and other forms of support. Nevertheless, despite these various constraints, the new Agreement seems to bode well for the future of this troubled international environmental governance process.

4 Mainstreaming Human Rights into the Global Climate Regime: Relevant Approaches

Human rights-based approaches are a conceptual framework that is normatively based on international human rights standards, and operationally directed to promoting and protecting human rights. These approaches seek to analyze obligations, inequalities and vulnerabilities and to redress discriminatory practices and unjust distributions of power that impede progress and undercut human rights. Under this approach, plans, policies, and programmes are anchored in a system of rights and guided by principles and standards derived from international human rights treaties. This helps promote the sustainability, empowering people themselves (*right-holders*) – especially the most marginalized – to participate in policy formulation and hold accountable those who have a duty to act (*duty-bearers*). These approaches provide benchmarks against which states' actions can be assessed, and offer additional criteria for the interpretation of applicable principles and obligations that states have to each other, to their own citizens, and to the citizens of other states in relation to climate change.

Human rights-based approaches, taken in their entirety, have the potential to bring much needed attention to individual welfare as well as to provide ethical moorings in the climate regime which is still generally characterized by self-interested deal-seeking. There may be various ways to bring human rights to bear on climate protection, but two key approaches deserve special focus in this analysis: either bringing climate protection within the context of the existing 'human right to the environment', litigated in several national and international fora, and thus enforced in discrete cases (Sect. 4.1); or applying a human rights perspective to climate impacts (Sect. 4.2). The latter has been considered by some studies (i.e. Rajamani 2010) as an ambitious approach because: it seeks a reframing of the climate problem in terms of urgency that would provide nations with a 'compass for policy orientation' and draw them towards ever more stringent multilateral actions; and it focuses on the broader range of human rights placed at risk by climate impacts and the ethical influence this might create, rather than exclusively on a human right to the environment and the litigation possibilities that might flow from it.

4.1   The Global Climate Regime from a 'Human Right to the Environment' Perspective

The multilateral environmental dialogue, in its anthropocentricity, has always held the human being firmly at its center, but it is only in the last few decades that environmental protection, which by logical extension encompasses climate protection, has been articulated in the language of human rights. Several international soft[33] law instruments recognize and protect environmental rights whether procedural, derivative or stand-alone ones. In addition, a significant number of national constitutions recognize an environmental right. Indeed, the Ksentini Report (1994), commissioned by the UN Sub-Commission on Prevention of Discrimination and Protection of Minorities, recorded 'universal acceptance' of environmental rights at the national, regional and international levels more than three decades ago.

Even if a few international treaties recognize a stand-alone or explicit human right to the environment, international judicial fora – such as the European Court of Human Rights – have increasingly recognized that environmental harms lead to human rights violations (Ksentini 1994). For Rajamani (2010), such recognition does not provide the evidence of the evolution of a distinctive autonomous right to a healthy environment (which is stuck in controversy), but it is simply an emerging understanding that environmental harms have the potential to impact human rights, such as the rights to health and life or even to respect for private and family life. The UN HRC, as well as the Inter-American Commission and Court on Human Rights,

---

[33] See, for instance, UNGA Res 2398 (XXII) (1968); Preamble, Declaration of the United Nations Conference on the Human Environment, A/CONF 48/14/Rev. 1 (1972), reprinted in (1972) 11 ILM 1416; Hague Declaration on the Environment, 1989, reprinted in (1990) 28 ILM 1308; and UNGA Res 45/94 (1990).

have also considered cases based on environmental harms in the absence of specific environmental rights[34] – however, not without dispute.

Despite the widespread recognition that environmental harms may impact human rights, an explicit human right to the environment – its scope, content, and justiciability, as well as the wisdom of pursuing a human rights path to environmental protection – remains controversial (Boyle 1996; McGoldrick 1996). Rajamani (2010) claims that the scope and content of a human right to the environment are, by their very nature, indeterminate. When applying this right to climate change, it raises more questions than it answers: *first*, to what qualitative level should the environment be protected? In the climate context, what degree of temperature increase would be acceptable, some temperature increase being inevitable? Even limiting global warming to below 2 °C temperature increase would result in serious climatic changes; *second*, against which of the numerous existing standards or benchmarks should the qualitative level of protection be assessed? How, for instance, do we determine the climate impacts that are acceptable and those that are not, given that impacts differ between people and communities?; *third*, even if determinable, to what extent, if at all, should the level of protection be uniform across states regardless of their specificities? Should all states have uniform obligations to protect the climate system, or should their obligations be tailored to their level of economic well-being or their historical responsibility?; *fourth*, who should bear the burden of the correlative duties – states (by themselves or collectively), private actors such as multinational corporations, NGOs, and/or individuals?; and *fifth*, to what extent should these rights be justiciable?

In addition to these pragmatic difficulties in conceptualizing an effective human right to the environment, there is the ethical concern that a focus on environmental protection from a human rights perspective may be excessively anthropocentric, and not accord due consideration to the intrinsic value of the environment, as for instance, the many species that are likely to face extinction as a result of global warming.

Given these numerous uncertainties and complexities in shaping and implementing a workable human right to the environment, the role that such a right can directly play in guiding climate policy and governance seems limited (Rajamani 2010). To the extent that climate litigation, based on a human right to the environment, however, gains ground in national and regional fora, it can serve to shape Parties' positions in the future climate negotiations.

---

[34] See Organization of American States, Inter-Am Commission on Human rights, 'Report on the Situation of Human Rights in Ecuador' (24 April 1997) OEA/Ser L /V/II 96, Doc 10, Rev. 1; Case of the Mayagna (Sumo) Indigenous Community of Awas Tingni, Judgment, (31 August 2001) Inter-Am Ct HR (Ser C) No 79, cited in Rajamani (2010).

4.2 The Global Climate Regime from a Human Rights Perspective

The second approach to mainstream human rights into the climate regime is by applying a broader human rights lens to address climate change. Two elements shape this approach: first, the focus is on those established human rights rather than a disputed human right to the environment, considered as limited, less workable, and vulnerable to the challenge of anthropocentricism; *second*, this approach rests on a broader ethical conception of human rights rather than solely on a legal construction of them. This approach is justified by the fact that numerous internationally protected rights, along with their progressive realization, are and will be severely affected by climate impacts (Rajamani 2010).

In addition, the majority of states, which are now Parties to the core human rights treaties, have obligations to respect, protect and fulfill the rights contained in these treaties, each of these requiring different degrees of public intervention. These obligations are binding on every Party, and must be implemented in good faith. Besides, once a state has ratified the ICCPR, it is not permitted to denounce or withdraw from it. It is true that the extent of enforcement of protected human rights differs from a state to another depending on national circumstances, constitutional culture, political will, judicial creativity, civil society advocacy, and governance mechanisms, but at a minimum, the core human rights instruments set standards and benchmarks in place, and impose process obligations – obligations to integrate human rights concerns into policy planning (Rajamani 2010).

Most of the Parties to the core human rights instruments are also Parties to the current climate regime (mainly UNFCCC, Kyoto Protocol, and Paris Agreement). Therefore, these Parties are in principle obliged to approach climate change not just as a global environmental problem, but also as a human rights concern. To do so, states are obliged to identify and explore the human rights that might be undermined by climate impacts and risks, and take the needed anticipatory and preventive actions in that regard. They are also required to design climate policies and build governance frameworks that are sensitive to the climate impacts that could have implications for the progressive enforcement of protected human rights. Certainly, the language of obligations in the absence of mechanisms for oversight and enforcement may be of limited use; however, to the extent that Parties recognize the existence of such obligations, their positions in the climate negotiations should be tailored to and informed by them (Rajamani 2010).

For a coherence purpose, it would be advisable for states to take their obligations under human rights instruments into account while shaping the future climate regime. If they do not, the performance of obligations under the regime may be inconsistent with, interfere or even impact the performance of obligations under certain human rights instruments. For instance, in designing policy approaches and framing positive incentives within the REDD framework, it is important to take into account the recognition, protection and specific dimensions of the rights of indigenous peoples and communities under human rights treaties.[35] It is on this basis that the OHCHR recommended to the COP16 that REDD and REDD+ programs should adopt a more rights-based approach, create legal awareness programs along with other support programs for indigenous peoples affected by REDD activities, and improve participatory and access to justice provisions (OHCHR, n.d.).

In addition to the obligations that stem from a legal construction of human rights, an ethical conception of human rights can also provide considerable value to the climate debate. The international human rights discourse is premised on: the notion of universality[36]; the notion that universally valid rights are located 'beyond law and history' (Koskenniemi 1999); and the belief that their protection transcends cultural, social, religious, economic, and political context. As such the institution of human rights combines 'law and morality, description and prescription'. This has the potential, as Douzinas (2007) argued, to lead to 'confusion and rhetorical exaggeration'. However, such confusion can be limited as long as a conceptual distinction is maintained between legal and moral human rights.

The content of legal human rights is dependent on the legislative, judicial, and executive bodies that maintain and interpret the laws in question. The validity of moral human rights, on the other hand, is independent of such governmental bodies, and it is indeed respect for moral human rights that imparts legitimacy to the acts of governmental bodies. While legal human rights derive their legitimacy from consent-based sources of international law (such as multilateral treaties), moral human rights derive their validity and rhetorical force primarily from natural law, and only secondarily from consent-based sources of international law. The recognition of moral human rights is significant because it creates the space for a critical assessment of existing international law, free from the narrow formalistic confines of consent-based mechanisms. The recognition of moral human rights may also serve as a catalyst to the legalization of these rights.

It is worth noting in this context that as international lawyers begin to engage with issues such as climate change and human rights, a tension will likely emerge between the formalistic consent-based renderings of international law and the ethically anchored revisionist renderings of international law, and this tension needs to be acknowledged and accommodated. The efforts of ALBA countries to tailor the post-2012 climate regime to fit their alternative ecocentric rights-based and socialistic vision expose such a tension.

---

[35] Bolivia had proposed that the REDD provisions operate within the UN Declaration on the Rights of Indigenous Peoples; see UNFCCC/AWGLCA/2008/16/Rev. 1 (n 25).

[36] See the Preambles of the UNDHR (1948) and the African Charter on Human and Peoples' Rights (1981). See also the Art. 1(1) of the Vienna Declaration and the Programme of Action (1993).

5   Mainstreaming Human Rights into the Global Climate Regime: Why and How?

Given the complexity of the ongoing climate negotiation process, and the low interest from many states regarding the integration of a human rights-based approach in this process, what space can this approach constructively occupy, and what role can it creatively play? (Sect. 5.1). In addition, since the linkages between climate change and human rights are currently more documented and recognized, what should be done on the ground to effectively consider a rights-based approach when developing responses to address climate change? (Sect. 5.2).

5.1   The Advantages of Mainstreaming a Human Rights–Based Approach

During the last decade, many official and non-official statements and declarations converge and confirm the relevance of mainstreaming a rights-based approach into the climate regime. For instance, the International Council on Human Rights Policy (ICHRP) suggested in its 2008 report (*Climate Change and Human Rights: A Rough Guide*) that human rights offer "a shared and codified moral language around which consensus can be built" (Humphreys 2008:22). Elsewhere, the report noted that the human rights discourse can add "considerable normative traction to arguments in favor of strong mitigation and adaptation", and that "human rights provide a legitimate set of guiding principles for global policy because they are widely accepted by societies and governments everywhere" (Humphreys 2008:20). This argument is premised on the notion that the institution of human rights offers a universally shared value system (understood here as a relatively permanent framework that shapes and influences actors' perceptions and behaviors). The conscience-affirming, but self-denying actions, that tackling climate change will require, can only occur if it is predicated on and spurred by a powerful value system. This is an intuitively persuasive argument, in particular if viewed in conjunction with the suggestion made by Argentina in 2008[37] that a rights-based approach could "provide us with a compass for policy orientation".

Along with preventing or minimizing the human implications of climate change, mainstreaming a rights-based approach into the current climate regime has the potential to ensure effective outcomes in managing the climate challenge. This is in line with the statement made by the UNHRC in its resolution 10/4 (2008), "human rights obligations and commitments have the potential to inform and strengthen international and national policymaking in the area of climate change, promoting policy coherence, legitimacy and sustainable outcomes". Indeed, a rights-based approach is critical to achieving the expected outcomes both in the climate

---

[37] UNFCCC/AWGLCA/2008/MISC5(n61)14.

negotiating process and in the development and implementation of response mechanisms on the ground (CIEL and CARE International 2015). Through a focus on empowerment, participation and transparency, this approach helps mobilize individuals and communities in all governance levels and create support for the needed actions. Basic human rights – such as the rights to access to information and full and effective participation in decision making – may increase support for, and public ownership of, climate policies.

A rights-based approach also helps clarify who is responsible for the delivery of key changes and, thus, who can be held accountable where this does not happen or when people are harmed by the actions of those in power. As already recognized by the UNHRC (2008), the "effects of climate change will be felt most acutely by those segments of the population who are already in vulnerable situations owing to factors such as geography, poverty, gender, age, indigenous or minority status and disability"[38]. This approach permits to focus on those populations who suffer the disproportionate impacts of climate change and helps bring attention to systemic issues like inequality, discrimination, and exclusion.

More practically, a rights-based approach can be used to guide climate policies and assess the implementation outcomes. It can inform assessments, strengthen processes, and ensure access to essential information, effective participation, and the provision of access to justice (remedies). Focusing on the rights of those who are already vulnerable and marginalized due to poverty and discrimination, the rights-based approach can be a useful tool to complement international efforts aimed at tackling the adverse effects of global warming.

5.2 What Should be Done to Effectively Mainstream a Rights-Based Approach?

Given the serious human rights ramifications of climate change, states are obliged to take all appropriate means to avoid and mitigate climate change, as well as assist vulnerable communities in adapting to its harmful consequences. States are also required to ensure that their responses to climate change are consistent with their human rights obligations under domestic and international law (Jodoin and Lofts 2013). These obligations are derived from the UN Charter, the UN Declaration on Human Rights (UNDHR), and other international human rights instruments (such as the ICCPR and the ICESCR), according to which states have a duty to cooperate internationally to prevent and address threats to human rights, and this extensively includes the duty to take effective actions such as the adoption and implementation of an agreement that adequately protects against the harms to human rights resulting from climate change.

---

[38] *Outcome of the work of the Ad Hoc Working Group on Long-term Cooperative Action under the Convention*, adopted by the Conference of the Parties to the UNFCCC, Decision 1/CP16, (4 December 2010), preamble [*Cancun LCA outcome*].

Increased attention to human implications of climate change both internationally (climate regime), regionally and nationally (climate policies) may improve the likelihood that climate-related policies contribute to the protection of human rights. Moreover, linking the climate negotiations and structures to existing human rights norms has the potential to enable states to use indicators and mechanisms anchored in the well-established human rights system to address the challenges posed by a changing climate (CIEL 2009, 2013).

In line with the OHCHR' submission (2015) to the COP21, mainstreaming human rights into the climate regime requires various actions to be taken on many levels. The guiding objective is to ensure that concrete policies and measures aimed at addressing climate change enhance, rather than undermine, internationally protected rights such as rights to life, work, family, health, food, water, shelter, education and culture (Jodoin and Lofts 2013). Such an objective is based on existing human rights and climate change instruments to which the majority of states are committed. Examples of what should be done in this regard are provided below:

5.2.1 Mitigating Climate Change, Preventing its Negative Human Rights Impacts, and Enhancing All Persons' Adaptive Capacity

States have an obligation to respect, protect, fulfill and promote all human rights for all persons without discrimination. Failure to take affirmative measures to prevent human rights harms caused by climate change, including foreseeable long-term harms, breaches this obligation. The Fifth Reports of the IPCC (2013, 2014) confirm that climate change is caused by anthropogenic carbon emissions and that its impacts negatively affect, among others, people's fundamental rights. These negative impacts will increase exponentially according to the degree of climate change that ultimately takes place and will disproportionately affect individuals, groups and peoples in vulnerable situations. Therefore, a rights-based approach should be integrated in any response mechanism, such as the promotion of alternative energy sources, forest conservation or tree-planting projects, resettlement schemes and others. More precisely, states must act to limit anthropogenic carbon emissions, including through regulatory measures, in order to prevent, to the greatest extent possible, the current and future negative human implications of climate change.

Similarly, states must ensure that appropriate adaptation measures are taken to protect and fulfill the rights of all persons, particularly those most endangered by negative climate impacts (e.g. individuals and communities living in vulnerable areas such as small islands, riparian and low-lying coastal zones, arid regions, and the poles). Actions should be taken to build adaptive capacities of vulnerable communities, including by recognizing the manner in which factors such as discrimination and disparities in education and health affect climate vulnerability, and by devoting adequate resources to the realization of the economic, social and cultural rights of all persons, particularly those facing the greatest risks.

In the context of mitigation policies for instance, the protection of cultural rights requires that states avoid or minimize policies that could impact these rights. To this end, the protection of cultural rights should be mainstreamed into mitigation policies, and appropriate scoping and risk assessment activities that take cultural rights into account should be undertaken in the context of mitigation actions.

5.2.2 Ensuring Accountability and Effective Remedy for Human Rights Climate–Related Harms

According to many human rights instruments, states are required to guarantee effective remedies for human rights violations. Negative climate change impacts have already inflicted human rights harms on millions of people. For states and communities on the frontline, survival itself is at stake. Those affected, now and in the future, must have access to meaningful remedies, including judicial and other redress mechanisms. The obligations of states in the context of climate change and other environmental harms extend to all rights-holders and to harm that occurs both inside and beyond boundaries. States should be accountable to rights-holders for their contributions to climate change, including for failure to adequately regulate the emissions of businesses under their jurisdiction regardless of where such emissions or their harms actually occur.

5.2.3 Protecting Human Rights from Business Harms

The United Nations Guiding Principles on Business and Human Rights affirm that states have the obligation to protect human rights from harm by businesses, while businesses have a responsibility to respect human rights and to do no harm. Therefore, states must take adequate measures to protect all persons from human rights harms caused by businesses by: ensuring that their own activities, including activities conducted in partnership with the private sector, respect and protect human rights; and where such harms do occur to ensure effective remedies. As duty-bearers, businesses must be accountable for their climate impacts and participate responsibly and ethically in climate actions with full respect for human rights. Public climate policies must ensure, in particular, that climate-relevant decisions by corporations – and the consumers of their products – are guided by prices that take full account of the true present and future costs of GHG emissions (Jodoin and Lofts 2013). Also, where states incorporate private financing or market-based mechanisms within the climate regime (such the CDM, SDM, Joint Implementation, and carbon trading), the compliance of businesses with respective responsibilities is critical and should be permanently monitored.

5.2.4   Mobilizing Maximum Available Resources for Sustainable, Human Rights–Based Development

Under core human rights treaties, states acting individually and collectively are required to mobilize and allocate the maximum available resources for the progressive realization of civil and political, economic, social and cultural rights. The failure to adopt reasonable measures to mobilize available resources to prevent foreseeable human implications of climate change breaches this obligation.

The mobilization of resources to address climate change should complement and not compromise other governments' efforts to pursue the full realization of all human rights for all, including the right to development. Innovative measures such as carbon taxes, with appropriate safeguards to minimize negative impacts on the poor, can be designed to internalize environmental externalities and mobilize additional resources to finance mitigation and adaptation efforts that benefit the poorest and most marginalized.

5.2.5   Ensuring Equity in Climate Action

The Rio Declaration on Environment and Development, the Vienna Declaration and Programme of Action, and '*The Future We Want*' all call for the right to development, which is articulated in the UN Declaration on the Right to Development, to be fulfilled so as to meet equitably the developmental and environmental needs of present and future generations. The UNFCCC calls for states to protect future generations and to take action on climate change on the basis of equity and in accordance with their common but differentiated responsibilities and respective capabilities (Art. 3/1). While climate change affects people everywhere, those who have contributed the least to carbon emissions (i.e. the poor, children, and future generations) are those most affected. Equity in climate action requires that mitigation and adaptation efforts should benefit people in developing and least developed countries, indigenous peoples, communities in vulnerable situations, and future generations.

5.2.6   Guarantying that Everyone Enjoys the Benefits of Science and its Application

The ICESCR states that everyone has the right to enjoy the benefits of science and its applications. All states should actively support the development and dissemination of new climate mitigation and adaptation science and technologies, including those related to sustainable production and consumption. Environmentally clean and sound technologies should be accessibly priced, the cost of their development should be equitably shared, and their benefits should be fairly distributed between and within countries. Science and technology transfers between states should take place as needed and appropriate to ensure a just, comprehensive and effective international response to climate change.

States should also take steps to ensure that global intellectual property regimes do not obstruct the dissemination of mitigation and adaptation science and technology, while at the same time ensuring that these regimes create appropriate incentives to help meet sustainability objectives. The right of indigenous peoples to participate in decision making related to and benefit from the use of their knowledge, innovations and practices should be widely protected.

5.2.7 Mainstreaming Gender Quality in Climate Response Mechanisms

Care should also be taken to ensure that a gender perspective, including efforts to ensure gender equality, is included in all planning for climate mitigation, adaptation, finance, etc. Future climate negotiations should consider further work on the subject of gender equality and equity. For instance, it will be necessary to clarify how to proceed on gender issues when the existing work program (Draft decision -/ CP.20, Lima work programme on gender[39]) expires. The climate conference in Doha in 2012 had called on the parties to staff the committees under the UNFCCC, as well as the national delegations, with women and men in a more balanced manner. Indeed, the delegations in particular, but also the technology and finance expert bodies, remain dominated by men. In the future, gender mainstreaming should not be treated as a separate work program, but integrated into all sectors of the climate negotiations and provided with core UNFCCC Secretariat financial resources to ensure continuous funding of such efforts (Fuhr et al. 2016).

5.2.8 Guarantying Equality, Non–Discrimination, and Meaningful and Informed Participation

States have committed to guarantee equality and non-discrimination. Efforts to address climate change should not exacerbate inequalities within or between states. For example, indigenous peoples' rights should be fully reflected in line with the UNDRIP, and actions likely to impact their rights should not be taken without their free, prior and informed consent. Likewise, the rights of children, older persons, minorities, migrants and others in vulnerable situations must be effectively protected.

Furthermore, many human rights instruments guarantee all persons the right to free, active, meaningful and informed participation in public affairs. This is critical for an effective rights-based climate policy and governance and requires open and participatory institutions and processes, as well as accurate and transparent measurements of carbon emissions, climate change and its impacts. States should make early-warning information regarding climate effects and natural disasters available

---

[39] https://unfccc.int/files/meetings/lima_dec_2014/decisions/application/pdf/auv_cop20_gender.pdf

to all sectors of society. Adaptation and mitigation plans should be publicly available, transparently financed and developed in consultation with affected groups.

Particular care should be taken to comply with relevant human rights obligations related to participation of persons, groups and peoples in vulnerable situations in decision-making processes (including the involvement of local organizations in developing objectives, policies and strategies for the negotiations as well as for ensuring the success of longer-term strategies), and to ensure that adaptation and mitigation efforts do not have adverse effects on those that they should be protecting. Human rights impact assessments of climate actions should be carried out to ensure that they respect human rights. Further, states should develop and monitor relevant human rights indicators in the context of climate change, keeping disaggregated data to track the varied impacts of climate change across demographic groups and enabling effective, targeted and human rights compliant climate action.

5.2.9   Promoting International Cooperation

The UN Charter, the ICCPR, and other human rights instruments impose upon states the duty to cooperate to ensure the realization of all human rights. As a human rights threat with transboundary causes and consequences, climate change requires a global response, underpinned by international solidarity. Therefore, states should share resources, knowledge and technology in order to address this challenge. International assistance for climate change mitigation and adaptation should be additional to existing official development assistance commitments. Pursuant to relevant human rights principles, climate assistance should be adequate, effective, transparent, and administered through participatory, accountable and non-discriminatory processes; it should be targeted toward persons, groups, and peoples most in need. Similarly, states should engage in cooperative efforts to manage climate-induced population displacements and address climate-related conflicts and security risks.

6   Conclusion

In this chapter, I attempt to highlight the relevance and usefulness of mainstreaming a rights-based approach into the climate regime given the strong linkages between climate change and fundamental human rights. An overview of the progress made so far regarding the integration of a rights-based approach into the development and implementation of effective and equitable solutions to climate change has been provided. In addition, actions and measures to be adopted by countries on the ground have been highlighted in line with the claims made by researchers, UN official bodies, and civil society actors.

To conclude, it is worth recalling that climate change has the potential to undermine many achievements made in the areas of human rights, human security, and development. Therefore, considering human rights by the climate regime will not be effective if rights are simply added to the existing frameworks, especially as another preambular recital; they would need to be genuinely integrated and mainstreamed. Existing climate instruments would need to be reinterpreted in a fashion not envisaged at the time when they were negotiated, and the ongoing climate negotiations need to take these into account.

Currently, the existing climate regime, instead of being rights-focused, is still primarily concerned with inter-state burden sharing for a global environmental problem. Any attempt to reshape it and reinterpret its components in a rights-focused manner will not be easy and substantial. Therefore, the role that a human rights approach could play in this setting would have to be carefully tailored to the needs and constraints of the climate regime (Rajamani 2010).

Indeed, while most climate negotiators consider the human rights approaches valuable insofar (especially during the last COPs) as they present a useful complement to the intergovernmental climate process, within the process itself, other negotiators still reject a structural reform of the process based on these approaches. This is due to: the complex and overloaded agenda of the negotiations; the limited space during previous negotiations for new methodological or conceptual approaches; and the reluctance to import the many differences among states on human rights issues into the climate negotiations.

Nevertheless, the recent progress made during the last climate negotiations and the actions taken so far (such as the human rights referential integrated in Paris Agreement) show that the receptiveness to the human rights discourse and related advocacy are evolving; hence it seems realistic to expect a continuing trend towards the full mainstreaming of a human rights-based approach in the future climate regime.

Acknowledgement I would like to express my sincere thanks and appreciation to Ms. Siham Marroune, a junior researcher at the Faculty of Arts and Humanities of Sais-Fes (Morocco), for revising and refining the language of the work.

References

Boyle A (1996) The role of international human rights law in the protection of the environment. In: Boyle A, Anderson M (eds) Human rights approaches to environmental protection. OUP, Oxford, p 43

Bultheel C, Morel R, Alberola E (2016) "Climate governance and the Paris Agreement: the bold gamble of transnational cooperation", Climate Brief, n°40, available at: http://www.i4ce. org/wp-core/wp-content/uploads/2016/11/16-11-03_I4CE-Climate-Brief-40_Cooperative-approaches-Paris-agreement.pdf

Center for International Environmental Law (CIEL) (2008a) Feasibility study: climate change, technology transfer and human rights (prepared for the international council on human rights policy)

Center for International Environmental Law (CIEL) (2008b) A rights-based approach to climate change mitigation: the clean development mechanism of the kyoto protocol (prepared for the International Union for the Conservation of nature)

Center for International Environmental Law (CIEL) (2009) Human rights and climate change: practical steps for implementation., available at: www.ciel.org/Publications/CCandHRE_Feb09.pdf

Center for International Environmental Law (CIEL) (2011) Analysis of human rights language in the cancun agreements., available at: http://www.ciel.org/Publications/HR_Language_COP16_Mar11.pdf

Center for International Environmental Law (CIEL) (2013) Climate change and human rights: a primer., available at: www.ciel.org/Publications/CC_HRE_23May11.pdf

Center for International Environmental Law(CIEL) and CARE International (2015) Climate change: tackling the greatest human rights challenge of our time- recommendations for effective action on climate change and human rights., http://www.carefrance.org/ressources/themas/1/4566,CARE_and_CIEL_-_Climate_Change_and_.pdf

Douzinas C (2007) Human rights and the empire. Routledge-Cavendish, New York

Fuhr L, Schalatek L, Ilse S (2016) 'Morocco must breathe life into the Paris Agreement', (31 Oct), available at <https://translate.google.com/#auto/en/Lili%20Fuhr%20Liane%20Schalatek%20Simon%20Ilse> (visited 1 Nov 2016)

Gupta H. (2015) Women and climate change: linking ground perspectives to the global scenario. Indian J Gend Stud 22(3):408–420. https://doi.org/10.1177/0971521515594278

Höhne N, Kuramochi T, Warnecke C, Röser F, Fekete H, Hagemann M, Day T, Tewari R, Kurdziel M, Sterl S, Gonzales S (2016) The Paris agreement: resolving the inconsistency between global goals and national contributions. Clim Pol:1–17

Human Rights Committee (HRC) (1982) Covenant on economic, social and cultural rights, General Comment No. 6, Article 6, Sixteenth session, 1982

Human Rights Committee (HRC) (2000) Covenant on economic, social and cultural rights, General Comment 14, The right to the highest attainable standard of health, E/C 12/2000/4

Humphreys S (2008) The human rights dimensions of climate change: a rough guide, international council on human rights policy, Geneva

IPCC (2007) Climate change 2007: impacts, adaptation and vulnerability, contribution of working group II to the fourth assessment report. CUP, Cambridge, p 18

IPCC (2013) Summary for policymakers. In: Stocker TF, Qin D, Plattner G-K, Tignor M, Allen SK, Boschung J, Nauels A, Xia Y, Bex V, Midgley PM (eds) Climate change 2013: the physical science basis. Contribution of working group I to the fifth assessment report of the intergovernmental panel on climate change. Cambridge University Press, New York

IPCC (2014) Climate change 2014: synthesis report. Contribution of working groups I, II and III to the fifth assessment report of the intergovernmental panel on climate change [Core writing team, Pachauri RK, Meyer LA (eds)]. IPCC, Geneva, Switzerland, 151 pp

Jagers SC, Stripple J (2003) Climate governance beyond the state. Glob Gov 9(3):385–400

Jane J (2006) ICC climate change petition rejected, NUNATSIAQ NEWS. Available at: http://www.nunatsiaq.com/archives/61215/news/nunavut/61215_02.html

Jodoin S, Lofts K (2013) Economic, social, and cultural rights and climate change: a legal reference guide. CISDL, GEM and ASAP, New Haven

Koskenniemi M (1999) The preamble of the universal declaration on human rights. In: Alfredsson G, Eide A (eds) The universal declaration of human rights: a common standard of achievement. Kluwer Law International, The Hague

Ksentini F (1994) Human rights and the environment, final report, E CN 4/Sub 2/1994/9, Annex III, 81-9

Lewis D, Ballard K (n.d.) Disability and climate change - understanding vulnerability and building resilience in a changing world. CBM available at: https://www.cbm.org/article/downloads/54741/Disability_and_Climate_Change.pdf

Limon M (2009) Human rights and climate change: constructing a case for political action. Harvard Environmental Law Review 33:439–476 available at: http://www.law.harvard.edu/students/orgs/elr/vol33_2/Limon.pdf

Mary Robinson Foundation (2014) Climate justice, incorporating human rights into climate action, available at: http://www.mrfcj.org/pdf/2014-10-20-Incorporating-Human-Rights-into-Climate-Action.pdf?v=2

McGoldrick D (1996) Sustainable development and human rights: an integrated conception. ICLQ 45:796–811

McInerney S, Lankford MD, Lavanya R (2011) Human rights and climate change. In: A review of the international legal dimensions. The World Bank, Washington, DC

Meridian Institute (2009) Reducing emissions from deforestation and forest degradation (REDD): an options assessment report, www.REDD-OAR.org (prepared for the Government of Norway)

Office of the UN High Commissioner for Human Rights (OHCHR) (2015) Understanding human rights and climate change, submission of the OHCHR to the 21st conference of the parties to the United Nations framework convention on climate change, available at: http://www.ohchr.org/Documents/Issues/ClimateChange/COP21.pdf

Office of the UN High Commissioner for Human Rights(OHCHR) (n.d.) Applying a human rights-based approach to climate change negotiations, policies and measures., available at: http://www.ohchr.org/Documents/Issues/ClimateChange/InfoNoteHRBA.pdf

Organization of American States (OAS) and Inter-American Commission on Human Rights (IACHR) (2005) Petition seeking relief from violations resulting from global warming caused by acts and omissions of the United States, available at http://www.inuitcircumpolar.com/files/up-loads/icc-files/FINALPetitionICC.pdf

Rajamani L (2010) The increasing currency and relevance of rights-based perspectives in the international negotiations on climate change. J Environ Law 22(3):391–429

Sachs W (2007) Climate change and human rights, World economy and development in brief, WDEV special report 1/200, 3 5, http://www.wdev.eu/4

Savaresi A (2016) The Paris agreement: an equity perspective., available at: http://www.benelex-blog.law.ed.ac.uk/2016/01/29/the-paris-agreement-an-equity-perspective/

UN Human Rights Council (UNHRC), Res. 7/23 (2008) in U.N. Human rights council, report of the human rights council on its seventh session, at 65, U.N. Doc. A/HRC/7/78 (July 14)

UN News Centre (2008) Secretary-General Ban Ki-moon: opening remarks at news conference, <http://www.un.org/apps/news/infocus/sgspeeches/search_full.asp?statID=176>

UN Practitioners' Portal on Human Rights Based Approaches to Programming (2003) The human rights based approach to development cooperation: towards a common understanding among UN agencies, online: HRBA portal <http://hrbaportal.org/the-human-rights-based-approach-to-development-cooperation-towards-a-common-understanding-among-un-agencies>

UNEP, The Adaptation Gap Report (2014) A preliminary assessment report (Nov 2014), available at: http://www.unep.org/climatechange/adaptation/gapreport2014

UNFCCC (2014) Elements for a draft negotiating text, as contained in COP20 decision "Lima Call to Climate Action", available at: http://unfccc.int/les/meetings/lima_dec_2014/application/pdf/auv_cop20_lima_call_for_climate_action.pdf

UNFCCC AWG-LCA (2008) Ideas and proposals on the elements contained in paragraph 1 of the Bali Action Plan, Submissions from Parties', UNFCCC/AWGLCA/2008/ MISC 5, 13

UNFCCC AWG-LCA (2009) 'Revised Negotiating Text', UNFCCC/AWGLCA/2009/Inf 1, 8

UNFCCC AWG-LCA (2010) Submission by the plurinational state of Bolivia to the AWG-LCA, 'Additional views on which the Chair may draw in preparing text to facilitate negotiations among Parties, Submissions from Parties' FCCC/AWGLCA/2010/MISC 2

United Nations (1948) United Nations Human Rights Declaration, General Assembly resolution 217 A (III) of 10 December.

United Nations Development Group (UNDG) (2009) Human rights-based approach to development programing (HRBA), online: United Nations Development Group <http://www.undg.org/?P=221>

United Nations Development Group (UNDG) (2010) Integrating climate change considerations in the country analysis and the United Nations Development Assistance Framework (UNDAF) -

A guidance note for United Nations country teams, available at: https://undg.org/wp-content/uploads/2014/06/UNDG-GuidanceNote_ClimateChange-July2011.pdf
United Nations Framework Convention on Climate Change (UNFCCC) (2015) Paris agreement, adopted in 12 Dec 2015, in force 4 Nov 2016, https://unfccc.int/resource/docs/2015/cop21/eng/l09r01.pdf
United Nations Permanent Forum on Indigenous Issues (UNPFII) (2008) Report on the seventh session, economic and social council, E/2008/43, e/c.19/2008/13 (Apr. 21-May 2) at 3-4
Watt-Cloutier S (n.d.) Global warming and human rights, EarthJustice and the Center for International Environmental Law (CIEL), available at: http://srenvironment.org/wp-content/uploads/2013/05/CIEL-and-Earthjustice.pdf
World Health Organization (WHO) (2011) World report on disability, available at: http://www.who.int/disabilities/world_report/2011/en/index.html
World Health Organization (WHO) (2013) The world health report 2013: research for universal health coverage, available at: http://apps.who.int/iris/bitstream/10665/85761/2/9789240690837_eng.pdf?ua=1

Chapter 2
Risks to Human and Environmental Security, Well-Being and Welfare: What are the 'Right' Indicators, How Are They Measured and Why Are They Only Rarely Used to Guide Policies?

Hartmut Sommer

Abstract This work starts with the classification of different forms of individual and systemic risks and their basic measurement. Thereafter, perceptions and attitudes towards risks will be analyzed. Reasons for the over- or underestimation of risks, basic risk behavior (risk seeking versus risk avoiding personalities and strategies) as well as recommendations to overcome possible misperceptions will be forwarded. The actual problems of non-sustainable development in the world are partially due to ill-conceptualized and misleading indicators for risks. The Gross Domestic Product (GDP) is used as one key objective for government policies in nearly all countries. Its deficiencies are well known and scientific consensus exists that other indicators are needed. Such sets of sustainability-oriented indicators and conformingly derived proposals for change exist since many years and will be resumed. Reasons why these indicators are only rarely used in official politics need to be discussed: errors, scandals, biased approaches ('wishful thinking'), complex models, false forecasts or simply missing knowledge about the future and bad data analysis. Examples from economy, ecology and demography are presented. Measures needed to overcome at least some of these deficiencies will be proposed claiming for risk management systems, robust decision making, and ethical principles guiding individual and collective behavior.

Keywords Risks · Well-being · Climate change · Forecasts · Scenarios

H. Sommer (✉)
Department of Life Sciences and Engineering, University of Applied Sciences Bingen, Bingen, Germany

© Springer International Publishing AG, part of Springer Nature 2019
M. Behnassi et al. (eds.), *Human and Environmental Security in the Era of Global Risks*, https://doi.org/10.1007/978-3-319-92828-9_2

1   Classification and Measurement of Risks

This work starts with the classification of the different forms of individual and systemic risks and their basic measurement. Thus, the first question to be asked is 'what can happen?'; the second 'what is the probability of the risk?'; and the third question concerns the possible effects of a risk or the gravity of the risk.

1.1   The Different Types of Risks

Probably the word 'risk' was created in ancient times when ships crossed the Mediterranean Sea. The Latin word 'resecum' means the cliff that has to be avoided when sailing on the sea. Probably one of the first certificates of insurance covered maritime risks (Nguyen and Romeike 2013:51). In common language, there is no fundamental difference made between the terms 'risk', 'hazard', and 'uncertainty'.

However, science in general, and statistics in particular, mean by 'uncertainty' events without being able to indicate the probability of the event. Events where such a probability exists are called 'risks'. While uncertainty and risk are neutral terms with possible positive or negative outcomes, 'hazard' characterizes the danger of an event or an object having negative effects (i.e. a substance causing cancer). As the calculation of probabilities is based on the frequency of past events, the derived probability for future events is only reliable if there are no fundamental changes in factors influencing or controlling events (e.g. the probability of heavy rainfalls or floods may increase due to climate change making past counts obsolete). Thus, it would be from a terminological point of view more appropriate to use the term 'uncertainties' instead of 'risks'. Being scientifically very correct would mean to reserve the term 'risks' to lotteries and games with clear rules where, by means of mathematical probability analysis, in fact probabilities are valid. However, many areas in practice show sufficient regularities (e.g. in biology, medicine, agriculture, and sports) that justify the (prudent!) use of past frequencies of events for calculating probabilities and confidence intervals for future events (see Sect. 1.2). Therefore, mostly in this chapter, the term 'risk' in the sense of possible, but 'uncertain events' will be applied in accordance with common language and the topic of this volume.

In this chapter, it will be argued from a human anthropocentric perspective. It does not deal with risks to other species or nature, however not excluding ecological aspects as biodiversity that are considered to be important also for humans and their survival.

Literature contains voluminous catalogues of risks in different areas. Useful classifications might be the separation of natural and manmade risks on the one hand, and individual and collective/systemic risks on the other hand. Natural risks comprise earthquakes, tsunamis, volcano eruption, sun-activity, meteorites, floods, droughts, weather anomalies like El Niño, etc. Manmade risks, probably much more frequent, include political, ecological, technical, health, economic, financial, demographical, social, personal security, gender and privacy risks, risks to justice, risks for minorities and others (Proske 2008; Renn 2014).

Some of these risks are caused by individual behavior (as smoking or obesity), others by the behavior of groups of humans or the setting of society (collective or systemic risks). Also, the effects of risks are either borne by individuals, groups of people, nations or even the whole mankind. A further distinction might be made according to the gravity of the risk from influencing mostly the comfortableness of one or a few people (i.e. noise in the neighborhood of wind turbines, airfields or factories) to affecting the future welfare and well-being of whole nations or mankind (climate change, biodiversity...).

1.2   The Probability of a Risk

After having answered the question 'what event or risk can happen', the next question to ask is 'how high is the probability or chance that a risk comes true'. For many events or risks, statistical information is available based on the frequency of events or calculations (Gigerenzer 2013; Gigerenzer 2014; Proske 2008; Renn 2014). However, a possible and rather common misunderstanding of such statistical data is – especially concerning man-made risks – that for example a very low risk for instance of a nuclear meltdown means that such an event will practically never take place in human's lifetime. The catastrophes of Chernobyl and Fukushima within a period of 25 years only show that such calculations might be wrong and misleading. Also, the fact that it took nearly 200 years between two very big earthquakes in Haiti does not exclude the possibility that there will be an earthquake in Haiti in the near future, thus justifying – eventually – even rather costly preventive measures or construction standards.

1.3   The Effects and Measurement of Risks

After knowing the possible events, respectively risks and their probability, the negative and/or positive effects of such events must be estimated, if possible and calculated in monetary or other concrete terms. Basically, the questions asked are: 'What is at risk and how big are the possible damages and losses?'. Standard financial analysis proposes the concept of actuarial expectations or values where the probability of an event is multiplied by its possible monetary positive or negative effect and summed up for all possible events. Respective values are used in cost-benefit analysis discounting future payments and earnings into net present values. Thus, the gravity of a risk can be measured and precautionary expenses can be evaluated. However, such calculations are rather inappropriate when dealing with human life, injuries or biodiversity. In those cases, other methods and indicators are requested as for example absolute life expectancy, days of lost life expectancy, mortality, number of years of 'healthy' life, quality adjusted life years, number of species at risk of extinction, etc. Especially the mortality indicator is widely used for measuring the risk exposure of different activities or events. Proske (2008:177–180) has established a very detailed catalogue of about 150 situations/events or risks ranging from the mortality of a new born in Mali which was about 10% in 2004 p.a. until the chance being killed by a meteorite (6× 10–11 p.a.). For illustrating certain risks, the indicator 'days of lost life expectancy' is also very impressive: according to Proske (2008:234), alcoholism costs 4000 days and poverty in France between 2555 and 3650 days. Not very convincingly – because positive health effects are not counted – jogging takes 50 days and hiking 0,9 days.

If human life is threatened, the upper tolerated limit of a risk per year, the so-called de minimis risk, for a breakdown of any kind of buildings or other technical devices usually is defined as 10–6 (Proske 2008:188). As the exposure time to a certain risk is quite different, the so-called 'fatal accident rate' calculates the risk assuming an identical exposure time. For example, the risk of a jockey to have a serious, even deadly accident is rather high, but the time he is riding the horse is very short, thus leading to a still acceptable fatal accident rate.

Risks of groups of people are called 'collective' risks. For example, the number of people being killed in a plane crash is usually bigger than in a car crash. The concept of cumulative Γ-N-diagrams gives an idea of risks of different events, and was widely used in designing and comparing technical devices including nuclear reactors (Proske 2008).

Even if risks are evaluated by indicators like mortality, cost considerations come into the focus of decision makers when deciding about preventive measures (i.e. appropriate construction methods regarding the prevention of earthquakes versus financing of development activities).

2 Perceptions and Attitudes Towards Risks

Perceptions and attitudes towards risks will be analyzed as they vary considerably on the individual and on the systemic level. Reasons for the over- or underestimation of risks, basic risk behavior (risk seeking versus risk avoiding personalities and strategies), as well as recommendations to overcome possible misperceptions, will be forwarded.

As was discussed above in different areas of science and practice, the term 'risk' is defined quite differently, usually with negative connotations as fear, danger or the (mostly unknown) probability of loss or damage, but also positive meanings as 'opportunity', 'chances' or 'adventure', nicely expressed by the slogan 'no risk – no fun' of mostly young and hedonistic people (Nguyen and Romeike 2013). Not only in the Middle Ages, risk and adventure were considered by parts of the population as a mean to increase honor and self-esteem (for example when knights were fighting against each other for winning the hearts of beautiful ladies). It seems to me that modern followers or adepts of these medieval knights can be found easily nowadays in international politics.

Risks are quite often perceived differently from one person to another. For example, some kinds of acoustic waves might be classified by one individual as noise, and by somebody else as music, wind turbines as positive instruments to fight against climate change or a risk to the natural environment and as bird killers.

Risk attitude depends on the possibility to decide on risk exposure: voluntarily people take and accept rather high individual risks (e.g. in sports or when smoking), but are much more sensitive about collective risks (Proske 2008). According to Renn (2014), individual risks are influenced to 2/3 by individual behavior and 1/3 by the environment. People generally underestimate systemic risks, for example financial crisis, migration, and climate change effects.

People overestimate the capability to deal with risk (for example car drivers, where a very big majority thinks that they are better drivers than the average driver) combined with the underestimation of individual risks (e.g. concerning exposure to accidents and diseases) and overestimation of collective completely or nearly completely uncontrollable natural or other risks (tornados, flooding, contaminated 'poisoned' food, risk of vaccination). However, this doesn't necessarily influence the behavior of people: regions with natural hazards (e.g. California with its high risk of earthquakes) are very much appreciated by people as working and living environment (Proske 2008). People ignore such risks because social risks (like losing a job, income or friends) strongly influence the decisions of people rather than very abstract natural risks.

Basic individual alternatives in dealing with risks are (Nguyen and Romeike 2013): acceptance (e.g. not changing driving style); avoidance (not driving with cars); reduction (driving prudently); limitation of possible damage (using a car that protects passengers); and transfer of risks to others (car and health insurance).

Risk reduction measures (e.g. better roads) do not necessarily lead to a lower rate of accidents as people adapt their behavior and e.g. drive faster on roads or drive faster in winter due to ABS[1] (Schmidt-Bleek 2014; Paech 2013).

The perception of risks depends heavily on the presence of the risk in the media. For example, at the beginning of the century the fear of the 'Creutzfeld-Jacob' disease due to beef consumption was shared by many people and governments. Imports of meat were blocked between many countries. However, in Europe only 150 people died from this 'Mad Cows Disease' within 10 years, approximately the same number dying from drinking perfumed lamp oil (Gigerenzer 2014:300). Another example of a problematic perception of risk is even more alarming: after the terrorist attack on the twin towers on September 11, 2001, many Americans (and not only Americans!) were afraid of flying, and decided to take instead the car as transport medium. According to Gigerenzer (2014:22), this decision produced very probably 1.600 additional deaths because taking a car is much more risky and causes more accidents and higher mortality rates than going by plane.

---

[1] ABS is the acronym of the German word 'Antiblockiersystem'.

Also, the time since the last accident of the same type counts. The perception of the risk diminishes as longer as a disaster attributed to a certain risk took place (Proske 2008:174).

On one hand, it seems natural that individuals and humanity try to minimize risks. On the other hand, taking risks is the very essence of an individual's life when choosing a partner, selecting a job, developing hobbies or investing money. Also, societies as a whole took big risks in the past – unaware of consequences and the gravity of their decisions – as they developed nuclear reactors or even conventional coal industries. Up to now, probably many more people died in coal mines and from the induced air pollution than from nuclear energy, but the potential and fear of much bigger catastrophes with nuclear reactors than thermal power plants motivated the German government to close down the nuclear reactors rather quickly, but only in the medium future, conventional thermal power plants will follow.

From the perspective of climate change, the whole process of industrialization can be seen as a very risky endeavor producing more and more greenhouse gases each year, demonstrated by the annual report of the World Meteorological Organization (WMO) on greenhouse gas concentrations (WMO 2015b). However, this process has led to huge improvements in the living conditions of many people as measured by conventional indicators, thriving development all over the world to new records and motivating the countries of the Southern hemisphere to copy the industrialization and modernization process of the so-called 'developed' countries. As the thriving factor of this development is the desire to achieve well-being and welfare everywhere, indexes and indicators for measuring well-being and welfare, but also the risks to human and environmental security, will be discussed in the next section.

3 Indexes and Indicators for Human and Environmental Security Well-Being and Welfare

The actual problems of non-sustainable development in the world are partially due to ill-conceptualized and misleading indexes and indicators measuring human objectives. For example, in conventional economic theory and politics Gross Domestic Product (GDP) is still used as a proxy for welfare (3.1). Sets of sustainability oriented indicators and conformingly derived proposals for change exist since many years and will be discussed (3.2).

3.1 The Gross Domestic Product

A negative or only small growth rate of the GDP per capita, respectively the Gross National Income (GNI) per capita, is still seen as a major risk for welfare, and therefore GDP/GNI is used as one key objective for government policies in nearly all countries. Its deficiencies are well known. Some of them are (Diefenbacher and Zieschank 2011:13ff):

- Many products are not accounted for as part of GDP such as domestic labor, activities of volunteers, black market or informal market activities.
- Costs are accounted for goods or services – as in the case of expenditures for hospitals, prisons or defense – disregarding the fact that these costs are the result of accidents, diseases, crime, wars, etc. that in general cause a lower welfare for the big majority of the population.
- Long-term effects of economic activities, as the use of raw material and effects of climate change, are not considered. Implicitly, it is taken for granted that scarcity of resources and negative environmental effects will result in price increases of scarce and polluting material, thus leading to their replacement or at least more efficient usage. The price evolution of raw materials shows clearly that market mechanisms are not working in this sense. Also, efficiency gains are often outweighed by rebound effects, for example lower consumption of fuel stimulates sales of Sport Utility Vehicle (SUV) with a quite high consumption (Paech 2013).
- The (unequal) distribution of income, fortune and welfare is not taken into account by the GDP concept neither within nations nor between them. To compensate the unequal distribution and satisfy poorer sections of an economy, growth of the GDP is necessary as a vehicle for generating a 'trickling down effect' of income, and thus reducing poverty. However, the result of this process is often an even more unequal distribution.
- The growth rate of the GDP is by itself problematic in a finite world. For example, the mathematical effect of a constant growth rate of 3% is equal to a doubling of GDP within 23.5 years. The actual growth rate goals, and also the real growth rates in many developing and emerging countries, were at least in the past much higher (common political strategies suggest that this is also necessary for reducing poverty – see above). As a consequence, huge ecological, social, economic, and also political problems have evolved in many countries. After the financial crisis, many countries have stimulated growth of the GDP by deficit spending programs for generating demand, thus increasing public debts to very high levels. This raises doubts if those levels are still sustainable or will lead at least in some cases (the case of Greece for example) to bankruptcy of whole states.

So, the growth ideology has been criticized heavily asking for a more balanced growth or even replacing 'quantitative' non-sustainable growth (this means GDP growth) by 'qualitative' growth with a constant or even decreasing GDP where economies are restructured for becoming sustainable (Miegel 2011). Alternative indicators were proposed for example by the Stiglitz-Sen-Fitoussi Commission appointed by the French government. They presented their report in 2010 suggesting that indicators and indexes should cover the following domains: material living standards (income, wealth), health, education, personal activities, political voice and governance, social relationships, and environmental sustainability (Stiglitz et al. 2010). Such indicators exist in the framework of the so-called happiness indicators which will be introduced and discussed in the next section.

3.2 Gross National Happiness, Human Development, Legatum Prosperity and Happy Planet Index

Many different alternative indicators have been developed since 1970 when Bhutan introduced its first Gross National Happiness indicator (GNH) (Centre for Bhutan Studies and GNH Research 2015). Recently, other countries followed Bhutan and created their own indexes, e.g. the Canadian index of well-being (2011) (University of Waterloo 2015), the Australian unity wellbeing index (Australian Unity 2015), and the United Kingdom (UK) national well-being index (2014) (Office for National Statistics 2015; Self 2014). The advantage of the national indexes is that they can be tailored ideally for the local situation reflecting cultural, ethnical or even religious characteristics. The disadvantage is the non-comparability on the international level.

Thus, in this analysis the objective is to present and discuss three international indexes: Human Development Index (HDI) (UNDP 2015a); Happy Planet Index (HPI) (Centre for Well-being at NEF 2015); and Legatum Prosperity Index (LPI) (Legatum Institute 2015). These indexes represent rather different approaches and illustrate the different dimensions of happiness on the international level. There are many more other indexes with more or less the same objective, but also very different approaches like the OECD Better Life Index (2011)[2], World Happiness Report (2012)[3], and the Quality of Life Index[4].

This rather big confusing number of different happiness measurement approaches is similar to other domains, as poverty analysis, where at least seven different concepts are used by statistical authorities in one country (Benin), leading often to logically non-convincing results (Sommer et al. 2013, 2014) (see below). The same might be true for the happiness indicators where the results are presented in the form of rankings of the rated countries (Table 2.1)[5].

As can be seen in Table 2.2, there is a very strong correlation between GNI and HDI. Probably not much astonishingly, it can be concluded that high income countries are also those with a rather long mean period of schooling and high life expectancies, the opposite being true for the countries with a low GNI. The same is true for the LPI, only to a slightly lower extent.

More astonishingly, the correlation between HDI and LPI is also very high (0,929). Taking into account the fact that LPI is calculated from 89 indicators and

---

[2] http://www.oecdbetterlifeindex.org/.

[3] http://worldhappiness.report/.

[4] http://www.numbeo.com/quality-of-life/rankings_by_country.jsp

[5] The detailed calculation of the correlation of these rankings can be forwarded by the author to everybody on demand.

Table 2.1 HDI, HPI and LPI

Index	Human development index (HDI)	Happy planet index (HPI)	Legatum prosperity index (LPI)
First year	1990	2006	2010
Domains	economy, health, education	health, well-Being environment	prosperity, economy, entrepreneurship and opportunity, governance, Education, health, personal freedom, social capital, safety and security
Indicators	GDP / capita, Life expectancy, mean years of schooling, average years of schooling	Life expectancy, experienced well-being, ecological footprint.	89 variables out of 200 were selected
Organization	UNDP	New Economics Foundation	Legatum Institute
Weighting method	Geometric mean of normalized indices for each of the three dimensions,	Individual formula: footnote (b)	Weights were determined by regression analysis at the start (2010)
Data sources	UNDP Human Development Report, UNESCO	Gallup World Poll, UNDP Human Development Report 2011, WWF	Official sources and "commercial providers", details not provided
Number of countries	185	151	142
Gender issues	No	No	–
Sources	http://hdr.undp.org/en/content/human-development-index-hdi	http://www.happyplanetindex.org/about/	http://www.prosperity.com
Time series analysis	Yes	No	No
Normalization of variables / sub-indicators	All: upper and lower limits defined by UNDP Footnote (a) GNI: logarithm of income, to reflect the diminishing importance of income with increasing GNI	Footnote (b)	Variables are standardized by subtracting value from the mean and dividing by the standard deviation.
Latest update	2013	2012, Footprint 2007 Footnote (c)	2013

[a]Reference values: life expectancy: lower limit: 20 and upper limit: 85 years; mean years of schooling 18, average years of schooling: 15, GNI: upper limit 75.000 USD, lower limit: 100 USD

[b]Complete calculation of HPI: Happy Planet Index $= \Omega \dfrac{(\text{Ladder of life} + \alpha \text{ Life expectancy}) - \pi}{(\text{Ecological Footprint} + \beta)}$

where: $\alpha = 2.93$, $\pi = 4.38$, $\beta = 73.35$, $\Omega = 0.60$, Ladder of Life (individual subjective classification from 1 to 10): researched by surveys from the Gallup Institute: http://www.happyplanetindex.org/assets/happy-planetindex-report.pdf

[c]Newest footprint data (2011): http://www.footprintnetwork.org/ecological_footprint_nations/

Table 2.2 Correlations between GNI, HDI, HPI, LPI and ecological footprint

	HDI	HPI	LPI	Ecological footprint
GNI	0,968	0,258	0,894	−0,884
HDI		0,332	0,929	−0,863
HPI			0,328	−0,001
LPI				−0,846

Table 2.3 HDI time series of selected countries

Growth rank	HDI rank	Country	1980	1990	2013	Growth rate
1	136	Cambodia	0,251	0,403	0,584	2,59%
2	169	Afghanistan	0,230	0,296	0,468	2,18%
3	176	Mali	0,208	0,232	0,407	2,06%
7	91	China	0,423	0,502	0,719	1,62%
16	129	Morocco	0,399	0,459	0,617	1,33%
69	6	Germany	0,739	0,782	0,911	0,64%
73	20	France	0,722	0,779	0,884	0,62%
131	5	USA	0,825	0,858	0,914	0,31%
132	57	Russia		0,729	0,778	0,28%
139	133	Tajikistan		0,610	0,607	−0,02%

Data source: UNDP (2015a)

HDI only from 4, this shows that indeed life expectancy, mean years of schooling and GNI are decisive factors for happiness (as measured by LPI). A detailed cluster analysis study by Otoiu et al. (2014:581) reveals that the best explanatory variables, that is the indicators with the strongest influence on happiness indexes are: CO_2 emissions per capita; GNI per capita; life expectancy at birth; mean years of schooling; total participation rates; share of fossil fuels; income Gini coefficient; and well-being.

On the other hand, the correlations between HPI and HDI are rather small. Also the correlations between HPI and LPI and HPI and GNI are low. This is due to the influence of the ecological footprint that is strongly negatively correlated to GNI: the richer the country the higher the footprint.

As HPI and ecological footprints are practically not correlated, the message of the HPI is not so clear. Rich countries reach medium ranks while medium countries as Costa Rica and Vietnam reach the highest ranks due to their lower footprints. Countries with lower ranks in HDI normally don't change their low positions due to low sub-indicators for wellbeing and for life expectancy.

It results from the discussion that HDI together with the ecological footprint are theoretically the best indexes and indicators for describing the situation of a country[6]. The HDI indicator allows intertemporal comparisons of countries (Table 2.3),

---

[6] As the latest available ecological footprint indicator dates from 2007 with an update in 2010, it is proposed to replace this indicator by an indicator measuring the CO2 emissions. Both indicators are strongly correlated.

Table 2.4 Comparison of seven selected countries

	Germany	Saudi Arabia	Russia	Costa Rica	South Africa	Vietnam	Morocco
HDI rank	6	34	57	68	118	121	129
GNI per cap.	43,049	52,109	22,617	13,012	11,788	4892	6905
Male life expectancy	78,3	73,9	61,8	77,8	54,7	71,3	69,1
Femal life expectancy	83,1	77,6	74,4	82,2	58,8	80,5	72,7
Adult literacy rate		87,2	99,7	96,3	93	93,4	67,1
Under 5 mortality	3	7	9	9	33	18	27
Obesity	14,7			8,1	19,2	4,4	10,7
Overall life satisfaction index	6,7	6,5	5,6	7,3	5,1	5,5	5
Footprint (2010?)	4,57	3,99	4,40	2,52	2,59	1,39	1,32
Female participation rate	53,5	18,2	57,0	46,4	44,2	72,8	43,0

Data sources: UNDP (2015a) and Statistisches Bundesamt (2014)

as well as comparisons across countries (Table 2.4). Probably not very astonishingly, developing and emerging countries are showing the highest annual growth rates. However, at least in the case of Afghanistan, this might be a rather unsustainable situation produced by the support of the allies due to the war against the Taliban. Also, the situation in Mali is not stable as it is still under attack of terrorists. Probably more astonishingly, the USA and Russia are nearly at the end of the ranking showing only half of the growth rate of Germany and France.

The comparison of seven very different countries from Germany to Morocco (Table 2.4) reveals that countries may show very different performances. Sometimes high GNI per capita hides shortages and deficiencies in other areas: Saudi Arabia with one of the highest GNI per capita in the world has a very low female labor participation rate and a lower literacy rate than South Africa. Only in South Africa, male life expectancy is lower than in Russia. Overall life satisfaction in Costa Rica is significantly higher than in Germany with less than one third of the German GNI per capita and similar life expectancies. Morocco shows only slightly lower values of the life satisfaction index than countries like South Africa and Russia with 50% or less of the footprint of South Africa and Russia.

In general, the detailed comparison of countries shows that high ranks in GNI do not necessarily mean higher life standards of the population: multiple reasons from gender issues to human rights and freedom of speech would probably motivate many people to live in Costa Rica (HDI rank 68) or in Morocco (rank 129) than in Saudi Arabia (rank 34). Thus, indexes as HDI, LPI and others implicitly hide tradeoffs between important aspects. The same happens when methods like cost-benefit analysis are used to support decision makers in ecological decision-making. Implicitly, such methods apply equal weightings for different costs and benefit items. Thus, a tradeoff between nature and social and economic aspects take place. For example, the loss of biodiversity that is the threat of extinction of species is accepted in exchange for increased agricultural production or measures against climate change like increased biogas production[7]. Such an approach means always applying a weak instead of a strong concept of sustainability where for example biodiversity or other social or ecological values must be protected and tradeoffs are not allowed.

If composite indexes like HDI are used, distribution effects (who has to bear the costs? who will get the benefits?), human security, empowerment, gender issues, regional, ethnical or local differences are not considered.

4   Why Politicians Still Use the GDP-Concept and Mistrust Science

Reasons why HPI and LPI indexes are often neglected in politics need to be discussed. Also, it should be investigated which indexes or indicators additionally to GDP/GNI are really used and why they are used.

4.1   Measurement Problems and Interpretation of Data

The great diversity of approaches reflects different perceptions of the world and its values by the index users. But, as shown in the previous section, HDI and LPI are closely correlated and they are also strongly correlated with GDP (Table 2.2). It follows that the error of using for example, the HDI instead of the LPI will not result in big differences. Even the GDP/GNI are rather strongly correlated with HDI and LPI. However, there are exceptions (as the case of Saudi Arabia shows) from this rule that were analyzed above. Thus, in summary the mistake of continuing to use GDP, instead of for example HDI, might be acceptable for a politician preparing economic programs and for investors deciding on fund applications.

However, there are some other problems with indexes and indicators: the values of – for example GDP – are frequently revised ex post. These revisions, due to statistical errors and corrections, reached in the case of Germany a magnitude between −1.0 and 0.9 percent points for specific years even quite some time later than the first values were published (Döhrn 2014:34).

Also, the magnitude of growth rates of GDP is a highly political and economic issue demonstrated by the Chinese GDP growth rates. The suspicion that China falsifies growth rates has a long tradition and continued until recently. In an article published by the Economist (2015), it was noted that: "all seems a little too perfect to be true. The Chinese government set a growth target of 'about 7%' this year. And

---

[7] In cost-benefit analysis also problematic discounting of future benefits and costs when calculating net present value takes place: for justifying discounting, it is implicitly assumed that in the future technological progress will solve problems, higher incomes will make it easier to bear costs.

for a second consecutive quarter, despite ample evidence of stress in its industrial sector, it managed to hit that right on its head". Thus, doubts that the data were 'smoothed' or even 'fabricated' are openly expressed. In its article, the Economist concludes that the first alternative is more likely, referring to examples from the past when China most probably underestimated intentionally the GDP growth rate for political reasons. The theory of 'smoothing' is supported by the fact that the Chinese government propagates actually a new target for economic growth of about 6.5% instead of 7%. So the publications earlier this year might be seen as a preparation of the market for the new target.

The same problem of uncertainties about actual figures can be observed with demographic data. In 2006, a (worst case) scenario was presented where the German population would decrease from about 82 million (2002) to roughly 62 million until 2050 (Schwägerl 2015). The resulting demographic gap alarmed the German public and was widely discussed by many authors (Birg 2005; Schirrmacher 2006). The tendency of a shrinking and older population seemed to be confirmed when officially the population number in Germany was reduced to 80.2 million (Census 2011). In the census 2011, it was found out that the resident's registration offices contained 1.5 million only nominal citizens (Statistische Ämter des Bundes und der Länder 2014; Der Spiegel 2013).

Due to the influx of many foreigners – 860.000 migrants from Non-EU-countries arrived in 2015 (plus 382.500 migrants from EU-countries) (Bundesamt für Migration und Flüchtlinge 2016) – the German government changed its projections for the coming years and expects actually 300.000 migrants per year until 2030 instead of 100.000 until recently. As a result, it is estimated that the German population will not shrink, but increase until 2030 and will only decrease afterwards (Creutzburg 2015; Schwägerl 2015).

In other countries like Nigeria, an ethnically and religiously very diverse country, the census is an even more political issue than in Germany since it has links with the borders of electoral districts and the composition of parliament. In the past most probably, the outcome of the census was frequently falsified leaving many doubts about the actual number of people living in Nigeria (Odunfa 2006; Population Reference Bureau 2006). Even officially, it was acknowledged: "the usual problems of under-coverage common to most censuses are compounded by those of over-counting, as states, ethnic groups, and religions compete for demographically based claims on power and resources" (Population Council 2007:206).

For checking especially urban population figures in West Africa, the French development agency (AFD) commanded the so-called 'Africapolis' study (AFD undated, 2011). By using satellite imaginary complimentary to census data, the study found out that especially in Nigeria, but also in other West African countries, the number of urban population is considerably smaller as assumed until now by the census. The results of the Africapolis methodology for Nigeria demonstrate that many settlements have either scarcely grown since the 1960s or were simply grossly over-estimated at that point. Astonishingly, it was found that the populations of nearly half of the smaller urban settlements assessed via their methodology were less than those recorded in the 1963 census (Potts 2012). "At the national level, the outcome of the Africapolis re-evaluation of the 2006 census in Nigeria is that its urbanization level is 30%. Yet the UN Department for Economic and Social Affairs reported a level of 49% in 2006" (Potts 2012:1386). Thus, a very critical view must be observed on statistical data produced by national statistical offices, especially in developing countries.

Also, a research we have undertaken in the Republic of Benin revealed many inconsistencies. For example, the incidence of income poverty in Benin in 2011 for households under three persons is published as 1.0%, that is to say almost zero, while non-monetary poverty in the same group is 39.6% (Sommer et al. 2013). Therefore, despite many efforts and initiatives in the past for improving the performance of statistical offices, much needs to be done to improve data quality. This might be often difficult or impossible on account of political motives. The reason for at least some of these errors may be even sometimes simple calculation errors, facts that even happen to very prominent scientists working in prominent universities and financed by prominent institutions like the IMF. One of the IMF's most prominent researchers, Reinhart and Rogoff (Chief Economist of the IMF between 2001 and 2003), committed obviously such (EXCEL) errors when calculating the threshold for country's debt ratios (Reinhart and Rogoff 2009). Especially, they tried to analyze the problem of how (higher) debt ratios might influence economic performance. Their results were understood especially by conservative politicians all over the world that there exists a threshold of 90% debt ratio to GDP from whereon GDP growth suffers from higher debts.

Herndon et al. (2013) found out, when replicating Reinhart and Rogoff's analysis, that coding errors, selective exclusion of available data, and unconventional weighting of summary statistics lead to serious errors that inaccurately represent the relationship between public debt and GDP growth among 20 advanced economies in the post-war period. Their finding is that when properly calculated, the average real GDP growth rate for countries carrying a public-debt-to-GDP ratio of over 90 percent is actually 2.2 percent, not 0.1 percent as published in Reinhart and Rogoff (2009). According to Herndon et al. (2013), differences in average GDP growth in the categories 30–60 percent, 60–90 percent, and 90–120 percent cannot be statistically distinguished. The Journal Newyorker asked thereafter: "after this new fiasco, how seriously should we take any economist's policy prescriptions, especially ones that are seized upon by politicians with agendas of their own?" (Cassidy 2013). However, this result should not be misunderstood: according to the German institute for economic research (Deutsches Institut für Wirtschaftsforschung, DIW), early indicators of the last financial crisis were high growth rates of credits in relation to GDP, and on the micro level high debt ratios and low equity ratios of the affected banks (Beck and Bremus 2014). Also, the example of Greece will probably show in future that it may be virtually impossible for this country to reduce its debt ratio to a sustainable level without debt reliefs.

In the case of the controversy on the debt threshold, calculation errors, and perhaps the misinterpretation of data, explain the deviations. In the case of 'Climategate' in November 2009, the email server of the Climate Research Unit of the University of East Anglia in the UK was hacked. This Unit is a member of the Intergovernmental

Panel on Climate Change (IPCC) and collaborating closely with the British National Weather Service. Especially, the emails of its Director Phil Jones contained rather problematic content, recommending for example manipulation of data. The report of an independent government committee found out later that there was at the end no manipulation of data. However, the very problematic communication policy only publishing affirmative data was heavily criticized (Fiaromonti 2014). Especially, the attempt to hide data from publication has to be considered an ethically very problematic behavior, because it provokes mistrust of people and politicians in the results of climate models and calculations. Probably, Climategate was one of the causes why the Copenhagen climate summit in 2009 failed.

In the climate change debate, another example of problems of really knowing 'the true and real' actual situation is the discussion about the so-called 'Climate Hiatus'. Climate Hiatus means the fact that according to many researchers and institutions, including IPCC (since 2013) and the British meteorology services, the temperature didn't increase as expected in the climate models during the last decade (Allan and Loeb 2015). IPCC states in its latest Summary report: "The rate of warming over the past 15 years (1998-2012; 0.05 [-0.05 to 0.15] °C per decade), which begins with a strong El Niño, is smaller than the rate calculated since 1951 (1951–2012; 0.12 [0.08 to 0.14] °C per decade)" (IPCC 2015). Another study summarizes: "despite ongoing increases in atmospheric greenhouse gases, the Earth's global average surface air temperature has remained more or less steady since 2001" (England et al. 2014:227). The IPCC forwards as an explanation "substantial decadal and inter-annual variability", and also the fact that "ocean warming dominates the increase in energy stored in the climate system, accounting for more than 90% of the energy accumulated between 1971 and 2010 (high confidence), with only about 1% stored in the atmosphere" (IPCC 2015:3). Other possible explanations are forwarded in the literature, for example a "recent intensification of wind-driven circulation in the Pacific" (England et al. 2014:227)

The publication of a study in June 2015 claiming that there never was a global warming hiatus was under these circumstances a big surprise: "The apparent hiatus, first reported by the IPCC in 2013, resulted from a shift during the last couple of decades to greater use of buoys for measuring sea surface temperatures. Buoys tend to give cooler readings than measurements taken from ships" (EOS Earth and Space Science News 2015). It would mean that the measurements of world average temperature were wrong, measuring too high values in the past and too low values now. Thus, according to these researchers no climate hiatus exists! This finding would be perfectly in line with many other observations confirming climate change, from average regional temperatures beating 1 year after the other all time high records, sea level rising until natural disaster statistics and extreme events (IPCC 2015; WMO 2015a, 2015b; Munich Re 2015:58). On a worldwide scale, the year 2015 was the warmest year ever observed (NOAA National Centers for Environmental Information 2016).

As a result, we may conclude that there is quite often only a low confidence in statistics about 'facts' of the past because they are wrongly measured, falsified or manipulated, badly ascertained, respectively researched or misinterpreted. However,

Table 2.5 Comparison of naïve forecasts of the German GDP with official forecasts

	2004	2005	2006	2007	2008	2009	2010	2011	2012	2013	Score
Official forecast	1,7	1,5	1,2	1,4	2,2	0,2	1,2	2	0,8	1	
Reality	1,2	0,7	3,7	3,3	1,1	−5,1	4	3,3	0,7	0,4	
Naive forecast		1,2	0,7	3,7	3,3	1,1	−5,1	4	3,3	0,7	
Deviations											
Off. forecast – reality		0,8	−2,5	−1,9	1,1	5,3	−2,8	−1,3	0,1	0,6	5
Naive forecast - reality		0,5	−3	0,4	2,2	6,2	−9,1	0,7	2,6	0,3	4

Data source: Statista (2015)

indexes and indicators as GDP/GNI, population figures or CO_2 are not only analyzed for comparing a country with other countries or evaluating past successes in development, but also for deriving recommendations and planning or even 'knowing' the future.

4.2 Forecasts, Projections and Scenarios

All indexes show only past performances of countries, but don't help to produce forecasts that show how the indicators will develop in the future. For doing this, a lot of efforts were and still are regularly undertaken for knowing better future developments in many fields, especially economy, demography, and ecology.

4.2.1 Economic Forecasts: The Example of GDP

One of the key indicators in economy is still GDP/GNI. At least in the case of industrialized countries, as for example Germany, many institutes publish every year a series of forecasts for the next year. It begins with the IMF in September, followed by OECD, the Council of independent experts nominated by the German government, several research institutes, and at the end of January finally the last forecast is made public by the German government. The question is how reliable such forecasts are (Döhrn 2014). A very simple comparison of naïve and scientific forecast is reproduced in Table 2.5.

As can be seen in the time series of the years 2005–2013, official forecasts were in 5 years more precise than naïve forecasts; naïve forecasts were better in 4 years. The impression of the relatively weak quality of forecasts is partially confirmed by the literature. For example, the OECD has detected in its forecasts a positive bias, forecasting constantly better values as realized (Lewis and Pain 2015). Other authors found out an acceptable degree of precision of German GDP forecasts (Döhrn 2014), if the time horizon doesn't exceed very much 1 year. However, the (rather low) precision of GDP forecasts decreases rapidly for forecasts exceeding 1 year.

According to Dühr, the 70% confidence interval has a width of 4 percent points for such forecasts in the case of Germany. Also, Tol (2008) confirms that GDP forecasts exceeding 18 months are not reliable. Julio and Esperanza (2012) evaluated the forecast quality of GDP components and claimed, "the overall accuracy of component predictions...is substantially low, meaning that GDP forecasts are assembled with rather inaccurate component predictions". As expected, the lowest forecast quality occurred in the financial crisis. Ikka Korhonen (Bank of Finland) and Maria Ritola investigated the success of forecasting economic developments in major emerging markets and conclude: "our results suggest that relying on any one forecast methodology is not prudent" (Korhonen and Ritola 2014).

4.2.2 Population Projections

Concerning population forecasts, it has to be repeated that there are in certain countries, as Nigeria, high uncertainties about the correct actual figure. Thus, it is widely accepted that it is impossible to forecast the future population of a country or the world, but to do instead 'projections'. According to the Merriam-Webster Dictionary (2015), a projection is "an estimate of what might happen in the future based on what is happening now". According to the German Gabler's economic dictionary, projection is not like forecasts since it is based only on observations from the past and objective methods, but uses also subjective guesses for example by experts (Gabler's Wirtschaftslexikon 2015).

The UN has updated recently its so-called 'population estimates and projections' applying statistical methods. The United Nations Development Program (UNDP) projections for the world population vary between 7.3 billion (low variant) and 16.6 billion people (high variant) for the year 2100 with an 80% confidence interval between 9.6 and 12.3 billion people (UNDP 2015b). However, all projections start with the same values in the year 2015, not acknowledging any doubt on the measurement of the existing world population. So, in all projections a population of 182 million people for 2015 is supposed for Nigeria.

According to other demographers, the UN's projection exaggerates heavily. In an article at Global Environmental Change, Wolfgang Lutz and his colleagues at the International Institute for Applied Systems Analysis (IIASA) in Vienna, Austria, use a very different method – one that involves canvassing a large group of experts – to argue that population is likely to peak at 9.4 billion in 2075 and fall to just under 9 billion by 2100 (Samir and Lutz 2017; Kunzig 2014).

Another study provides simulations showing what global and regional population sizes would actually be if the rest of the world would have experienced similar population growth patterns as what was observed in Europe during the demographic transition. The implications of differences in population growth patterns are large. If Nigeria would have followed the French population trajectory, it would grow from 38 million in 1950 to only 72 million in 2100, while the UN projections suggest it would reach 914 million people by 2100. Thus, population growth is more influenced by cultural variations causing uneven growth rather than by any universal population growth trajectories over the demographic transition (Skirbekk et al. 2015).

Therefore, the only possible conclusion of the different approaches from an observer's point of view and a concerned world citizen is that there is high uncertainty about future world population as well as national populations.

4.2.3 Climate Change Scenarios

Instead of projection in climate science, scenarios are widely used. Scenarios are descriptions of different plausible futures. According to Tol (2008:22), they combine "art and science". They are based on empirical regularities (as the assumption of dependencies between temperature and CO_2 levels) and theoretical understanding (the greenhouse effect). The basis of climate change scenarios are projections of emissions of greenhouse gases and their supposedly (close) correlation with average world temperature, sea level or other meteorological phenomena as the frequency of storms, droughts, floods, etc. For deriving the scenarios, assumptions must be made on the behavior of people, respectively socio-economic systems including population growth, technological progress and capital growth (Tol 2008:23). For Storch (2008:5), "scenarios are not predictions, but 'storyboards' (used usually for producing movies), a series of visions of futures which are possible, plausible, internally consistent but not necessarily probable". Even one might say the purpose of many scenarios is to motivate politicians to implement adequate measures that undesired scenarios become self-defying prophecies, that is they don't come true.

However, the IPCC scenarios are often presented in the form of forecasts. According to Tol (2008:18), "this mistake is often made by journalists and policy makers". But, "although scenarios cannot be falsified or validated, they can be tested against data nonetheless" (Tol 2008:32). Thus, if the temperature hiatus – if there exists any at all (!)– would continue for some years, the scenarios developed by the IPCC need to be adapted and the models 'calibrated'.

Last but not least, an important reason for politicians not following scientific recommendations is very often the complexity of the texts published by scientific bodies. An example is the Summary for Policy Makers (SPM) of the IPCC. A linguistic analysis of IPCC summaries for policymakers was undertaken recently by researchers from the university of Bonn in Germany that resulted in the evaluation that "IPCC SPMs clearly stand out in terms of low readability, which has remained relatively constant despite the IPCC's efforts to consolidate and readjust its communications policy" (Barkemeyer et al. 2015).

4.3 Data Politicians (and People) are Really Interested in

Beside GDP growth, unemployment and debt ratios, politicians are interested in data like polls (especially measuring their own popularity and of their parties!). Such data is provided worldwide by the Gallup[8] or many national institutes, researching the opinion of population or specific target groups. For example, since 1973 the EU publishes two times a year in spring and fall a survey in all member countries called 'Eurobarometer'. One standard question since the beginning is as follows: 'On the whole, are you satisfied with the life you lead?'. The possible answers are: very satisfied; fairly satisfied; not very satisfied; and not at all satisfied. Taking all countries together, about 80% of the EU citizens are very satisfied or fairly satisfied with no big variations in time since the last 40 years (European Commission 2015).

At the first glance, it seems that this result confirms the so-called set point theory in psychology that happiness and satisfaction are mostly constant factors in the life of a person and, as a consequence, also have characteristic values in specific countries. However, a more detailed analysis reveals that rather big deviations exist between countries which may be explained by national factors. But, there are also important fluctuations in time: between 2007 and 2014, the percentage of people satisfied with their life increased in Germany from 82% to 87%, and decreased in Italy from 76% to 63% and in Spain from 90% to 70%. It seems that the most important reason for these changes is the economic evolution in the Eurozone since 2009 (European Commission 2015).

The big problem with information contained in polls, GDP growth rates, and also well-being indicators is its short mindedness, which is the unwillingness to look or care about long-term implications. However, as these implications are highly uncertain, forecasts are impossible, communication policies are inappropriate, and scenarios are not sufficiently plausible or really understood or accepted by people and politicians, this attitude is perhaps understandable.

5 What Needs to be Done – Risk Management and How to Find the Right Decisions for Reaching Sustainability

Despite all problems of measuring, analyzing and forecasting indicators and risks, one of the most important risks is population growth, especially in poor countries. A growth rate of about 3% p.a. can still be observed in many sub-Saharan African countries. This rate means that the population will double within 23.5 years. High population growth provokes high economic growth necessary for reducing poverty, but also increases in resource consumption, pollution and CO_2 production. Therefore, climate change, depletion of resources and migration will be intensified.

---

[8] http://www.gallup.com/home.aspx

Thus, the fight against population growth must become a priority, especially in Sub-Saharan Africa.

By using the words of Keynes (who meant them for defending his mostly short-term oriented economic policy proposals), there is only one certainty in life: 'in the long run we are all dead'. Not only for religious people and parents and grandparents caring for their children and grandchildren this doesn't mean that we should live following the principle 'après nous le deluge' ('after me, the deluge'). Since the Brundland Commission published its report 'Our common future' in 1987 defining 'sustainability' as "not compromising the ability of future generations to meet their own needs" (UN 1987), basically all nations accepted this principle.

Concerning short and medium term risks, risk management systems exist in many areas from fire brigades, medical care to management of any kind of disasters as accidents, floods, hurricanes or earthquakes. In all these areas, prevention plays nowadays an important role. It may reduce the risk to life dramatically as is shown by the low dead rates in countries having implemented adequate risk prevention measures (for example Japan and Chile regarding earthquakes). The development of risk management in organization is supported by principles and guidelines laid down in ISO 31000:2009: "It can be used by any organization regardless of its size, activity or sector. Using ISO 31000 can help organizations increase the likelihood of achieving objectives, improve the identification of opportunities and threats and effectively allocate and use resources for risk treatment" (ISO 2015). "Organizations using it can compare their risk management practices with an internationally recognized benchmark, providing sound principles for effective management and corporate governance" (ISO 2015).

These risk management systems are limited up to now to disasters and many types of prevention are only possible for rich countries. It may be hoped that the creation of additional funds may improve the situation of poorer countries. The actual behavior and mobilization of funds concerning refugees in Europe raise many doubts on the ability of the international community to increase solidarity. At the end, this is a question of ethical principles guiding individual as collective behavior including ethically sound scientific research.

One approach that is basically, and in general also accepted by many, is the robust decision making. Policy options must be identified to prevent unacceptable outcomes in as many scenarios as possible. In practice, it means realizing co-benefits in as many areas as possible like for example from poverty reduction to climate change.

6   Conclusion

As was shown previously, richer countries are in general also the happier countries. The indicators with the strongest influence on happiness are: CO_2 per capita; GNI / GDP per capita; life expectancy at birth; mean years of schooling; total participation rates; share of fossil fuels; income Gini coefficient; and well-being. This is underlined by the fact that poor countries show in general a low performance in most if not all these areas. Thus, regarding the importance of CO_2 per capita and resource consumption there seems to be no chance for sustainable development. However, the example of Costa Rica shows that alternatives are possible; people there have a higher score in life satisfaction than people of much richer countries as Saudi Arabia and even Germany.

Despite all uncertainties demonstrated above regarding the correct description and analysis of the actual situation, that is the selection of the 'right' indicators and the impossibility to forecast long-term developments, there are some fundamental robust strategies that should be promoted. These include the fight against high population growth rates, mainly in poor countries, the introduction of clean technologies and the promotion of a more dematerialized world by fostering the insight that happiness can (also) be reached when changing the actual patterns of consumption.

However, it should not be underestimated that there is a long way to go. Especially the following questions need to be answered when preparing and deciding about concrete measures for reducing risks to human and environmental security, well-being and welfare: How to define and achieve sustainable long-term solutions? How to change the individual perception of risks from short term to long term? How to develop better early warning systems? How to deal with worst-case scenarios? How to value non-monetary items (for example in the case of biodiversity)? And how to overcome free rider situations?

References

Allan RP, Loeb N (2015) *Warming over the last decade hidden below ocean surface.* Retrieved at November 8, 2015. https://www.ncas.ac.uk/index.php/en/climate-science-highlights/284-warming-over-the-last-decade-hidden-below-ocean-surface

Australian Unity (2015) *The Australian unity wellbeing index.* Retrieved at November 7, 2015. http://www.australianunity.com.au/about-us/wellbeing/auwbi

Barkemeyer R, Dessai S, Monge-Sanz B, Napolitano G (2015) *Linguistic analysis of IPCC summaries for policymakers and associated coverage.* Retrieved at November 8, 2015. Nature Climate Change : http://www.nature.com/nclimate/journal/vaop/ncurrent/full/nclimate2824.html

Beck A, Bremus F (2014) *Wie kann systemisches Risiko beschränkt werden?* Retrieved at November 9, 2015. DIW Roundup 36: https://www.diw.de/documents/publikationen/73/diw_01.c.481461.de/diw_roundup_36_de.pdf

Birg H (2005) Die demographische Zeitenwende : der Bevölkerungsrückgang in Deutschland und Europa. Beck, München

Bundesamt für Migration und Flüchtlinge (2016) *Wanderungsmonitoring: Erwerbsmigration nach Deutschland - Jahresbericht 2015.* Abgerufen am 24. 11 2016 von http://www.bamf.de/SharedDocs/Anlagen/DE/Publikationen/Broschueren/wanderungsmonitoring-2015.pdf?__blob=publicationFile

Cassidy J (2013) *The Reinhart and Rogoff controversy: a summing up.* 26. 4, Retrieved at November 7, 2015. http://www.newyorker.com/news/john-cassidy/the-reinhart-and-rogoff-controversy-a-summing-up

Centre for Bhutan Studies and GNH Research (2015) *Gross National Happiness*. Retrieved at November 8, 2015. http://www.grossnationalhappiness.com/home/media/

Centre for Well-being at the New Economics Foundation (NEF) (2015) *Happy planet index*. Retrieved at November 7, 2015. http://www.happyplanetindex.org/

Creutzburg D (2015) *Arbeitskräftemangel: Zuwanderer entlasten Deutschland*. Frankfurter Allgemeine Zeitung, 5.2.2015, Retrieved at November 8, 2015. http://www.faz.net/aktuell/wirtschaft/wirtschaftspolitik/arbeitsmarkt-prognose-2030--andrea-nahles-13411792.html

Der Spiegel (2013) *Zensus 2011: Deutschland hat weniger Einwohner als angenommen*. Der Spiegel, dated 31.5.2013, Retrieved at November 8, 2015. http://www.spiegel.de/politik/deutschland/zensus-2011-in-deutschland-leben-80-2-millionen-menschen-a-902992.html

Diefenbacher H, Zieschank R (2011) Woran sich Wohlstand wirklich messen lässt : Alternativen zum Bruttoinlandsprodukt. Oekom Verlag, München

Döhrn R (2014) Konjunkturdiagnose und -prognose: Eine anwendungsorientierte Einführung. Springer Gabler, Berlin Heidelberg

Economist (2015) *Whether to believe China's GDP figures*. Economist dated 15 July 2015, Retrieved at November 7, 2015. http://www.economist.com/blogs/freeexchange/2015/07/chinese-economy

England MH, McGregor S, Spence Paul MG, Timmermann A, Cai W, Gupta AS et al (2014) Recent intensification of wind-driven circulation in the Pacific and the ongoing warming hiatus. Nat Clim Chang 4:222–227

EOS Earth and Space Science News (2015) *Global warming "Hiatus" never happened, study says*. Retrieved at November 8, 2015. https://eos.org/articles/global-warming-hiatus-never-happened-study-says

European Commission (2015) *Eurobarometer surveys*. Retrieved at November 8, 2015. http://ec.europa.eu/public_opinion/index_en.htm

Fiaromonti L (2014) How numbers rule the world: the use and abuse of statistics in global politics. Zed Books, London/New York

Gabler's Wirtschaftslexikon (2015) *Projektion*. Retrieved at November 8, 2015. http://wirtschaftslexikon.gabler.de/Definition/projektion.html

Gigerenzer G (2013) Risk savvy: how to make good decisions. Penguin, New York

Gigerenzer G (2014) *Risiko : wie man die richtigen Entscheidungen trifft*. München: btb

Herndon T, Ash M, Pollin R (2013) *Does high public debt consistently stifle economic growth? A critique of Reinhart and Rogoff*. University of Massachusetts Amherst, Political Research Institiute working paper series Number 322

IPCC (2015) *Climate change 2014 - synthesis report - summary for policymakers*. Retrieved at November 7, 2015. http://www.ipcc.ch/pdf/assessment-report/ar5/syr/AR5_SYR_FINAL_SPM.pdf

ISO (2015) *ISO 31000 - Risk management*. Retrieved at November 7, 2015. http://www.iso.org/iso/iso31000

Julio P, Esperança PM (2012) *Evaluating the forecast quality of GDP components - An application to G7*. Retrieved at November 7, 2015. Econ papers: http://econpapers.repec.org/paper/mdewpaper/0047.htm

Korhonen I, Ritola M (2014) An empirical note on the success of forecasting economic developments in major emerging markets. Asian Econ Pap 13(1):131–154

Kunzig R (2014) *A world with 11 billion people? New population projections shatter earlier estimates*. Retrieved at November 8, 2015. National Geographic: http://news.nationalgeographic.com/news/2014/09/140918-population-global-united-nations-2100-boom-africa/

Legatum Institute (2015) *2014 Legatum Prosperity Index™*. Retrieved at November 7, 2015. http://www.li.com/activities/publications/2014-legatum-prosperity-index

Lewis C, Pain N (2015) Lessons from OECD forecasts during and after the financial crisis. OECD J Econ Stud 2014:10–39

Merriam Webster (2015) *"projection"*. Retrieved at November 8, 2015. http://www.merriam-webster.com/dictionary/projection

Miegel M (2011) *Exit : Wohlstand ohne Wachstum*. Berlin: List

Munich Re (2015) *Natural catastrophes 2013*. Retrieved at November 11, 2015. http://www.munichre.com/site/corporate/get/documents_E1043212252/mr/assetpool.shared/Documents/5_Touch/_Publications/302-08121_en.pdf

Nguyen T, Romeike F (2013) Versicherungswirtschaftslehre : Grundlagen für Studium und Praxis. Springer Verlag, Wiesbaden

NOAA National Centers for Environmental Information (2016) *State of the climate: global analysis for annual 2015*. Abgerufen am 23. 11 2016 von https://www.ncdc.noaa.gov/sotc/global/201513

Odunfa S (2006) *Nigeria's counting controversy*. BBC, 21.3.2006, Retrieved at November 7, 2015. http://news.bbc.co.uk/2/hi/africa/4512240.stm

Office for National Statistics (2015) *National wellbeing - measuring what matters*. Retrieved at November 7, 2015. http://www.ons.gov.uk/ons/guide-method/user-guidance/well-being/index.html

Otoiu A, Titan E, Dumitrescu R (2014) Are the variables used in building composite indicators of well-being. Ecol Indic 46:575–585

Paech N (2013) Grünes Wachstum? Vom Fehlschlagen jeglicher Entkoppelungsbemühungen: Ein Trauerspiel in mehreren Akten. In: Sauer T (ed) Ökonomie der Nachhaltigkeit. Metropolis Verlag, München, pp 161–181

Population Council (2007) Report of Nigeria's national population commission on the 2006 census. Popul Dev Rev 33(1):206–210

Population Reference Bureau (2006) *In the news: The Nigerian census*. Retrieved at November 8, 2015. http://www.prb.org/Publications/Articles/2006/IntheNewsTheNigerianCensus.aspx

Potts D (2012) Challenging the myths of urban dynamics in sub-saharan Africa: the evidence from Nigeria. World Dev 40(7):1382–1393

Proske D (2008) Katalog der Risiken - Risiken und ihre Darstellung. Saechsische Landesbibliothek-Staats- und Universitaetsbibliothek Dresden, Dresden

Reinhart CM, Rogoff K (2009) This time is different: eight centuries of financial folly. Princeton University Press, Princeton/New Jersey

Renn O (2014) Das Risikoparadox : warum wir uns vor dem Falschen fürchten. Fischer, Frankfurt

Samir K, Lutz W (2017) The human core of the shared socioeconomic pathways: population scenarios by age, sex and level of education for all countries to 2100. Global Environ Change 42 181–192. https://www.sciencedirect.com/science/article/pii/S0959378014001095

Schirrmacher F (2006) Das Methusalem-Komplott. Blessing, München

Schmidt-Bleek F (2014) Grüne Lügen : nichts für die Umwelt, alles fürs Geschäft ; wie Politik und Wirtschaft die Welt zugrunde richten. Ludwig, München

Schwägerl C (2015) *Deutschland will einfach nicht veröden*. FAZ, 23.9.2015, Retrieved at November 7, 2015. https://www.genios.de/document?id=FAZ__FNUWD12015092346785 69IFAZT__FNUWD1201509234678569

Self A (2014) *Office for national statisitics*. Retrieved at January 7, 2015 Measuring National Wellbeing: Insights: http://www.ons.gov.uk/ons/dcp171766_371427.pdf

Skirbekk V, Stonawski M, Alfanim G (2015) Consequences of a universal european demographic transition on regional and global population distributions. Technol Forecast Soc Chang 98:271–289

Sommer H, Houngbo NE, Montchowui E (2013) Mesurer la Pauvreté et la Sécurité Alimentaire au Bénin : Concepts, Pratique, Résultats et Défis. Document de recherche de l'Université Agricole de Kétou / Bénin and FH Bingen, Cotonou

Sommer H, Houngbo NE, Montchowui E (2014) Messung Armut und Nahrungsmittelsicherheit in Benin - Konzepte, Ergebnisse und Herausforderungen. In: Kunz FR (ed) Konfliktfelder und Perspektiven im Umweltschutz. oekom Verlag, München, pp 239–259

Statista (2015) *Das Statistik Portal*. Retrieved at November 13, 2015. http://de.statista.com/infografik/1541/bip-prognose-versus-reales-bip/

Statistische Ämter des Bundes und der Länder (2014) *Zensus 2011* Retrieved at November 7, 2015. https://www.zensus2011.de/DE/Zensus2011/zensus2011_node.html

Statistisches Bundesamt (2014). *Nachhaltige Entwicklung in Deutschland - Indikatorenbericht 2014*. Retrieved at November 13, 2015. https://www.destatis.de/DE/Publikationen/Thematisch/UmweltoekonomischeGesamtrechnungen/Umweltindikatoren/IndikatorenPDF_0230001.pdf?__blob=publicationFile

Stiglitz JE, Sen A, Fitoussi JP (2010) Mismeasuring our lives: why GDP doesn't add up. The New Press, New York

Storch Hv (2008) Climate change scenarios - purpose and construction. In H. Storch, R. Tol, and G. Föser, Environmental crises (S. 5–16). Berlin/Heidelberg/New York: Springer

Tol RS (2008) Economic scenarios for global change. In: Storch H, Tol R, Flöser G (eds) Environmental crises. Springer, Berlin/Heidelberg/New York, pp 17–36

UNDP (2015a) *Human development index (HDI)*. Retrieved at November 7, 2015. http://hdr.undp.org/en/content/human-development-index-hdi

UNDP (2015b) *World population prospects - the 2015 revision*. Retrieved at November 8, 2015. http://esa.un.org/unpd/wpp/DVD/

United Nations (UN) (1987) Report of the world commission on environment and development: our common future. Retrieved at November 8, 2015. http://www.un-documents.net/our-common-future.pdf

University of Waterloo (2015) Canadian index of wellbeing. Retrieved at November 7, 2015. https://uwaterloo.ca/canadian-index-wellbeing/

WMO (2015a) New report on extreme events. Retrieved at November 8, 2015. https://www.wmo.int/media/content/new-report-extreme-events

WMO (2015b) Greenhouse gas concentrations hit yet another record. Retrieved at November 10, 2015. https://www.wmo.int/media/content/greenhouse-gas-concentrations-hit-yet-another-record

Chapter 3
Building Socio-Ecological Coviability: An Efficient Way to Combat Poverty, Reduce Inequality and Address Insecurity Risks

Amine Amar

Abstract The proliferation of conflicts throughout the world, increasing poverty, rising inequality, and the necessity of a social cohesion require adherence to the concept of socio-ecological systems (SES). This nexus constrains policies to be conceptualized in order to face, jointly, social and ecological dimensions and coerce identified decisions to spur mutual progress towards interlinked goals. To address such challenges with the required level of comprehensiveness and urgency, researchers and decision makers must first acknowledge the linkages between social and ecological problems within a coviability perspective. In addition, they are required to offer both empirical and analytical insights to these linkages in order to implement efficient policies, which are able to tackle major problems such as poverty, inequality, and insecurity risks. Traditional approaches to poverty eradication, for instance, often disregard environmental degradation and biodiversity loss as externalities, whereas the proposed SES approach considers not only the social but also the environmental cause, thereby providing a new redefinition of wealth and its components. Thus, efforts should be complemented by implementing initiatives aimed at enhancing social well-being for all people, and by addressing the root causes of poverty through coherent, coordinated and coviable strategies. To address the aforementioned issues in this chapter, some tools adapted from economic analyses and development economics, in addition to some success stories and empirical cases, are presented.

Keywords Coviability · Development economics · Differential and social-ecology vectors · Inequality · Insecurity risks · Poverty · Socio-ecological systems

A. Amar (✉)
National Institute of Statistics and Applied Economics (INSEA), Mohamed V University of Rabat, Rabat, Morocco

Researcher at the Center for Research on Environment, Human Security and Governance (CERES), Rabat, Morocco

© Springer International Publishing AG, part of Springer Nature 2019   65
M. Behnassi et al. (eds.), *Human and Environmental Security in the Era of Global Risks*, https://doi.org/10.1007/978-3-319-92828-9_3

1   Introduction

The majority of debated environmental challenges in the recent years can be attributed to many social problems, which arise from income and power inequality. According to this point of view, inequalities (social, income or other kinds of inequalities) can be conceived as an environmental issue and, in a reverse sense, the environmental degradation can be treated as a social question. This paradigm forms what we call a socio-ecological nexus. It constrains the policies to be conceptualized in order to face jointly social and ecological dimensions and coerce identified decisions to spur mutual progress towards interlinked goals.

Therefore, to address such challenges with the required level of comprehensiveness and urgency, researchers and decision makers must acknowledge the interrelatedness of social and ecological problems within a coviability context. In addition, they are required to offer both empirical and analytical insights of this relationship to implement efficient policies, which are able to treat many common risks, pertained to poverty, inequality, and insecurity.

In this regard, sustained, inclusive and equitable growth can be apprehended as a key requirement for achieving the Sustainable Development Goals (SDGs), and especially for eradicating poverty. Thus, efforts should be complemented by implementing initiatives aimed at enhancing social well-being for all people, and by addressing the root causes of poverty through coherent, coordinated and coviable strategies. In this sense, many projects related to agriculture, fisheries and aquaculture, tourism, transportation and mobility, water and energy, can be developed.

For agriculture, the sustainability of production can be assured through improvement of the market functioning, international cooperation, strengthening of urban-rural linkages, maintenance of natural ecological processes that support food production systems, and the preservation of marine ecosystems. For water resources, projects should concern the progressive realization of basic sanitation and access to safe and affordable drinking water given their positive impacts in terms of poverty eradication and women's empowerment. Within the same perspective, it should be noted that tourism can make a significant contribution in regard to sustainable development, essentially through the creation of decent and green jobs and the generation of trade opportunities. Thus, efforts should be made to support sustainable tourism activities and relevant capacity building, which promote environmental awareness and improve the welfare and livelihoods of communities. Regarding the energy sector, there are many evidences on the critical role played by energy in the development process. Access to sustainable modern energy services often contributes to poverty eradication and helps provide basic human needs. Thus, policies and strategies should be implemented based on individual national circumstances and developmental aspirations, using an appropriate energy mix to meet developmental needs, including the use of renewable energy sources and other low emission technologies. Needs should also be recognized to address the challenge of access to sustainable modern energy services for the poor, who are unable to afford these services even when they are available.

In fact, there are many evidences, on the empirical level, about the existence of a strong, reciprocal, and complex relationship between ecological and social systems. On the one hand, social inequalities can drive ecological crises due to increasing ecological irresponsibility of the richest in society and among nations. Such inequalities can probably fuel the demand for economic growth among the rest of the population, increase social vulnerability, and hinder collective action. On the other hand, ecological crises can generate new forms of inequality such as structural environmental inequalities.

Furthermore, it can be shown in this analysis that the eradication of poverty presents a major obstacle to sustainability. In economic terms, the decline of poverty comes, in the majority of cases, at the cost of biodiversity degradation. The analysis of different statistics throughout much of the twentieth century reveals that as the human development indicators increased, the global biodiversity declined. Thus, from 1970 to 2010, the human development index improved on average by about 40%, while the global biodiversity index fell by 30–50% (UNDP 2010; WWF 2014). This can be explained by the fact that since the wealth of the world's poor lies in natural capital, due to the lack of access to other forms of capital, the depletion of such natural resources leads to further impoverishment. Thus, the eradication of poverty should not be considered only as a social cause, but also an environmental one, provided that it takes the form of a redefinition of comprehensive wealth and its components.

On the macro-ecological level, five channels, through which rich and poor interact in environmental degradation, crises, and policies, can be distinguished (Laurent 2015):

- Channel 1: inequality increases the need for environmentally harmful and socially unnecessary economic growth.
- Channel 2: inequality increases the ecological irresponsibility of the richest, within each country and among nations.
- Channel 3: inequality, which affects the health of individuals and groups, diminishes the social-ecological resilience of communities and societies and weakens their collective ability to adapt to accelerating environmental change.
- Channel 4: inequality hinders collective action aimed at preserving natural resources.
- Channel 5: inequality reduces the political acceptability of environmental preoccupations and the ability to offset the potential socially-regressive effects of environmental policies.

Thus, a socio-ecological approach is needed to address the knowledge gap between social and ecological systems, by considering the reciprocal relationship between the two systems, demonstrating how social logics determine ecologic damages, and exploring the consequences of these damages on social inequality. The integrative social-ecology represents the first vector of causality; it shows that the gap between the rich and the poor and the interaction of the two groups often lead to the worsening of environmental degradation and ecological crises. On the contrary, the second vector, labeled differential social-ecology vector, shows that the social impact of ecological crises is not the same for different individuals and groups given their socio-economic status (Laurent 2015). The two vectors should be supported by stressing the importance of stronger inter-linkages among disaster risk reduction and long-term development planning, which integrate climate change adaptation and humanitarian considerations in a coviability context.

To address the aforementioned issues, this chapter presents in the following section, an overview on poverty, inequality and insecurity risks and trends. The third section is dedicated to some dimensions of socio-ecological systems and coviability concepts. The fourth part analyzes the impact of coviability considerations on poverty and inequality reduction and insecurity risk management, with some illustrations and case studies. The last section concludes by some key messages and recommendations.

2 Overview on Poverty, Inequality, and Insecurity Risks and Trends

Wealth inequalities tend to dominate recent discussions across many circles of economists. This interest can be justified by the fact that such disparities trap large pockets of society in poverty and exclusion and shape the opportunities that people have in life as well as their children's future (UNDP 2013). At an international level, inequality trends have not followed a universal pattern and the distribution of income remains very uneven. In fact, some countries of Africa and Latin America have been able to reduce economic inequalities, while income disparities have increased within many other countries over the world (UNCTAD 2014). Thus, in 2010 for example, high-income countries, which accounted for only 16% of the world's population, were estimated to generate 55% of global income while the low-income countries created just above 1% of global income even though they represented 72% of the global population (UNESCAP 2014). In addition, available statistics prove that, as countries have grown and developed, inequalities have increased, the relative situation of the poor has not improved, and rich individuals have become wealthier (Dabla-Norris et al. 2015). All these facts induce a growing recognition among stakeholders, that economic growth is not sufficient to sustainably reduce poverty if it is not inclusive.

The urgency to address inequalities and their consequences requires, at first step, the quantification of income distribution and the level of economic inequality. In this regard, different metrics can be used (20:20 Ratio, Palma Ratio, Hoover index, coefficient of variation, Theil index or others). All available indicators have strengths and limitations, but their appropriateness can be assessed against a number of criteria, especially scale-invariance, principle of transfers and the symmetry or anonymity axiom. The scale-invariance, as the first criterion, signifies that the values of indicators should not change when all incomes change proportionally. The second criterion – principle of transfers – means that the transferring of income from a richer to a poorer person should result in a reduction in inequality as measured by the indicator and the reverse should also hold. Regarding the third criterion – symmetry or anonymity axiom – it signifies that the index must depend only on the income values used to construct it and it should not take into account additional information.

The most widely used indicator is the Gini coefficient, which ranges from 0 (perfect equality) to 100 (complete inequality). It has clear graphical representation and it is easy to interpret but, like other measures of inequality, it suffers from a number of limitations. *First*, it is not additive across groups, which means that the total Gini for a society is not equal to the sum of the Gini for its sub-groups (Galbraith 2012). *Second*, it doesn't identify whether rises or falls in inequality were triggered by changes at the bottom, middle or top of the income distribution ranking. *Third*, the Gini coefficient itself is more responsive to changes in the middle of the income distribution ladder than to changes at the very bottom, or at the very top (Palma 2011).

For this coefficient, two variants can be conceptualized. The first one, named non-weighted Gini coefficient, is obtained by taking each country's GDP per capita as one observation or data point. The second variant refers to weighed coefficient, obtained by weighting each national GDP per capita by each country's population to account for example, the fact that, China's economic growth should have affected more people than growth in a smaller country.

The monitoring of the two coefficients indicates that international inequality has been declining since the early 1980s. Most of this decline can be explained statistically by the rapid growth of China. However, between 1980 and 2000, the international income inequality increased quite sharply, if China is excluded. This increase is due to several factors, particularly declining incomes in Latin America during the 'lost decade' of the 1980s, and the prolonged economic implosion of countries in sub-Saharan Africa, as well as the economic collapse of transition countries in the late 1980s and 1990s. However, since about 2000, the decline in international inequality has been observable even without China. Stronger economic growth in all three major developing regions – Asia, Africa, and Latin America – has contributed to this trend (United Nations (UN) 2013a, b) (Fig. 3.1).

Other discrepancies and facts can be identified. Figure 3.2 for example, shows the national GDP per capita, as well as average GDP per capita of the top and bottom 10% of the population of some selected countries, where the top and bottom of each bar represent the average GDP per capita (PPP in constant 2005 international dollars) of the top and bottom 10% of the population of each country, respectively and the marker in between represents national average GDP per capita. Since this measure of global inequality among individuals reflects, in principle, inequalities within and across countries, and since inequalities across countries did not increase in this period, the rise in global inequality must be due to increased inequalities within countries.

Figure 3.2 shows that the mean income of a resident in Albania or Russia was lower than that of an individual in the lowest 10% of the distribution in Sweden, who also earned 200 times more than an individual in the bottom decile in the

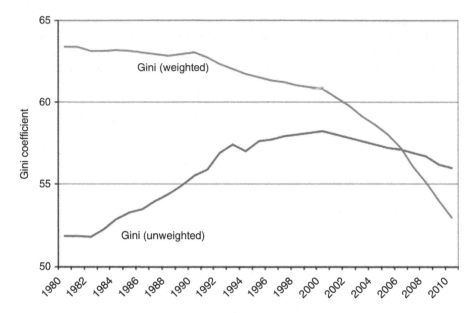

Fig. 3.1 International income inequality (1980–2010)
Source: United Nations (2013a, b)

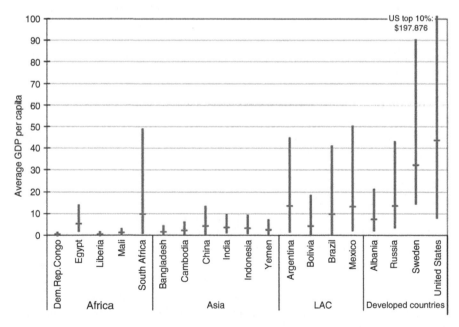

Fig. 3.2 Average income per capita of the top and bottom 10% of the population and of the total population in some selected countries (late 2000s)
Source: United Nations (UN) (2013a, b)

Democratic Republic of the Congo in the late 2000s. In addition, poor people in countries with more inequality can have lower living standards than poor people in countries with lower average incomes but less unequal distribution. Thus, individuals in the bottom 10% earned less in the United States than in Sweden, in Brazil than in Indonesia, and in South Africa than in Egypt in the late 2000s. Historically, Asia has experienced lower inequality than other developing regions. However, despite remarkable growth and impressive declines in extreme poverty, the region has seen widespread increases in income inequality at the national level.

In the opposite side, two key factors can be identified based on their contribution in countries where inequality has declined: the first one is the expansion of education; and the second is related to public transfers to the poor. However, the redistributive impact of social transfers depends on the degree of progressiveness of the tax system (income and property taxes are usually progressive while indirect taxes are regressive). In fact, the increase in public expenditure on education throughout Latin America and the Caribbean in the early 1990s led to rising secondary enrolment and completion rates, and this became a major determinant of the fall in income inequality (Ferreira and Gignoux 2007; Lopez et al. 2012).

Kuznets (1955) noted that inequalities tend to be low at the early stages of development when societies are mostly agricultural, and inequalities increase as industry is developed, countries urbanize and economies grow faster. Thus, increasing inequality has been assumed as a cost of the development process. However, regional and national trends in economic inequality suggest that there is no clear relationship between inequality and development. In addition, the empirical evidence of such a relationship is still ambiguous.

In fact, a comparison of income distribution across countries by GNI per capita in 2012 (Fig. 3.3) shows a slightly inverted U-shape, but country observations are significantly scattered and the correlation between the two variables is small.

Furthermore, the situation regarding poverty over the world is not better than inequality and the number of people living in extreme poverty remains unacceptably high. In fact, according to the most recent estimates of the World Bank (World Bank 2016), 10.7 percent of the world's population in 2013 (=767 million) lived on less than US$ 1.90 a day. This means that poverty reduction may not be fast enough to reach the target of ending poverty by 2030. However, statistics proves that overall prevalence of poverty is in retreat. Based on monetary terms, which capture levels of income or consumption per capita or per household, the last 20 years have seen significant reductions in the depth and severity of extreme poverty in the developing world. In absolute terms, extreme income poverty has fallen substantially, with the number of people living on less than 1.25$ a day having declined from a high of 1.9 billion in 1981 to a low 1.4 billion in 2005. The dramatic drop in poverty levels has been attributed to many factors, especially the improved rates of economic performance and higher wages, as well as the provision of social protection systems.

Statistics also show that the global poverty landscape is changing. In fact, poor people are increasingly found in middle-income countries and in fragile states, with the ascertainment that the number of countries classified as low income has fallen by two fifths and the number of middle-income countries has ballooned to over 100.

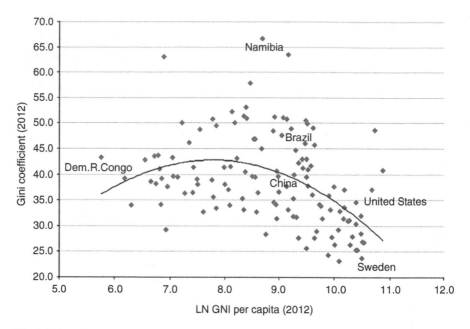

Fig. 3.3 Gini coefficient and Gross National Income (GNI) per capita by country (Source: UNDP (2013))

In addition, the number of fragile states across the world has risen from 28 in 2006 to 37 in 2011 (Chandy and Gertz 2011). The combination of the two trends mentioned above illustrates the evolution of the global poverty landscape. Thus, each country can be measured along three dimensions (Fig. 3.4): the number of poor people (circle area); the degree of fragility (y-axis); and the real gross national income per capita (x-axis). According to this representation, countries that score 90 or above on the Fund for Peace's Failed States index are classified as fragile and countries with real GNI per capita above 1000$ are classified as middle income (Chandy and Gertz 2011).

The analysis of statistics between 2005 and 2010 (Table 3.1) shows that at the start of the series, the majority of the world's poor are accounted for by stable low-income countries. But over time, countries change category and poverty occurs at different speeds in different countries.

These facts induce a need for a new thinking and differentiated approaches for poverty reduction, especially an approach which integrates the link between fragility (conflicts risk), inequality, and poverty. In this context, notwithstanding comparability and inconsistency problem of collected data, statistics show that poorer countries tend to have more conflicts than wealthier countries (Collier 2007). In addition, horizontal inequalities between ethnic groups and states can promote conflict (Cederman et al. 2011).

In fact, there is a growing body of empirical research investigating the link between conflict, inequality, and poverty, which has occupied the minds of thinkers

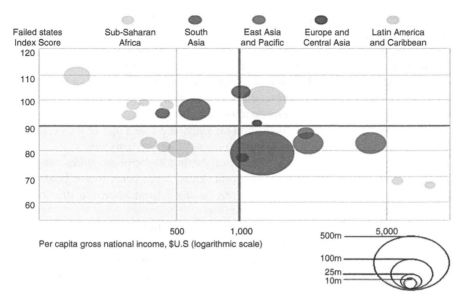

Fig. 3.4 Poverty over the world according to level of income and fragility (2010)
Source: Chandy and Gertz (2011)

Table 3.1 Share of world's poor by country category (Source: Chandy and Gertz (2011))

	2005		2010	
Fragile	19.6%	0.9%	23.7%	17.1%
Stable	53.9%	25.6%	10.4%	48.8%
	LIC	**MIC**	**LIC**	**MIC**

and practitioners for many years. While there is consensus among researchers on the economic consequences of conflict, there is less agreement regarding economic causes of conflict. Regarding consequences, Stewart (1993) explains that the cost of conflicts (especially armed conflicts) appears at three levels:

- Macro: destruction of infrastructure, disruption of markets, etc.
- Meso: influence on the level of public resources for health and education facilities or food subsidies, etc.
- Micro or household: household disintegration, decline of individuals' health and education, and psychological shock.

The costs are often immediate and dramatic. Thus, in a survey of 78 conflicts between 1950 and 2000, Pottebaum (2002) found that real GDP per capita at the end of conflict in Cambodia, Liberia, and Lebanon was more than 50% less than at the beginning. Real GDP per capita declined annually by about 13% during the conflict in Rwanda, and by more than 5% in Sierra Leone and Haiti (USAID 2005).

Regarding the causes, the conflicts may emerge from the resource scarcity due to overuse or depletion, which are often exacerbated by the social and economic repercussions of environmental degradation, unsustainable increase in demand due to population pressures, distributive inequities or increased per capita consumption. Many examples of social-ecological systems (SES) unpleasant dynamics can be evoked such as depletion of water resources, growing interference in ecosystems from forests to wetlands and coral reefs, overexploitation of fisheries, and degradation of arable land. Apart from these, the sea-level rise, the decline of natural habitats, change in precipitation patterns, and the occurrence of more frequent extreme events, such as intense storms, floods, and droughts, are also a part of the problem. All these climate change facts are majorly a result of human activities.

Contestations related to access, control or sharing lucrative resources can also lead to tensions and violence over the world. Thus, if the benefits and burden of extractive projects such as mining, logging, and oil production are distributed unequally, the result may be protracted conflicts.

Poor economic performance increases poverty and reduces government legitimacy and ability to provide services. Inequality among groups also tends to worsen in poor economic environments and during violent conflicts. Thus, researchers perceive poverty as a factor that can fuel grievances. In the reverse, conflicts are recognized as a cause which deepens poverty. In this regard, Collier and Hoeffler (2002) use data from 161 countries and 78 civil wars over the period 1960–99 to investigate two alternative hypotheses for conflict causes. The first one is related to grievances (inequality, political oppression, and ethnic and religious divisions) and the second is related to greed or sources of finance (income from natural exploitation, diaspora, and hostile governments), which largely cause conflicts. The authors conclude that a combination of the two models is most useful in determining the risk of conflict, and argue that, in addition to level and growth of income, the abundance of natural resources is a factor which may increase conflict risk, but the relationship is non-linear.

Furthermore, a growing number of practitioners and researchers indicate that inequality is among the most important causes of violent conflicts. Remembering in particular, the study of Alesina and Perotti (1996), which proves that inequality, entered into their regressions using income shares of the five quintiles of population, leads to an unstable sociopolitical environment and fuels social discontent which is conductive to violent conflict. Besides, Nafziger and Auvinen (2002) conclude that inequality is an important factor in rising conflicts. Based on Gini coefficients, they find that income inequality contributes to humanitarian emergencies and argue that high-income concentration increases the perception of relative deprivation.

Considering all these facts, policy makers must recognize that security issues are complex and conflicts can arise due to various causes that can meet at different levels while reinforcing each other. Interventions cannot focus on a single cause or level, so policy makers must seek solutions for each underlying cause at each particular level. Thus, they must encourage broad and innovative approaches, which integrate socio-ecological systems (SES), defined as integrated systems of human society and ecosystems, with reciprocal feedback and interdependence. SES should be apprehended in a context of coviability that concentrates on the identification of factors which are important in dynamics of the analyzed system. In this context, the coviability analysis can be applied to a complex SES in order to assess the possible reconciliation between human needs, conservation of environment, and economic requirements.

3 SES Characteristics and Coviability Concepts

The interconnection between social and ecological dimensions defines SES. This essential concept is used since it emphasizes the humans in the environment perspective and highlights the impact on ecosystem from local areas to the biosphere as a whole, in providing the biophysical foundation and ecosystem services for social and economic development. Besides, it is worth noting that ecosystem have been shaped, directly or indirectly, by human decisions throughout history. Thus, social and ecological systems are not just linked, but truly interconnected and co-evolving across spatial and temporal scales.

However, it is difficult and even impossible to truly understand ecosystem dynamics and their ability to generate services without accounting for the human dimension. Thus, if the focus is only on the ecological side, results become incomplete and the conclusions partial. In this context, studies have often focused on investigating processes within the social domain, assuming that if the social system performs adaptively or is well organized institutionally, it will also manage ecosystems sustainably. Some key literature can be provided (Table 3.2).

Similarly, focusing on the ecological side only, as a basis for decision making for sustainability, leads to wrong conclusions (Folke 2006). To manage this interaction between social and ecological systems, analytical approaches are needed. One approach is building a common framework and using it to conduct research related to performance of SESs regarding governance, productivity, resilience, equity, etc. This common framework is essential because it helps identify variables that affect the structure of 'Action-Situations', leading to interactions and outcomes (Fig. 3.5). It also helps study similar systems that share some variables while differing in others and, finally, it allows understanding why some systems are not resilient.

Resilience refers to the capacity of an ecosystem to recover from disturbance or withstand ongoing pressures. It is a measure of how well an ecosystem can tolerate disturbance without collapsing into a different state that is controlled by a different set of processes. However, ecosystem resilience is complex to understand and assess because a number of factors can affect it (GBRMPA 2009). For example, an ecosystem's ability to absorb or recover from impacts, and its rate of recovery, depend on the inherent biology and ecology of its component species or habitats, the condition of these individual components, the nature, severity and duration of the impacts, and the degree to which potential impacts have been removed or reduced.

However, identifying the boundaries, components, and relationships between the apparently different parts of this system is not an easy task. In addition, valid knowledge to cope with global problems of SES sustainability are not purely scientific, or

Table 3.2 Some social dimensions of environmental sustainability (Source: Bacon et al. (2012))

Themes	Selected social dimensions
Human health	Food security, hunger, nutrition, wellness, morbidity and mortality from pesticide exposures, food contamination, livestock to human diseases, drinking water contamination, obesity.
Democracy	Participation (voice and vote), decision making, rural associations/ cooperatives, social capital and community cohesion, inequalities in social power, representation, accountability mechanisms, food sovereignty, social movements, governance and government policy (overlaps with equity and justice).
Work	Paid and unpaid agricultural and food system labor (within and beyond households). Employment, wages, changing labor routines, injuries, migration/ immigration, discrimination, collective bargaining (overlaps with equity and justice, health)
Quality of life and human well-being	Income, economic poverty, education, employment, housing conditions, security, life expectancy, as well as subjective perceptions.
Equity, justice, and ethics	Procedural and distributional dimensions of environmental and food justice. Environmental and food access inequalities. Influence of geography, race, class, gender and other markers of social identity upon the distributions of environmental benefits and burdens in Agri-food systems. Ethics of eating, farming, food systems, and intergenerational ecosystem stewardship.
Cultural diversity	Cultural practices, languages spoken, indigenous and hybrid ecological knowledge systems, diet, planned and associated diversity in farms and forests, oral traditions (overlaps with resiliency and vulnerability)

Fig. 3.5 Action-situations embedded in broader SESs Source: Ostrom (2007)

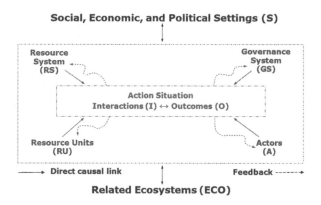

at least not in traditional sense. Thus, to understand how multiple forms of governance influence resource users of various scales and background, and how they affect resource systems that have diverse characteristics, researchers need to draw on multiple scientific disciplines. Ideally, a framework helps policymakers accumulate knowledge from empirical studies and assessments of past efforts and reforms and organize their analytical, diagnostic, and perspective capabilities.

Fundamental to the SES framework is the presumption that humans can make conscious choices as individuals or as members of collaborative groups, and that these individual and collective choices can, at least potentially, make a significant difference in outcomes. These choice processes are not required to follow any specific model of decision or policy making, nor are all outcomes observed required to have been intended by participants in that process.

Processes in SESs are dominated by non-linear phenomena and an essential quality of uncertainty. These observations have led to the notion of complexity, developed through the work of many academicians (SFI 2012). Earlier challenges to the idea of linear causality and reductionistic science go back to general systems theory developed in the 1930s and 1940s (Bertalanffy 1968). In fact, SES need to be sufficiently predictable that users can estimate what would happen if they were to establish particular harvesting rules or no entry territories. It should be noted that unpredictability at a small scale may lead users of pastoral systems to organize at larger scales to increase overall predictability.

A simple system is a system which can be adequately captured using a single perspective and a standard analytical model. By contrast, a complex system often has a number of attributes not observed in simple systems, including non-linearity, uncertainty, emergence, scale, and self-organization.

Non-linearity is related to inherent uncertainty. Mathematical solutions to non-linear equations do not give simple numerical answers, but instead produce a large collection of values for the variables that satisfy an equation. Complex systems organize around one of several possible equilibrium states or attractors. When conditions change, the system's feedback loops tend to maintain its current state. Resilience may be considered an emergent property of a system, one that cannot be predicted or understood simply by examining the components of the system. Resilience absorbs change and provides the capacity to adapt to change. Scale is important in dealing with complex systems, because this kind of systems should be analyzed or managed simultaneously at different scales. In addition, complex systems are also characterized by self-organization, which means that open systems will reorganize at critical points of instability. These characteristics of complex systems have a number of rather fundamental implications for resource and environmental management, as the necessity of integrating coviability dimension.

In this context, the analysis of coviability is important because it defines more precisely the current and desired condition of system targets. It permits to identify priorities, get insight into the nature of the threats to be targeted, and identify the most scientifically-appropriate indicators. The output of coviability is a realistic plan for SES monitoring to assess the impact of actions. In simple words, the coviability of SES aims to provide the right analytical framework, to address sustainability and to avert catastrophic outcomes, by remaining in the system within the realms of safety or acceptability (Krawczyk et al. 2013). To achieve this, the coviability analysis starts with a very general question: *is the system threatened, and if so, why?* In this initial step, the coviability analysis concentrates on the identification of factors that are important in the dynamics of the system. In the following advanced stage, the coviability combines mathematical and algorithmic tools to investigate the adaptation to coviability constraints of evolutions, governed by complex systems under uncertainty. However, the coviability does not require just mathematical tools, but involves interdisciplinary investigations spanning fields that have traditionally developed in isolation. But, it is important to take into consideration, when combining interdisciplinary fields, data availability, consistency of all the aspects of used coviability approaches and finally, to establish a tradeoff between reality and simplicity.

4   The Impact of Coviability Considerations on Insecurity Risks, Poverty and Inequality Reduction

The dysfunction of SES mechanisms can lead to various deleterious effects. In fact, the wealth generated from environmental resources can contribute to the increase of inequality between individuals, groups, and nations. In this context, inequality generates a less stable and less efficient economic system, which stifles economic growth and the participation of all members of society in the labor market (Stiglitz 2012).

Inequality has snowball impacts. First, unequal concentration of income and wealth reduces aggregate demand and slows down economic growth. In the next step, inequalities pose a serious barrier to social development by slowing the pace of poverty reduction and limiting opportunities for social mobility. This situation leads to intergenerational transmission of unequal economic and social opportunities. As a result, it creates poverty traps, wastes human potential, and leads to less dynamic and less creative societies. Intermixing of these varied impacts can generate economic crises, social tension, and provide fertile grounds for political and civil unrest heightening human insecurity. Impacts of all these aforementioned facts can be aggravated by global environmental hazards, such as climate change, stratospheric ozone depletion, biodiversity loss, changes in hydrological systems and the supplies of freshwater, land degradation, and stresses on food-producing systems (WHO 2005) (Fig. 3.6).

Development failures may increase people's exposure and vulnerability to natural hazards and disasters, thereby creating risk through reducing the effectiveness of established coping mechanisms and by generating new hazards. However, the poorest people and countries are often the most affected by environmental degradation. They have to eke out a living from marginal lands, forests, coastal waters or the peripheries of urban centers. The urban poor are the most exposed to severe air and water pollution and cannot escape the negative impacts of urbanization (Fig. 3.7).

In fact, the environment-poverty nexus is a two-way relationship. Environment can affect poverty situations in three distinct dimensions: by providing sources of livelihood to poor people; by affecting their health; and by influencing their vulnerability (Jehan and Umana 2003). On the other hand, poverty also affects environment in various ways: by forcing poor people to degrade environment and overuse its resources; by encouraging countries to promote economic growth at the expense of environment; and by inducing societies to downgrade environmental concerns,

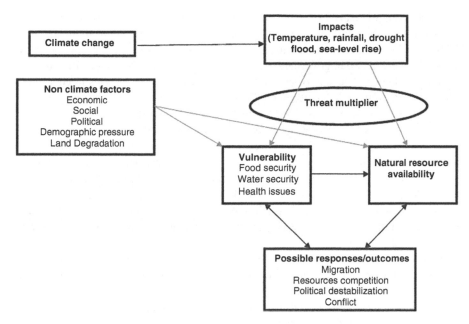

Fig. 3.6 Climate and non-climatic factors that can lead to violent conflicts
Source: United Nations (2012)

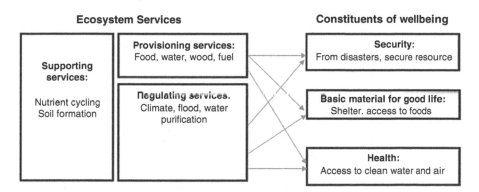

Fig. 3.7 Illustration of linkages between ecosystem services and human well-being
Source: Keese (2006)

including failing to channel resources to address such concerns. In the reverse, environment matters a lot to poor people since their well-being is strongly related to the environment in terms of, inter alia, health, earning capacity, security, physical surroundings, energy services, and decent housing. In rural areas, poor people may be particularly concerned with their access to and control over natural resources, especially in relation to food security. For poor people in urban areas, access to a clean environment may be a priority.

In addition, impacts can be very deleterious when observing fragile states which are defined by their failure to deliver security and basic services to their citizens. Thus, fragile states refer to countries that are suffering from severe development challenges such as weak governance and political legitimacy, limited administrative capacity, violence or legacy of conflict, complex array of weaknesses in economic management, regulatory quality, social inclusion, and institutional effectiveness. All these aspects of weakness can lead to violent conflicts, but the precise mechanisms are frequently less explored. Many examples can be evoked. The first one concerns conflict in Sudan, where the environmental issues and competition over agricultural land use were an important causative factor in the instigation and perpetuation of conflict. The second example concerns the challenge of rising land and water scarcity in Rwanda, where high population pressure and acute land scarcity in rural areas have resulted in land fragmentation. Another example is related to the increasing water scarcity in the Gaza Strip, driven by demand and supply factors (United Nations 2012).

Moreover, the environmental impacts of development projects can also create tensions if communities are not compensated for the damage and do not receive a share of the development benefits, financial or otherwise (United Nations 2012). Overlapping resource rights and discriminatory policies are contributing to violent conflicts in the highlands of Afghanistan, Bougainville and Ogoniland (Nigeria) and deleterious impact of water privatization without community consultation in Cochabamba (Bolivia).

Thus, there is a very strong link between conflict, natural hazards, and the governance of environmental resources and particularly, disagreements regarding statutory, customary, informal, and religious rules of environmental regulation are often at the heart of conflict (United Nations 2012) (Fig. 3.8).

It should be noted that non-violent resolution of conflicts is possible when the parties have trust in their governing structures and institutions to manage incompatible or competing interests. But conflict becomes problematic when mechanisms for managing and resolving them break down and give way to violence. In addition, while competing interests or opposing beliefs lie at the heart of any conflict, these also interact with the level of available information, the previous history, level of trust between concerned parties, and the prevailing system of decision-making (Sidaway 2013). In this violent context, institutional reform is often difficult or impossible and countries that fail to build legitimate institutions risk entering a vicious cycle of repeated violence and weak institutions (World Bank 2011).

The same lines of reasoning that connect poverty and environment also connects the inequality to the environment. In this regard, two hypotheses can be advanced. The first confirms that environmental harm is not randomly distributed across the population, but instead the distribution of wealth and power. The relatively wealthy and powerful tend to benefit disproportionately from the economic activities that generate environmental harm (Boyce et al. 2007). The second hypothesis states that the total magnitude of environmental harm depends on the extent of inequality. Thus, societies with wider inequalities of wealth and power will tend to have more environmental harm.

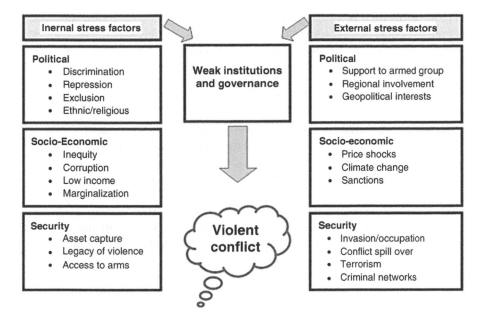

Fig. 3.8 Internal and external stress factors contributing to violent conflicts
Source: United Nations (2012)

The relationship between environmental resources, conflict, poverty, and inequality should convince decision makers to better manage SES and their mechanisms. However, large socio-ecological problems are difficult to model, analyze and mitigate because they involve interacting environmental, economic and social systems with complex, non-linear dynamics (Taylor 2014). At present, researchers from many disciplines are exploring ways to improve the resilience of SES, prevent failures, and support the transition to a viable system.

In this regard, natural resources can provide a stepping stone for moving to more sophisticated and higher-valued industries (Keese 2006). As example of management action which can assure coviability, the sound forest management has the potential to open the door to a range of industries linked to wood processing and generate considerable employment opportunities, including for skilled and semi-skilled labor. Concerning rich fisheries resources, we can for example move from fishing to caning and to more advanced agro-processing industries. Besides, eco-tourism is often presented as a profitable activity which employs skilled workers. In this concern, a study for the Kruger National Part in South Africa (UNDP et al. 2005) suggests that wildlife conservation is 18 times more profitable than options such as using the land for livestock and crops.

Many cases show that coviability of SES can have a positive impact on well-being. One such case concerns livelihood diversity and redundancy in coastal communities in East Africa. Households along the coast often engage in small-scale fisheries as part of a diverse livelihood portfolio which might include working in tourism, agriculture or casual labor. While households may maximize their total income by specializing in a single livelihood activity, households who have a portfolio of options tend to be more resilient, particularly if different livelihood activities are not affected by the same disturbances (SRC 2015). For example, in households with diverse livelihood portfolios, fishing activities can continue when the tourism sector suffers low number of tourists due to global perceptions of security.

The second case concerns avoiding poverty traps in Tanzania. In fact, while feedbacks can help keep a system in a desirable situation, they can also lock a system into an undesirable configuration (SRC 2015). The consequence is that they become trapped in a vicious cycle of poverty. In Tanzania, for instance, rainwater harvesting and conservation tillage can help restore soil fertility and reduce the impacts of drought.

The third case concerns the complex adaptive systems (CAS) thinking. The evolution of management paradigms in the Tisza River Basin in Europe, as an example of CAS thinking, has supported change in approaches to river management (SRC 2015). The network used participatory science to develop a CAS understanding that recognized cross-scale drivers, uncertainty, and the importance of incorporating multiple views into river management practices.

The last example is related to polycentric governance in southern Arizona, where a number of collaborations on environmental management and the promotion of ecosystem services were taken together. Thus, over 20 different groups and actors contribute to decision-making processes about pressing environmental challenges in the region.

Furthermore, improving environmental management goes hand in hand with economic modernization and diversification of income opportunities. This, in turn, goes in tandem with improved literacy and education, better access to communication technologies, and enhanced participation in global markets. So, environmental management, development, poverty education, economic growth, and maintenance of life-supporting environmental resources are closely linked. In this context, the links between poverty and the environment can be best understood by focusing on the diverse asset base of the poor, which ranges from housing to access to infrastructure, from livestock to tools and cash, from community relationships to soil quality, forests and clean air and water.

In fact to make the case for linked poverty and environment policy and planning, it is important to look at the economic and governance dimensions of sound environmental management as it relates to poverty reduction, examining the specific cases of mainstreaming conservation practices into the agricultural sector, using consumer choice to catalyze sustainable production patterns, and exploring the emergence of new market-based instruments for sustainability such as payments for ecosystem services.

Once various interactions are placed in a dynamic context, different factors in response to each other, the picture becomes even more complex. Rapid population change tends to make resource degradation worse, which, in turn, requires institutional adaptation to cope with the changes; however, adaptation itself depends on many factors and is not automatic.

While a great deal is known about coviability of SES through poverty-environment relationships, there remain large gaps in the information base required to effectively manage the environment for poverty reduction. Priorities for new information include (UNDP et al. 2005):

- Geographic data on the spatial overlay of poverty, environmental quality, and resource rights.
- Improved wealth accounting including a wider range of resources and ecosystem services, as well as household-level wealth accounts.
- More consistent evaluation of costs as well as benefits when assessing environmental investments.
- Empirical evidence of the links between biological diversity per se, ecological resilience, and poverty.
- Better understanding of how common property regimes evolve in the face of increasing resource scarcity.
- More analysis of the environmental impacts of subsidies for water, fishing, and land conservation.
- Better understanding of the social (poverty) impacts of environmental policy and investments at a sub-national level.
- Better understanding of the interactions and interdependencies among different forms of capital (social, natural, human, etc.).

Understanding the links between inequalities, poverty, environmental degradation, and potential negative spiral effects, and taking responsive actions, can accelerate positive dynamics and promote sustainability outcomes (AFDB 2007). The focus on these connections, as a means to inform development policies, builds upon political ecology (PE) analysis, which aims to influence policy development and investment programs by offering chains of explanations rather than single root causes. PE highlights the sociopolitical dimensions of natural resource access, control, and distribution (Robbins 2004). Hence, PE is founded on a coviability analysis which helps us rank a set of choices or alternative management measures and allows us to explore which policies will ensure strong sustainability. Nevertheless, the PE is not yet a part of mainstream political and economic practical actions, especially in developing countries.

5   Some Key Messages and Recommendations

The chapter confirms that ecological and social problems are tightly and strongly linked. In fact, environment represents different ecosystems which are characterized by diverse actors and multiple stressors and timescales. In this broader context, negative effects generated by some ecological and social mechanisms often degrade the well-being of populations and present a burden for groups that are particularly vulnerable to extreme dynamics such as climate extreme events, conflicts, and extreme poverty. Thus, a key challenge pointed out by prior research on SESs is how to find an approach which allows matching the best governance to the scale of problems and social-ecological mechanisms. Similar challenges have been exposed in ecology and conservation biology, where the importance of scale and the human dimension have been regularly underestimated in the past. Nevertheless, several concepts and initiatives, such as the SES coviability, have not gained prominence.

This chapter demonstrated how the coviability is essential to combat poverty, reduce inequality, and address insecurity risks. Coviability in this particular context moves beyond viewing people as external drivers of ecosystem dynamics, and rather looks at how they are a part of and interact with environment, through their use of different ecosystem services, regulation of the climate, and their spiritual and cultural connections to ecosystems. The coviability approach aims at investigating how these interacting SESs can be best managed to ensure a sustainable and resilient supply of the essential ecosystem services on which humanity depends. To enhance and build coviability, different factors can be proposed such as: maintaining diversity and redundancy; managing connectivity; fostering complex adaptive systems thinking; encouraging learning; broadening participation; and promoting polycentric governance systems.

Maintaining diversity and redundancy is based on evidence from several fields of study, which suggest that systems with many different components are generally more resilient than systems with few components. Thus, functional redundancy can provide insurance within a system by allowing some components to compensate for the loss or failure of others. This redundancy is even more valuable if diversity is present, which means that insurance is guaranteed, if the components providing it react differently to change and to disturbance. Furthermore, limited connectivity can sometimes boost the coviability by acting as a barrier to the spread of disturbances, but high levels of connectivity between different social groups can increase information-sharing and help build trust and reciprocity. Thus, in order to continue to benefit from a range of environment services, we need to understand the complex interactions and dynamics that exist between actors and ecosystems in a SES. This can be done by supporting long-term monitoring of key social and ecological components, providing opportunities for interaction that enable extended engagement between participants and by enabling people to network and to create communities of practices. Participation through active management is considered fundamental, because it helps build the needed trust and relationships and improve legitimacy of knowledge and authority during decision-making process. In this context, polycentricity governance system is highly recommended. This management approach represents a system in which multiple governing bodies interact to make and enforce rules within a specific policy arena or location.

The level of human development through reducing poverty, inequality, and conflict risks, can be raised and the environment can be protected if a concerted effort is made to align the aforementioned criteria into clear and consistent set of practices, policies, and behaviors. This can be done, essentially, by improving collaboration across institutions and scales and by implementing actions which target vulnerable population and area.

References

African Development Bank (AFDB) (2007) The poverty-environment Nexus in Africa. In: Gender, poverty and environmental indicators on African countries, vol 92, Tunis

Alesina A, Perotti R (1996) Income distribution, political instability and investment. Eur Econ Rev 40(6):1203–1228

Bacon N, Cochrane D, Woodcraft S (2012) Creating strong communities: how to measure the social sustainability of new housing developments. Berekeley Group, London

Bertalanffy LV (1968) General system theory: foundations, development, applications. Braziller, New York

Boyce JK, Narain S, Stanton EA (2007) Reclaiming nature: environmental justice and ecological restoration. Athem Press, London

Cederman LE, Nils BW, Kristian SG (2011) Horizontal inequalities and ethno nationalist civil war: a global comparison. Am Polit Sci Rev 105(3):478–495

Chandy L, Gertz G (2011) Poverty in numbers: the changing state of global poverty from 2005 to 2015. Brooking Institute, Washington, DC

Collier P (2007) The bottom billion, Oxford. Oxford University Press

Collier P, Hoeffler A (2002) Aid policy and growth in post-conflict societies. The World Bank, Washington DC

Economic and Social Commission for Asia and the Pacific ESCAP (2014) annual report, ESCAP, Bangkok, Thailand

Ferreira FHG, Gignoux J (2007) Inequality of economic opportunity in Latin America. The World Bank, Washington DC

Folke C (2006) Resilience: the emergence of a perspective for social-ecological systems analyses. Glob Environ Chang 16:253–267

Galbraith J (2012) Inequality and instability, Oxford\New York. Oxford University Press

Great Barrier Reef Marine Park Authority (GBRMPA) (2009) Ecosystem resilience. Great Barrier Reef Outlook report, Townsville

Jehan S, Umana A (2003) Environment-poverty nexus. Dev Policy J 3:53–70

Keese, M (Ed) (2006) Live longer, work longer. OECD, Paris

Krawczyk J, Pharo A, Serea OS, Sinclair SD (2013) Computation of viability kernels: a case study of by-catsh fisheries. Comput Manag Sci 10(4):365–396

Kuznets S (1955) Economic growth and income inequality. Am Econ Rev 65:1–28

Laurent E (2015) Social-ecology: exploring the missing link in sustainable development. Standard University

Lopez-Calva LF, Lustig N, Scott J, Castaneda A (2012) Cash transfers and public spending on education and health in Mexico 1992–2010: impact on inequality and poverty. Mimeo. The World Bank, Washington DC

Milanovic B (2012) Evolution of global inequality: from class to location, from proletarians to migrants. Global Pol 3(2):125–134

Nafziger EW, Auvinen J (2002) Economic development, inequality, war and state violence. World Dev 30(2):153–163

Norris ED, Kochhar K, Suphaphiphat N, Ricka F, Tsounta E (2015) Causes and consequences of income inequality: a global perspective. International Monetary Fund, Washington, DC

ParisOstrom E (2007) A diagnostic approach for going beyond panaceas. Natl Acad Sci Proc 104 (39): 15181:15187

Palma JG (2011) Homogeneous middles vs. heterogeneous tails and the end of the 'Inverted-U': the share of the rich is what it's all about. Dev Chang 42(1):105–118

Pottebaum D (2002) Economic and social welfare in war-affected societies. Cornel University, Ithaca/New York

Robbins P (2004) Political ecology. A critical introduction. Blackwell Publishing, Malden MA

Santa Fe Institute (SFI) (2012) www.santafe.edu/about, an independent, not-for-profit research and education center, Santa Fe, New Mexico

Sidaway JD (2013) Geography, globalization and the problematic of area studies. Ann Assoc Am Geogr 103(4):984–1002

Stewart F (1993) War and underdevelopment: can economic analysis help reduce the costs? J Int Dev 5(4):357–380

Stiglitz JE (2012) The price of inequality: how today's divided society endangers our future. Kindle edition

Stockholm Resilience Center (SRC) (2015) Applying resilience thinking: seven principles for building resilience in social-ecological systems, Stockholm University

Taylor G (2014) Viability: a priority criterion for the mitigation of climate change and other complex socio-ecological issues. J Futures Stud 19(1):77–96

United Nations (UN) (2012) Renewable resources and conflict, United Nations Interagency Framework Team for Preventive Action, New York

United Nations (UN) (2013a) The Millennium development goals report, Department of Economics and Social Affairs, New York and Geneva

United Nations (UN) (2013b) Inclusion Matters: The Foundation for Shared Prosperity-Overview, Report on the world social situation, Department of Economic and Social Affairs, Washington DC

United Nations Conference on Trade and Development (UNCTAD) (2014) Investing in the SDGS: An action plan. World investment report, New York and Geneva

United Nations Development Programme (UNDP) (2010) The real wealth of nations: Pathways to human development. Human development report, New York

United Nations Development Programme (UNDP) (2013) The rise of the South: Human progress in a diverse world. Human development report, New York

United Nations Development Programme (UNDP), United Nations Environment Programme (UNEP), International Institute for Environment and Development (IIED), International Union for Conservation of Nature (IUCN), World Resources Institute (2005) Sustaining the environment to fight poverty and achieve the MDGs: The economic case and priorities for action, prepared on behalf of the Poverty-Environment Partnership, New York

United States Agency for International Development (USAID) (2005) Conflict, poverty, inequality, and economic, growth. Human development report, New York

World Bank (2011) Conflict, Security, and Development Development Report. Washington, DC

World Bank (2016) Poverty and shared prosperity: taking on inequality, Washington, DC

World Health Organization (WHO) (2005) Multi-country Study on Women's Health and Domestic Violence against Women, Geneva

World Wildlife Fund (WWF) (2014) Living planet report: species and spaces, people and places. Switzerland

Chapter 4
'Whaling Out' the Climate Problem: A New Framework of Analysis?

Laurent Weyers

Abstract It is often considered that states may quite discretionarily define the course of their actions in order to combat global warming. While states do indeed enjoy discretion in this matter, it is not without limit and it can arguably be reviewed. Although the task of determining where the threshold lies is an arduous one, reference to the reasoning followed by the International Court of Justice in the *Whaling in the Antarctic case* can somewhat facilitate its accomplishment. This is due to the analogy that can be drawn between the kind of legal questions raised in the *Whaling case* and those pertaining to combating climate change: whether and how a state's discretionary power can be subject to judicial review; whether and how disputed scientific facts can be arbitrated upon; whether states, when determining the appropriate course of their actions, may be expected to give regard to certain elements of fact and law? From the *Whaling case*, it thus seems that a new framework of analysis can be drawn, allowing to better articulate the politics of climate change with the law and to transcend some of the main shortcomings of the international climate regime. This chapter is aimed at sketching out the main contours of this framework of analysis, the benefit of which, it is argued, is to uncover new ways around the complex issues that the climate problem encompasses, and to thereby help specify what obligations states have in this regard.

Keywords Climate change · Whaling case · Discretionary power · Scientific uncertainty · Duty to give due regard

L. Weyers (✉)
Université Libre de Bruxelles (ULB), Bruxelles, Belgium
e-mail: lweyers@ulb.ac.be

© Springer International Publishing AG, part of Springer Nature 2019
M. Behnassi et al. (eds.), *Human and Environmental Security in the Era of Global Risks*, https://doi.org/10.1007/978-3-319-92828-9_4

87

1   Introduction

According to a widespread opinion, the international climate regime fails, or is being much too slow, to address the climate problem adequately. This is due, as is often criticized, to the inability of the regime to legally constrain states' behavior. But pragmatics, on the other hand, highlight the numerous complexities involved and, in light of these, defend the view that crucial achievements have been made possible through, and thanks to the climate negotiations (Vihma and van Asselt 2012; Depledge and Yamin 2011:434; Gupta 2014: 194). In fact, these opposing views merely reflect different expectations that one can have towards the international climate regime. Indeed, it is often just seen as arranging the process by and through which a shared understanding of the problem can be sought, and hopefully found, but only gradually, one piece after another, allowing for a common plan to eventually unfold.

Yet, it is also sometimes more boldly expected from it, as would a magic wand, to provide concrete ways in which the problem can be resolved once and for all, disregarding states' dissent if and where need be. Gradual process or magic wand, the story so far credits the former rather than the latter perspective, and the Paris Agreement (PA) seems to again recently have confirmed it when definitively endorsing the shift to a bottom-up dynamic. Hence, trying to work things out from within the climate regime may raise concern as to the content and scope of states' obligations with regards to climate change, seemingly strictly dependent upon what states concede to and therefore likely to remain, for some more time to come, of a generally and persisting abstract nature.

Meanwhile, as is now well acknowledged, the warming of the climate does not wait, nor do the adverse consequences brought along with it. The question thus often arises whether and how the climate problem can otherwise be sorted out. In this regard, though many differents paths have been scouted, one has so far been left relatively unexplored. Could the climate problem not just simply be *'whaled out'*? By this, we mean that the kind of reasoning followed by the International Court of Justice (ICJ) in the quite recent *Whaling case* (ICJ 2014) uncovers interesting new ways around the complex issues that the climate problem encompasses, and it may thus possibly help specify what obligations states have in this regard. Indeed, the context in which judgment was passed in this case, as would undeniably be that of a climate ruling, was one marked with scientific uncertainty and ample state's discretion. Be this as it may, the ICJ allocated international responsibility to Japan, finding the latter, in light of the relevant circumstances, not to have reasonably behaved.

Now, can the same standard – reasonale behavior – be of equal relevance when it comes to allocating international responsibility for contributing to climate change? In our view, there is future in arguing in this sense – and others have even started to do so already (Young and Sullivan 2015:337; Maljean-Dubois et al. 2015:73) – but contestation is however likely to follow, as exemplified by a quite recent encounter between Sands (2016) and Bergkamp (2015). Whereas the *Whaling case* may offer novel grounds from which to tackle the climate problem, it also thus prompts some resistance along the way. To the best of my knowledge, these new grounds have not yet been fully explored, nor have the possible motives for resistance been fully addressed, which I here propose to do. For this purpose, I will first review the specifics of the *Whaling* case, with a particular focus on the reasoning which the Court therein followed. I will then reflect on the question of the relevance that this case has with respect to climate problem, relevance that, in my view, can neither be denied in terms of facts nor in terms of law. Finally, I will assess the use that can concretely be made of the teachings provided by the ICJ to overcome the long-lived indeterminacy of states' obligations with respect to climate change mitigation.

Despite remaining obstacles, I believe that a strong case can be made – and, the more embrace the view, the further it shall be strengthened – that the climate problem can be *whaled out*. From the *Whaling* case, a new framework of analysis can indeed be drawn, allowing to better articulate the politics of climate change with the law and to transcend some of the main shortcomings of the international climate regime, among which the loose sense of obligation that it permeates as well as its poor interaction with other relevant rules and institutions of international law.

2 The Whaling Case in (Very) Brief[1]

On 31 March 2014, the ICJ found that Japan, in granting permits to kill whales, had not complied with the requirements of Article VIII, par. 1 of the International Convention for the Regulation of Whaling (ICRW) – which allows only for such permits to be granted 'for purposes of scientific research' – and that it had therefore acted in contradiction with its obligations under the Convention (ICJ 2014: par 247). The case had been brought before the ICJ by Australia, with New-Zealand also later intervening in the proceeding. According to these two, the conduct of Japan in relation with its alleged scientific whaling programme in the Antarctic (JARPA II) was in breach of that state's obligations under the ICRW. More specifically, it was claimed that JARPA II could not be considered to have been designed and implemented 'for purposes of scientific research', and hence the killing of whales, on which Japan had quite heavily relied on for sample collection, was not justifiable under Article VIII, par. 1 of the ICRW. It was even quite straightforwardly argued that the real motives behind the killing of whales were lying, not in Japan's interest in science, but in that state's dissimulated desire to pursue commercial whaling notwithstanding the provisions of the Schedule to the ICRW. On its side, Japan had readily conceded that whales had been killed, and their meat been sold, throughout the implementation of JARPA II, but it had claimed a sovereign right to authorize these killings on the basis of the discretion left to all states parties by Article VIII, par. 1 of the ICRW. The main issue at hand was thus whether the killing

---

[1] The content of this section is inspired by a previous article in which I commented on the *Whaling* case (Weyers 2013).

permits had been issued in compliance with the requirements set out in the afore-mentioned article – which included a determination of what discretion there was for states parties to exercise when issuing scientific killing permits – and, if they had not, whether that constituted a breach of the ICRW.

Now, as the terms used unambiguously indicate, Article VIII, par. I, of the ICRW does indeed leave, to states parties to the ICRW, a certain room for discretion. Yet, this discretion is not unlimited. In fact, and in light of the relevant circumstances, it is subject to a certain control by the judiciary as to how it is being exercised, and if that is in a reasonable manner, as the judgment rendered by the ICJ on 31 March 2014 would reveal. Article VIII, par.1, of the ICRW reads as follows:

> 'Notwithstanding anything contained in this Convention, *any Contracting Government may grant* to any of its nationals a special permit authorizing that national to kill, take and treat whales for purposes of scientific research subject to such restrictions as to number and subject to such other conditions *as the Contracting Government thinks fit*, and the killing, taking, and treating of whales in accordance with the provisions of this Article shall be exempt from the operation of this Convention [...]' (ICRW 1948: Art VIII par 1)

The italics – which are mine – are applied to the terms from which it resulted, in Japan's opinion, that states parties to the ICRW discretionarily could decide to issue special permits to kill whales. In its counter-memoir (§222), Japan had indeed argued that Article VIII, par. 1 of the ICRW allowed for states parties to the ICRW to exercise nearly absolute discretion to issue those permits. For its part, the ICJ considered that states did indeed enjoy a certain discretion in this regard, but however decided that:

> 'whether the killing, taking and treating of whales [...] is for purposes of scientific research cannot depend simply on [the] State's perception' (ICJ 2014: par 61).

Hence the direction taken by the ICJ to engage in a review of Japan's decision to grant the killing permits. As we now know, the ICJ found Japan to not have complied with the requirements of Article VIII, par. 1 of the ICRW, but this conclusion was not so obvious from the outset. Indeed, there were elements likely to cast some doubts on what the outcome of this proceeding would be. Before all else, did Australia have locus standi to bring the case before the ICJ? Could the ICJ review a decision to grant a permit 'for purposes of scientific research' without having to take on the most delicate task of defining what qualifies as 'scientific'? And in the affirmative, which were then the elements that the ICJ could possibly take into account to evaluate the scientificity of what did very much look – at least at first sight – like a scientific research programme? This is only to name but a few of the elements that, I think, could have proven obstructive, yet did not obstuct much at all.

Firstly, the seemingly dubious question of standing turned out to be less problematic than could have been awaited. As has been elsewhere argued (Hamamoto 2014:1–6), this is probably somewhat related to the fact that, not long before judgment was passed in the *Whaling case*, the ICJ had also rendered its landmark decision in the *Obligation to Prosecute or Extradite case*. There, the ICJ had found that all states parties to the Convention against Torture had a common and legal interest in ensuring protection of the rights, and compliance with the obligations therein enshrined (ICJ 2012: par 67). It could surely be expected that the same kind of finding would hold in the *Whaling case*. It is therefore not so surprising that, in the *Whaling case*, the ICJ did not reverse a position it had upheld just about a year before; not more so that Japan did not try to oppose the admissibility of Australia's claim.

After this first hurdle, the question as to whether the ICJ would have to define what 'scientific research' was, and conversely what it was not, was next to come. It may seem counter-intuitive to even suggest that this would not necessarily have to be the case. If one is to decide if a whaling programme is 'for purposes of scientific research', well then one surely has to make plain what 'scientific research' is. In its Memorial (pp 120–126), Australia had therefore provided the ICJ with quite a detailed description of what it considered the 'essential characteristics of a programme for purposes of scientific research in the meaning of Article VIII' to be. However, the ICJ chose not to walk along that path. Not only did the ICJ dismiss the criteria advanced by Australia, it also clearly stated that it did not consider 'necessary to devise alternative criteria or to offer a general definition of scientific research' (ICJ 2014: par 86). Instead, the ICJ indicated that it would consider

'if the killing [...] of whales [was] *for purposes of scientific research* by examining whether, in the use of lethal methods, the programme's design and implementation [were] reasonable in relation to achieving its stated objectives' (ICJ 2014: par 67).

From that point on, the question thus became what reasonable behavior could be expected from Japan, and if Japan had indeed behaved in such a reasonable manner. One could say, in other words, that it all came down to a matter of the relevant facts being examined in the light of the relevant circumstances. As for the facts, the following was particularly looked into by the ICJ:

'decisions regarding the use of lethal methods; the scale of the programme's use of lethal sampling; the methodology used to select sample sizes; a comparison of the target sample sizes and the actual take; the time frame associated with [the] programme; the programme's scientific output; and the degree to which the programme coordinates its activities with related research projects' (ICJ 2014: par 88).

As for the relevant circumstances, the ICJ interestingly noted that there were reasons for Japan to consider the feasibility of non-lethal methods, and not to use lethal methods to a greater extent than was necessary. One such reason was that Japan, as a state party to the ICRW, had

'a duty to co-operate [...] and thus should give due regard to recommendations calling for an assessment of the feasibility of non-lethal alternatives' (ICJ 2014: par 83)

In addition, the ICJ had also carefully pinpointed several acknowledgements and recognitions made by Japan, namely that of being

'under an obligation to give due regard to [...] IWC resolutions and Guidelines call[ing] upon States parties to take into account whether research objectives c[ould] be achieved using non lethal weapons' (ICJ 2014: par 137)

And so, after carefully and lengthily reviewing all of the elements of JARPA II's design and implementation, the ICJ finally reached its conclusion. On the one hand, the ICJ did acknowledge the existence of a certain discretionary power to be exercised by Japan when granting whaling permits, and it also even recognized that JARPA II did comprise activities that could broadly be characterized as scientific research, yet decided, on the other hand, that the permits had not in fact been issued 'for purposes of scientific research' in the sense of Article VIII, par 1 of the ICRW (ICJ 2014: par 227). This was, as we have hereinabove explained, because of a range of facts and circumstances from the examination of which it resulted, in the eyes of the ICJ, that

> 'little attention [had been] given [by Japan] to the possibility of using non lethal research methods more extensively' (ICJ 2014: par 225)

As much discretion as Japan did arguably enjoy, the ICJ nonetheless considered that it had been abused or misused. More exactly, perhaps, somewhat reversing the burden of proof to that state's disadvantage, the ICJ considered Japan not to have provided convincing explanations as to why the killing of whales served the purpose of scientific research. From this, it would then logically follow that the killing of whales was contrary to the prohibition of commercial whaling contained in the Schedule attached to the ICRW.[2]

3   Searching for an Analogy

At first sight, one cannot help but wonder what relevance the *Whaling case* could possibly have in respect of problems and issues arising out of the warming of the climate. Can it be dealt with climate the same way it can with whales? If, by that, it is simply proposed to apply the provisions of the ICRW to legal issues arising out of the warming of the climate, the question is absurd and the answer, of course, negative. But if the question is whether it can be reasoned about duties and obligations resulting from climate treaties in the same way that it can with those of the ICRW, the answer might then just be a positive one. In international legal reasoning, recourse is indeed sometimes being made to the use of analogies, a technique that is meant to infer the unknown or indeterminate from what one knows and can observe (Salmon 1984:495–498). Not all analogies, of course, are equally convincing, nor has the recourse to such a method always been accepted (Anzilotti 1929; Strupp 1930). Here, it is therefore all the more necessary to provide elements that will make the claim for an analogy convincing, at best, or at least acceptable. Some such elements have elsewhere already been sketched out, as we mentioned in our introduction. At a very well-attended symposium held in September 2015 at the UK Supreme

---

[2] The ICJ distinguishes between three different categories of whaling: scientific whaling (permitted under Article VIII, par. 1); aboriginal subsistence whaling (permitted under paragraph 13 of the Schedule to the ICRW); and commercial whaling. Accordingly, whaling activities not falling into the first two categories necessarily fall into the third (ICJ 2014, par 230). For a critique, see the dissenting opinion of Judge Bennouna.

Court,[3] Sands had indeed referred to the *Whaling case* and expressed the view that it could be of some assistance to resolve climate change-related disputes. I will first concisely comment on Sands' assertion – as well as on the encounter which followed with Bergkamp – and give my reasons for agreeing with the former that the ICJ did somewhat clear the way as to how to take on the challenge of scientific uncertainty in the court room. I will then more personally reflect on the relevance the *Whaling case* can have in respect of climate change, with particular regard given to the approach followed by the ICJ to review how a state – in this case Japan – exercises its discretionary competences.

In his reply to Sands' contention on the role of the judiciary with regards to climate change, Bergkamp (2015: 87) wrote that:

> '[t]he relevance of this case to a climate science ruling [...] is tenuous and solely procedural'.

In so writing, Bergkamp seemed to want to oppose what a possible consequence would be of recognizing the relevance of the *Whaling case* to climate change-related disputes, that is, as he puts it, to see

> 'the ICJ [...] pontificate on climate science [...] so that dissent be squashed and ambitious climate policies [...] be put in place around the world' (Bergkamp 2015:83).

However, this does not exactly appear to reflect how Sands had envisioned the role of the judiciary in such matters, nor his appreciation of the extent to which the *Whaling* case could be of relevance to sort out the climate problem. Sands did say, that is true, that

> 'in the face of sharply differing [scientific] opinions [...] a less robust court might have concluded that [...] it was not for the court to take a view'.

And, having said that, he then seemed to rejoice that '[t]he ICJ [had] not follow[ed] that path' (Sands 2016:29). From such words, one might conversely be led to believe that the ICJ had there taken a firm stance on the state of science, and that, in Sands' opinion, a similar approach should now be replicated in climate change-related disputes. Yet, it seems, things are in fact more nuanced. As Sands later explains, what the *Whaling case* shows is that the ICJ can perhaps play a role

> 'in assisting in the resolution of legal disputes that involve competing contentions as to matters of [...] scientific facts [and in] recogniz[ing] that the room for real doubt on the central factual issues as to climate change has – *for the purposes of legal adjudication, and legal obligations* – disappeared' (Sands 2016:30) [The italics are mine].

In other words, what climate science is to climate scientists, who seek scientific truth, it must not necessarily be to lawyers, who serve a very different purpose. In the *Whaling* case, as would also be in that of climate change, scientific issues complexify the operation of determining whether a given conduct complies with the law. However, in neither cases should the judge necessarily refrain from adjudicating for

---

[3] Adjudicating the Future: Climate Change and the Rule of Law, London Symposium, 17–19 September 2015: http://www.kcl.ac.uk/law/newsevents/climate-courts/symposium-puts-focus-on-courts.aspx. Accessed on 26 January 2017.

what may remain unsure in the world of science might very well be sure enough in the court room, i.e. *for the purposes of legal adjudication*. The reasoning followed in the *Whaling case* can therefore prove particularly helpful as it provides guidance on how to deal with science – sometimes complex and contentious science – when doing law. And if the task is certainly an ambitious one, especially when it comes to climate change, its accomplishment does not imply that the Court would have to 'pontificate on science', as I will later further explain (see part 4).

The *Whaling* case is thus firstly relevant to the extent hereinabove described, and I will now argue that more elements of fact and law support my claim for an analogy. There is no need to recall that the warming of the climate impacts on whales, as it also affects most – if not all – of the other marine species.[4] As true as this may be, the argument that I here want to make goes somewhat deeper.

The atmosphere and whales are commonly shared natural resources. Because, in both cases, the *sharing* was not functioning so well, the ICRW and the United Nations Framework Convention on Climate Change (UNFCCC) were adopted, respectively in 1946 and 1992. Indeed, human activities were being carried out in such a manner that stocks of whales, on the one hand, and the climate system, on the other, were being endangered. It was overfishing doing the job in the first case, and uncontrolled economic development and industralization in the second. Whereas whales and climate surely have little in common, there are at least these similarities. And to some extent, thus, the ICRW and the UNFCCC departed from similar premises. As a result, their object and purpose were unsurprisingly resembling as well, e.g. to regulate human activities in order to conserve common natural resources. In both cases, though, it is only a sustainable conservation that is being pursued. This is very clear in the case of the UNFCCC. Indeed, the proposition that some countries had made at the time of its negotiation, being that its objective be formulated in terms of ensuring 'that human activities [...] d[id] not contribute to climate change' (Elements proposed by Vanuatu 1991:25) was not followed, and instead, as is well known, the Convention merely strives to achieve

'stabilization of greenhouse gas concentrations in the atmosphere at a level that would prevent dangerous anthropogenic interference with the climate system' (UNFCCC: Art 2).

Over the time, and I will later return to that, this has come to be understood as requiring that the increase of the Earth's surface temperature be kept below 2 °C above pre-industrial levels (Brunnée and Toope 2010:146–151). In the case of the ICRW, the object was at first that

'whaling [be] properly regulated [and to] achieve optimum level of whale stocks' in order not to 'endanger these natural resources [and] make possible the orderly development of the whaling industry' (ICRW: Preamble).

However, since 1946, the object of the treaty seems to have evolved in such a way as to practically translate into a prohibition of all non-scientific whaling activities.

---

[4] Suffice it to note, in this regard, that the International Whaling Commission (IWC) has considered the implications of climate change for its work since the early nineties. In 2014, the Commission Scientific Committee even established a Climate Change Steering Group. See IWC (2014).

What conduct is required from a state may thus vary over the time, as may the discretion left to that state in choosing its course of action. In the case of the ICRW, states parties may still carry out whaling activities, but only 'for purposes of scientific research', as Article VIII paragraph 1 of the ICRW demands. And when they do indeed use this faculty, states enjoy a certain discretion, yet, as mentioned above (see part 2), not an unlimited one, and one that is possibly subject to judicial review. Climate treaties similarly allow for states to exercise some discretion as to what actions should be taken to combat climate change. Perhaps, the words *some discretion* is here somewhat of an understatement but I do however see no reason, here again, why the discretion would be absolute. There are, for that, several reasons, on which I will further elaborate below (see part 4): an objective was agreed upon and, although it was formulated in quite vague terms in the text of the UNFCCC, its substance subsequently more clearly emerged; climate science progressed such as to render more precisely determinable, and less contestable, what is needed for the said objective to be attained; states have been negotiating for decades, making statements, declarations, acknowledgements, all of which may contribute, to a certain extent, to the emergence of concrete expectations; the existence of linkages between climate change, more precisely the effects thereof, and other regimes has been gradually recognized, providing legal reasons for states to take actions, and accordingly affecting the width of their margin of appreciation.

In respect of the ICRW, the only remaining discretion for states to exercise is through the *scientific research* window left open by Article VIII, par 1 of the ICRW. In respect of climate change, the discretion of states is certainly much wider. Yet, at least in some measure, it is today dependent on what can reasonably be considered consistent with the objective of keeping the increase of the Earth's surface temperature below 2 °C above pre-industrial levels – or even a bit colder than that, as more ambitious targets have gradually begun to gain grounds (SED 2015), certainly even more so now that the PA has entered into force – and on what assessment is made of measures reasonably necessary to that end. Now, as the ICJ demonstrated in the *Whaling case*, where a state abuses or misuses its discretion, a finding that the conduct of that state is contrary to the law may follow. And this, it is worth noting, was also the line of reasoning followed by the ICJ in other instances.[5] Indeed, the *Whaling case* only just provides one very telling example of a method that the ICJ has more generally been using to review the manner in which states exercise their discretionnary competences (Salmon 1981; Jovanovic 1988; Corten 1998). Therefore, a similar conclusion to that of the ICJ in the *Whaling case* – e.g. that the state does not comply with its obligations and therefore incurs international

---

[5] Without this list being exhaustive, suffice it here to mention the following cases: *Conditions of Admission of a State to Membership in the United Nations* (1948): p. 63; *Rights of Nationals of the United States of America in Morocco* (1952): p. 212. More recently: *Certain Questions of Mutual Assistance in Criminal Matters* (2008): par 245; Dispute regarding Navigational and Related Rights (2009): par 85. Also, jurisdictional bodies competent in the field of human rights are quite accustomed to the exercise of reviewing the use that states make of their margin of appreciation, particularly when it comes to assessing the legality of a state's interference with human rights and freedoms (Arai 2002).

responsibility – could be reached where a state contributes to the warming of the global climate as a result of activities that are not – or no longer – consistent with the extent of that state's remaining discretion under the relevant climate treaties. This is facilitated by the fact that, as I earlier explained, there seems to be a way to ensure that scientific uncertainty will not always obstruct the process of legal adjudication and the operation of determining legal obligations with regards to climate change.

Therefore, in our view, there are convincing, or at least acceptable analogies that can be drawn between the kind of legal questions raised in the *Whaling case* and those pertaining to the issue of combating climate change. The approach followed by the ICJ to resolve these legal questions may thus very well inspire a framework of analysis for determining what obligations states have in respect of climate change.

4   'Whaling Out' the Climate Problem

So then, remaining scientific uncertainty will not necessarily be obstructive, nor will the fact that states enjoy discretion as to what measures must be taken to combat climate change. This discretion, I claim, is indeed not absolute, and the exercise thereof may therefore be subject to judicial review, which implicitly amounts to affirming that a threshold could somewhere be crossed. But what would be the result of that? Would an obligation then be breached? Which obligation would that be? Also, as the scope of the review will quite logically be narrowly dependent on the extent of discretion there is to exercise, how much of the latter do states enjoy? Finally, if one assumes that such a determination can be made, what is then to be taken into account to establish, in a particular given situation, that a state's discretion has been abused? These are some of the questions that I will address in this last section, yet not in an exhaustive and systematic manner. For reasons of space, I will indeed limit myself to laying the main grounds of this new framework of analysis in favor of which I here advocate.

In 1992, the UNFCCC was adopted. Although this was a source of general satisfaction among states, they did not in fact commit to much of anything at the time. The objective of the Convention was not to halt the warming of the climate, rather to contain it within acceptable levels, and even that, as we explained earlier, was not very precisely phased in the Convention. Moreover, there was not yet an agreement on concrete mitigation actions that should be carried out.[6] If satisfaction there had been, then, the reason was surely not related to what states had managed to agree upon in substantial terms.[7] Arguably, in concluding this treaty, states had not renounced much of their discretion, and their freedom of doing would appear to

---

[6]There would be such an agreement in the years that followed, as is well known, with the adoption of the Kyoto Protocol, but one could surely dare to call the experience only half-convincing (Rayner and Prins 2007: v).

[7]To be sure, although taking delight in the adoption of the UNFCCC, many states did express disappointment about not being able to close a more ambitious deal (Report of the United Nations Conference on Environment and Development 1992: p. 14 (Malta); p. 26 (United Kingdom); p. 55 (Micronesia); p. 107 (Netherlands); p. 108 (Colombia); etc).

have remained quite unchanged at the time. Yet, one should look at the UNFCCC from a broader perspective, in light of what preceded it, which may help understand what state's discretion there was to begin with, and in light of what has happened ever since it was adopted, which might also be informative as to what state's discretion is remaining 25 years later. Depending on what finding follows from this assessment, it might well be that the discretion that states enjoy with regards to climate change is actually more limited than what seems at first sight and is generally thought to be.

It should firstly be stressed that the UNFCCC did not emerge out of a legal vacuum. Prior to negotiating for the adoption of a treaty on climate change, this had indeed quite generally been acknowledged. In June 1988, in Toronto, at the *World Conference on the Changing of the Atmosphere*, it had been stated that

'[t]he first steps in developing international law and practices to address pollution of the air ha[d] already been taken [and] should be actively used and respected by all nations' (WMO 1988: par 16)[8]

By these *first steps*, it was namely thought of, and referred to the *Trail Smelter case* and Principle 21 of the Stockholm Declaration. Regarding the Trail Smelter case, it is here of particular interest to recall the Tribunal's conclusions, that:

' […] under the principles of international law, as well as of the law of the United States, no State has the right to use or permit the use of its territory in such a manner as to cause injury by fumes in or to the territory of another or the properties or persons therein, when the case is of serious consequence and the injury is established by clear and convincing evidence'

And, as for Principle 21 of the Stockholm Declaration, as is well known, it is therein affirmed that:

'States have […] the responsibility to ensure that activities within their jurisdiction or control do not cause damage to the environment of other states or of areas beyond the limits of national jurisdiction'

Then, in 1989, the year that followed the Toronto Conference, when it officially convened the United Nations Conference on Environment and Development (UNCED), the General Assembly decided that one of its objectives, among others, would be

'[t]o promote the further development of international environmental law, taking into account the Declaration of the United Nations Conference on the Human Environment […] and relevant existing international legal instruments' (UNGA 1989: par 15d).

It also more precisely referred to Principle 21 of the Stockholm Declaration, though implicitly, when reaffirming

'the responsibility of states, in accordance with national legislation and applicable international law, for the damage to the environment and natural resources caused by activities within their jurisdiction or control through transboundary interference' (UNGA 1989: par 8).

---

[8] Similarly, in 1990, a report by the IPCC's Response Strategies Working Group referred to the 'general view that […] legal instruments and institutions with a bearing on climate [exist] [and] should be fully utilized and further strengthened' (IPCC 1990: 261).

Other such examples could be provided, but suffice it here to say that, before the UNFCCC was adopted, international law was not entirely silent as to the manner in which it should be dealt with the warming of the global climate. To be sure, the very preamble of the UNFCCC does recall

'the pertinent provisions of the Declaration of the United Nations Conference on the Human Environment, adopted at Stockholm on 16 June 1972 [and also recalls] that states have, in accordance with the Charter of the United Nations and the principles of international law, [...] the responsibility to ensure that activities within their jurisdiction or control do not cause damage to the environment of other states or of areas beyond the limits of national jurisdiction' (UNFCCC 1992: Preamble)

If anything, then, the rule enshrined in Principle 21 of the Stockholm Declaration, that the Tribunal quite centrally relied on in the *Trail Smelter case*, and that has later been referred to in other cases, reflected by state practice, reaffirmed in various declarations and international treaties, existed prior to 1992. It should also be noted, as has elsewhere convincingly and recently been argued, that the UNFCCC did not have the effect – by creating a *lex specialis* – to displace the application of this principle with regards to climate change (Mayer 2016:16–17). Quite the contrary, in fact, as it is very explicitly referred to in the preamble of the Convention. Therefore, the rule enshrined in Principle 21 surely remains of relevance, and the question is thus rather how it relates to the UNFCCC, more particularly its objective of stabilizing GHG concentrations at a level that would prevent dangerous anthropogenic interference, and whether the latter can possibly be seen as substantiating the meaning of this principle when it comes to what obligations states have in respect of climate change. Now, if this argument can be made – and I do think it can, although I have no room here to elaborate on the matter – legal consequences are likely to follow. And they will follow all the more since the objective of the UNFCCC, though not phrased in a very clear-cut manner, has now come to be much more precisely understood as meaning that the increase of the Earth's surface temperature should be kept below 2 °C above pre-industrial levels (Brunnée and Toope 2010:146–151). As we know, this evolutive interpretation was also recently, and more formally endorsed with the adoption of the PA (PA 2015: Art 2).

From what has hereinabove been explained, it then firstly results that the discretion of states in respect of climate change was never absolute, but always limited, at least in theory, by the effect of the rule enshrined in Principle 21 of the Stockholm Declaration. And, as a more precise definition of what constitutes a *dangerous anthropogenic interference* emerged, the discretion of states accordingly lessened over the time, arguably unveiling a clearer threshold for determining what damage to the climate system can acceptably be caused without infringing the rule enshrined in Principle 21. That is, we claim, a first important stage in the process of *whaling out the climate problem*. If a state's discretion is indeed to be exercised with consideration for an agreed objective, one that is – or has come to be – quite precisely determined, the width of that state's margin of appreciation accordingly lessens. In the *Whaling case*, the ICJ could not see how Japan's granting the permits could reasonably connect to 'purposes of scientific research'. That is not to say that Japan did not enjoy a certain deal of discretion in deciding to grant those permits, but it had to do so in due consideration of the agreed objective, and therefore within certain limits. And in respect of these limits, the more precise the objective to be pursued, the less discretionary the state's competence and freedom of doing. Indeed, as the ICJ made very plain in the *Navigational and Related Rights case*, the fact of

'expressly stating the purpose for which a right may be exercised [...] imposes the limitation thus defined on the field of application of the right in question' (ICJ 2009: par 61).

What freedom of doing is therefore left to states when it comes to global warming, what rights they may continue to exercise, is limited to what can be viewed as consistent, reasonably, with the objective that has been agreed upon.

So there is thus a limit to what states can do. Yet, to be frank, where lies this limit is still quite abstract, and that is to say the least. What does the agreed objective mean in terms of mitigating GHG emissions? How much has the climate warmed so far, and how much more warming is left to go? How can a state's carbon budget be determined? Undoubtedly, there do remain questions for the resolution of which scientific uncertainty seems to stand in the way. However, some answers might here again be found by taking on the approach of the ICJ in the *Whaling case*, or at least more precisely circumscribed – enough so, in any case, to help further determine where a state exceeds its margin of appreciation.

Differing scientific opinions certainly do exist in respect of climate change, even often *sharply differing* ones, to recycle Sands' words hereinabove referred to. In the face of these existing divergences, the only way that a state's actions in respect of climate change can legally be reviewed is, where it can, and to the maximum possible extent, by trying to objectify what, at first, might seem to be for the state to discretionarily decide. In other words, one has to objectify the subjective. In the *Whaling case*, the ICJ did indeed insist on the fact that the test to be applied was an objective one. According to the ICJ, it did not

'need [to] pass judgment on the scientific merit or importance of [Japan's research] objectives [...] nor to decide whether [...] [Japan had had recourse to] the best possible means of achieving its stated objectives' (ICJ 2014: par 88).

Rather, the test to be applied would only serve to determine

'whether [the means] [were] reasonable in relation to achieving the stated [...] objective' (ICJ 2014: par 97).

And for Japan to successfully pass the test, as all of the parties involved had agreed, the decision to grant permits had to be

'objectively reasonable, or supported by coherent reasoning and respectable scientific evidence and...in this sense, [be] objectively justifiable' (ICJ 2014: par 66).

The Court's review would thus consist in putting the integrity of Japan's decision making under scrutiny. And, of course, the more *respectable scientific evidence*, the thinner the margin of appreciation, and the closer the scrutiny. Now, as to what *respectable scientific evidence* means, one should note that the ICJ refrained from any final and peremptory determination where it considered a matter to fall within the ambit of scientific opinion and where Japan had not in any way conceded to

Australia's views on how scientific whaling had to be carried out. But – and that turned out to be key – Japan had precisely somewhat given in on certain issues, namely and most decisively with regards to the feasibility of non-lethal methods to obtain scientifically relevant data. Indeed, Japan 'ha[d] accepted that it [was] under an obligation to give due regard to [the] recommendations [of the IWC]' (ICJ 2014: par 137). And Japan had also 'state[d] that, for reasons of scientific policy, [i]t [did] not [...] use lethal means more than it considere[d] necessary' (ICJ 2014: par 138). Arguably, had Japan not given in to that, its decision to grant permits would have been reviewed less strictly by the ICJ for the matter would have then remained open to contestation on the basis of the exact state of science. But it had, and from that, it resulted that the coherence and reasonableness of that state's decision making were now dependent, in some measure, on how much consideration had been given to the feasibility of non-lethal methods to obtain data relevant to the research objectives and to the possibility of making a more extensive use of such methods. What is therefore meant by *respectable scientific evidence* can surely refer to uncontested science, but also to what may still be scientifically uncertain to some extent, yet no longer constitutes a point of divergence between the parties. That is indeed one way by which subjectivity can be objectified.

Now, when it comes to climate change, of course, such pieces of evidence will namely have to be looked for in the reports that have been published over the time by the IPCC. Hence the IPCC, for instance, considered in its fifth assessment report that

'limiting total human-induced warming to less than 2°C relative to the period 1861–1880 with a probability of [more than] 66% would require cumulative CO_2 emissions from all anthropogenic sources since 1870 to remain below about 2,900 GtCO_2 [whereas] [a]bout 1,900 GtCO had already been emitted by 2011' (IPCC 2014:10).

It was also asserted by the IPCC that

'[t]here are multiple mitigation pathways that are likely to limit warming to below 2°C relative to pre-industrial levels [but] these pathways would [all] require substantial emissions reductions over the next few decades and near zero emissions of CO_2 and other long-lived greenhouse gases by the end of the century' (IPCC 2014: 20).

This surely is informative of how much more GHG can globally be emitted, and of how mitigation strategies should best be designed and implemented. Earlier, being even more precise, the IPCC had expressed the view that an emission trajectory consistent with the 2 °C objective implied, for industrialized countries, that their GHG emissions be reduced by 25 to 40% in 2020, and by 80 to 95% in 2050, compared to the levels of 1990 (IPCC 2007:776- Box 13.7).

Apart from the IPCC, there have been, and are, other suppliers of scientific facts and data that can be relevant to determining the reasonableness of what states do in relation with the 2 °C objective. Since 2010, the UNEP has for instance annually been releasing reports in which it is closely looked into what states do, or say they will do, in the light of the emission trajectories consistent with the long-term objective of the UNFCCC. In the most recent of these reports, it was stated that '[e]ven if fully implemented, the unconditional [INDCs] are only consistent with staying below an increase in temperature of 3.2°C by 2100 and 3.0°C, if conditional [INDCs] are included' (EGR 2016: xvii).

Again, this gives an idea of what and how much effort is needed to stand a chance to meet the 2 °C target, and it is also illustrative of how reasonably consistent with this target the behavior of most states currently is.

Now, can these sources do more than give ideas, inform or illustrate? On a general basis, neither the IPCC nor the UNEP (nor, for that matter, any other supplier of climate science) produce what I earlier referred to as uncontestable science. This is very clear, in the case of the IPCC, for findings it makes are accompanied by an indication of the degree of confidence and agreement in their validity (IPCC Guidance Note 2010). But *respectable scientific evidence*, on the other hand, they might very well be, and we therefore do believe that a state's course of action cannot be decided without reasonable consideration for what work is done, and what findings and conclusions are reached, by the IPCC and other relevant institutions.

As far as the IPCC is concerned, one first reason for that is that some of the findings and conclusions reached are no longer seriously contestable, one very telling example being the near 100 percent certainty that 'humans are the main cause of current global warming' (IPCC 2014b:v).

Secondly, due to the nature and features of the IPCC's work - e.g. to assess the scientific basis of climate change, its impacts and future risks, and options for adaptation and mitigation, all of which with the contribution of hundreds of leading scientists and with multiple rounds of drafting and review to ensure comprehensiveness, objectivity, openness and transparency (IPCC 2013) -, those of the IPCC's findings marked with a degree of confidence ranging from high to very high, and a degree of likelihood, from likely to very likely, reach a level of credibility that may be considered high enough for the burden of proof – if not clearly and convincingly discharged or fulfilled beyond reasonable doubt – to be reversed. The latter assertion reflects a passage of the judgment rendered in the *Whaling case*, where the ICJ considers fit to

'look to the authorizing State, which has granted special permits, to explain the objective basis for its determination [that] the programme's use of lethal methods is for purposes of scientific research' (ICJ 2014: par 68).

Similarly, where preponderance of respectable scientific evidence exists as to what is necessary to attain the 2 °C objective, it could be argued that a state will either have to adjust its behavior accordingly, or be expected to explain the objective basis for the divergent course of action it has chosen to adopt. This argument could also be further strengthened by the effect that the precautionary principle has on the determination of which party carries the burden of proof (Foster 2011:240; Foster 2012:90).

Thirdly, the IPCC's assessment reports do not come out into the light of day without states carefully reviewing them, making observations and finally playing a part in their adoption. This is not to say that states necessarily agree to all of it, but the process is such that it must, at least in some measure, lead to scientific findings of a *respectable* quality.

Fourthly, as again reflected in the *Whaling case*, where the ICJ considered Japan to be under a duty to give due regard to some resolutions that had been adopted by the IWC (namely grounding the rationale for this duty in the obligation of cooperation incumbent upon states parties to the ICRW), it can surely be said that states, as they also have an obligation to cooperate under the UNFCCC, are under a duty to give due regard to a range of decisions that have been adopted under its auspices, particularly those adopted by Conferences of the Parties (COPs) (Young and Sullivan 2015:337). And these, if we keep on following the views expressed by the ICJ in the *Whaling case*, because they are adopted by consensus, are also relevant for the interpretation of the UNFCCC and other climate treaties (ICJ 2014: par 46; Maljean-Dubois et al. 2015:73). This is of importance for the COPs make constant reference to the work of the IPCC and have also explicitly endorsed some of its findings, thereby recognizing, at the very least, their status as *respectable scientific evidence*. This is true, for instance, of the view that the IPCC had taken, in its fourth assessment report, that industrialized countries should reduce their GHG emissions by 25 to 40% in 2020, and by 80 to 95% in 2050, compared to the levels of 1990, as it was later upheld by the COP in Cancun (Decision 1 / CP.16: par 37; Decision 1 / CMP.6: Preamble). Related and in addition to that, as had been the case of Japan in the *Whaling case*, some industrialized countries later acknowledged, in what has been referred to as the *Cancun pledges*, that the scale and pace of the reductions of GHG emissions needed to meet the 2 °C objective were indeed those referred to in the IPCC's fourth assessment report (Compilation of economy-wide emission reduction targets 2011). In so doing, these countries were not taking a legal commitment, but it surely is nonetheless an element that may be accounted for in order to assess the reasonableness of their now chosen course of action. Although, undoubtedly, there are more such elements to expose, it goes well beyond the scope of this chapter to precisely identify all IPCC's findings that deserve due regard from states, or of other relevant bodies and institutions, let alone to inventory all relevant state practices from which the duty to give due regard may be deduced, confirmed or reinforced. Suffice it here to say that such a duty exists, and that it may serve as an instrument to disclose what relevant facts and circumstances – even if they are of a scientific nature – can be examined for determining what is reasonably consistent with achieving the 2 °C objective, and therefore within the limits of a state's freedom of doing. A margin of appreciation will probably always remain, but the breadth of this margin will normally keep on diminishing. As we already mentioned, this is because a state's doing has to be supported by coherent reasoning and scientific evidence, the amount and respectability of which extend as years pass by.

This duty to give due regard is certainly an important piece of our puzzle. Not only can one rely on it in order to delimit the factual background in reference with which a state's decision making must reasonably be justifiable, and in light of which it may therefore be reviewed, but there is more. Indeed, it may also help uncover other rules of international law that are relevant for determining what obligations states have in respect of climate change. As we earlier briefly explained, the UNFCCC should not be seen as rendering Principe 21 of the Stockholm Declaration inapplicable to climate change. Rather, it provides guidance on what concrete substance and meaning the principle has when it comes to its application to the climate problem.

Now, apart from this principle, are there other sources in international law to which due regard should be given? Unlike Principle 21 of Stockholm Declaration, human rights law is not referred to in the UNFCCC, but that does not necessarily mean that human rights should not be considered by states in deciding upon the course of their actions. Indeed, human rights law is of very general application, either through treaties or on a customary basis, so that it must universally be complied with. This would be of little interest if there was no linkage to be drawn between the effects of climate change and human rights. Yet, there undoubtedly is, as has indeed been stressed by a number of institutions and bodies, most pioneeringly and persistently by the United Nations Human Rights Council (HRC 2008; HRC 2009; HRC 2011; HRC 2014; HRC 2015; HRC 2016). Quite notoriously, the work of the HRC was taken note of in the Cancun Agreements (2010a, b) in which it was also

'[e]mphasize[d] that Parties [to the UNFCCC] should, in all climate change related actions, fully respect human rights' (Decision 1 / CP.16: par 8).

Perhaps even more notoriously, in the PA, it was acknowledged that

'[...] Parties should, when taking action to address climate change, respect, promote and consider their respective obligations on human rights [...]' (PA, Preamble).

Nowadays, it can thus hardly be contested that the need to respect human rights falls within the ambit of the states' duty of due regard, the effect of which is manyfolds. *Firstly*, and quite importantly, the idea that a state's discretion in respect of combating climate change is absolute, should such claim be made, is of course no longer tenable. The duty that states have to respect human rights, and ensure their enjoyment, is of a binding nature, which consequently provides a stronger legal basis to affirming that a state's freedom of doing somewhere ceases by the effect of the law. *Secondly*, as to where that must be, and given the alleged fact that global warming by 2 °C would already lead to serious implications on the enjoyment of human rights, there is room for arguing that the states' freedom of doing is to be much more constrained than what has until now been the case, and their decision making accordingly much further reviewed. *Thirdly*, as to how this review should be carried out, and since it is thus clear that the climate problem encompasses human rights issues, a human rights-based approach could more heavily be relied on, from which there is much to learn, as several authors have already shown (Shelton 2009; Limon 2010; Knox 2015; Behnassi in Chap. 1 of this volume).

5 Conclusion

The argument that states may discretionarily define their course of action in taking, or refraining from taking measures to combat climate change, is no longer tenable. There might have been wide discretion in the beginning; yet, even then, it was not absolute. And it has been less and less so as time went by. Hence, there is room for review, which necessarily implies, through thorough examination of the relevant facts and circumstances, that a state's doing may be considered excessive or abusive. Now, when that happens, there is no falling into a legal vacuum. As in the *Whaling case*, where the prohibition or ban on commercial whaling had resurfaced, there are rules that can be found, allowing for a determination that a state's abusing or exceeding its discretion is contrary to the law. Principle 21 of the Stockholm Declaration is one example, and it had been there for quite some time even before the UNFCCC was adopted. As I briefly argued, human rights law has now also made a way into the toolkit of rules that can be considered of relevance, and as deserving due regard from states, when it comes to climate mitigation and adaptation. And this argument could surely also be made in respect of other rules and regimes of international law. To be sure, research has already been conducted in this sense, enlightening some more existing ways in which the climate problem can relate to other concerns and legal obligations (Ruppel et al. 2013; Hollo et al. 2013).

However, quite lesser attention has until now been accorded to how this could all be subsumed in an integrated framework of analysis. We here intended to do more of that, yet not in an exhaustive manner. We departed from the specifics of the *Whaling case*, from which we thought an analogy could convincingly be drawn, and thereafter tried to replicate what pieces of the ICJ's reasoning could be transposed to the climate problem, particularly with regards to whether and how a state's discretionary competences can be reviewed, disputed scientific facts arbitrated upon, and the so-called duty of due regard given a more precise meaning to when it comes to addressing and hopefully sorting out the climate problem. From that, we concluded that the room for absolute discretion, had it ever existed, has now surely disappeared. The view could even arguably be upheld that the discretion left to states has shrunk a great deal over the time, allowing for a very close scrutiny to now be exercised. The task, however, keeps on being an arduous one. This is, indeed, because some room for discretion will always remain.

Yet, in the meantime, the more due regard must be given, the less discretion states may exercise. This is why we have here claimed that a state's course of action could be reviewed on the basis of what is reasonably consistent with the 2 °C objective and with the scale and pace of GHG reductions necessary to attain it. Moreover, as a result of recent or plausible coming developments, there will also soon be more due regard to be given, and the scope of the review shall accordingly extend to what is reasonably consistent with attaining more ambitious objectives.

The picture can even be further embellished, should legal action be envisaged, by the prospect that a state, any state, as in the *Whaling case*, would most likely be recognized a legal interest in the standard of conduct to be complied with. Of course, in most cases, there would remain the problem of the absence of jurisdictional basis (Sands 2016:28). But that is also where other concerns and legal obligations, to which the climate problem relates, may prove helpful, by providing a basis on which to ground the jurisdiction of an international court, tribunal or arbitrator. In addition to that, if the proof can indeed be brought those international standards have not been complied with, as I believe our framework of analysis allows, the responsibility of states may also more easily be established at the domestic level. Indeed, there is more to international law than what can effectively be enforced through international courts and tribunals. To be sure, international law also may have relevance in the domestic scenery, as the Urgenda climate case[9] recently exemplified, and other such proceedings that are currently underway – in Belgium, Switzerland, etc.

Once one accepts that the indeterminacy of states' obligations in respect of mitigating climate change can be overcome, and I do hope that my framework of analysis will encourage such acceptance, there will be little left in the way of sorting out – or should I say *whaling out* – the climate problem.

References

Anzilotti D (1929) Cours de droit international. Sirey, Paris

Arai Y (2002) The margin of appreciation doctrine and the principle of proportionality in the jurisprudence of the ECHR. Intersentia, Antwerp

Bergkamp L (2015) Adjudicating scientific disputes in climate science : the limits of judicial competence and the risks of taking sides. In: Environmental liability – law, policy and practice. Lawtext Publishing Limited, pp 80–102

Brunnée J, Toope S (2010) Legitimacy and legality in international law. An interactional account. Cambridge University Press, Cambridge, pp 126–219

Cancun Agreements (2010a) Outcome of the work of the Ad Hoc working group on further commitments for annex I parties under the Kyoto protocol. Report of the conference of the parties serving as meeting of the parties to the Kyoto protocol on its 6th session, 1/CMP.6

Cancun Agreements (2010b) Outcome of the work of the Ad Hoc working group on long-term cooperative action under the convention report of the conference of the parties on its 16th session, decision 1/CP.16

Compilation of economy-wide emission reduction targets to be implemented by Parties included in Annex 1 to the Convention (2011) Subsidiary body for scientific and technological advice, 34th session, Bonn, 6–16 June 2011, FCCC/SB/2011/INF.1/Rev.1

Corten O (1998) Motif légitime et lien de causalité suffisant : un modèle d'interprétation rationnel du 'raisonnable. AFDI 1998:187–208

---

[9] The Urgenda Climate Case is the first case in which regular citizens have managed to hold their government accountable for taking insufficient action to keep them safe from dangerous climate change. On June 24 2015, the District Court of The Hague ruled that the Dutch government is required to reduce its emissions by at least 25% by the end of 2020 (compared to 1990 levels). This means that the Dutch government is now, effective immediately, forced to take more effective action on climate change. It is also the first case in the world in which human rights are used as a legal basis to protect citizens against climate change (for more details: http://www.urgenda.nl/en/climate-case/).

Depledge J, Yamin F (2011) The global climate-change regime, a defence. In: Helm D, Hepburn C (eds) The economics and politics of climate change. Oxford University Press, Oxford

Elements for a framework convention on climate change, proposed by Vanuatu on behalf of States Members of the United Nations and of the specialized agencies that are members of the Alliance of Small Island States (AOSIS) (letter of 5 June 1991), A/AC.237/Misc.1/Add.3

Emissions Gap Report (EGR) (2016) A UNEP Synthesis Report. UNEP. Nairobi

Foster C (2011) Science and the precautionary principle in international courts and tribunals. Expert evidence, burden of proof and finality. Cambridge University Press, Cambridge

Foster C (2012) International adjudication – standard of review and burden of proof: Australia-apples and whaling in the Antarctic. RECIEL 21(2.) 2012):80–91

Gupta J (2014) The history of global climate change governance. Cambridge University Press, Cambridge

Hamamoto S (2014) Procedural questions in the whaling judgment: admissibility, intervention and use of experts. ICJ judgment on whaling in the antarctic: its significance and implications, the honorable Shigeru Oda commemorative lectures, Japanese society of international law (19–21 September 2014 @ Niigata). http://www.edu.kobe-u.ac.jp/ilaw/en/whaling_docs/2014manuscript_Hamamoto.pdf. Accessed on 24 Jan 2017

Hollo E et al (eds) (2013) Climate change and the law. Springer, Dordrecht/London

Human Rights Council (2008) Human rights and climate change. Resolution 7/23. Adopted on 28 Mar 2008. A/HRC/RES/7/23

Human Rights Council (2009) Human rights and climate change. Resolution 10/4. Adopted on 29 Mar 2009. A/HRC/RES/10/4

Human Rights Council (2011) Human rights and climate change. Resolution 18/22. Adopted on 30 Sept 2011. A/HRC/RES/18/22

Human Rights Council (2014) Human rights and climate change. Resolution 26/27. Adopted on 27 June 2014. A/HRC/RES/26/27

Human Rights Council (2015) Human rights and climate change. Resolution 29/15. Adopted on 2 July 2015. A/HRC/RES/29/15

Human Rights Council (2016) Human rights and climate change. Resolution 32/33. Adopted on 1 July 2016. A/HRC/RES/32/33

Intergovernmental Panel on Climate Change (IPCC) (1990) Climate change: the IPCC response strategies 1990 (Working group III)

Intergovernmental Panel on Climate Change (IPCC) (2007) Climate change 2007 mitigation of climate change. Contribution of working group III to the fourth assessment report of the intergovernmental panel on climate change [Metz B, Davidson OR, Bosch PR, Dave R, Meyer LA (eds)], Cambridge University Press, Cambridge, United Kingdom and New York, USA

Intergovernmental Panel on Climate Change (IPCC) (2010) Guidance note for lead authors of the IPCC fifth assessment report on consistent treatment of uncertainties [Core writing team, Mastrandrea MD, Field CB, Stocker TF, Edenhofer O, Ebi KL, Frame DJ, Held H, Kriegler E, Mach KJ, Matschoss PR, Plattner G-K, Yohe GW, Zwiers FW (eds)]. IPCC, Switzerland

Intergovernmental Panel on Climate Change (IPCC) (2013) IPCC factsheet : what is the IPCC ? IPCC, Geneva, Switzerland: https://www.ipcc.ch/news_and_events/docs/factsheets/FS_what_ipcc.pdf. Accessed on 14 Jan 2017

Intergovernmental Panel on Climate Change (IPCC) (2014) Climate change 2014 synthesis report. Contribution of working groups I, II and III to the fifth assessment report of the intergovernmental panel on climate change [Core writing team, Pachauri RK, Meyer LA (eds)]. IPCC, Geneva, Switzerland

International Convention for the Regulation of Whaling (ICRW) (1948) Signed in Washington DC on 2 Dec 1946. Entered into force on 10 Nov 1948

International Court of Justice (ICJ) (2009) Dispute regarding navigational and related rights (Costa Rica v. Nicaragua). Judgment of 13 July 2009

International Court of Justice (ICJ) (2012) Questions relating to the obligation to prosecute or extradite (Belgium v. Senegal). Judgment of 20 July 2012.

International Court of Justice (ICJ) (2014) Whaling in the Antarctic (Australia v. Japan: New Zealand intervening). Judgment of 31 Mar 2014

International Whaling Commission (IWC) (2014) Report of the IWC climate change steering group meeting, 19 Aug, Glasgow, UK

Jovanovic S (1988) Restriction des compétences discrétionnaires des Etats en droit international. Pedone, Paris

Knox J (2015) Human rights principles and climate change. In: Carlarne C, Gray KR, Tarasofsky R (eds) Oxford handbook of international climate change law

Limon M (2010) Human rights obligations and accountability in the face of climate change, GA. J Int Comp 38(3):543–592

Maljean-Dubois S, Spencer T, Wemaere M (2015) The legal form of the Paris climate agreement: a comprehensive assessment of options. CCLR 9(1):68–84

Mayer B (2016) The relevance of the no-harm principle to climate change law and politics. Asia Pacific J Environ Law 19:79–104

Rayner S, Prins G (2007) The wrong trousers: radically rethinking climate policy. Institute for Science, Innovation and Society, Oxford

Report of the United Nations Conference on Environment and Development (1992) Statements made by heads of state or government at the summit segment of the conference, Rio de Janeiro, 3–14 June 1992, volume III, A/CONF.151/26/Rev.1

Report on the structure expert dialogue on the 2013–2015 review (SED) (2015) Subsidiary body for scientific and technological advice. forty-second session. Bonn, 1–11 June 2015. FCCC/SB/2015/INF.1

Ruppel O et al (eds) (2013) Climate change: international law and global governance, Volume 1: Legal responses and global responsibility

Salmon J (1981) Le concept de raisonnable en droit international. In: Mélanges offerts à Paul Reuter. Paris, Pedone

Salmon J (1984) Le raisonnement par analogie en droit international. In: Mélanges offerts à Charles Chaumont, Le droit des peuples à disposer d'eux-mêmes – Méthode d'analyse du droit international. Pedone, Paris, pp 495–525

Sands P (2016) Climate change and the rule of law : adjucating the future in international law. J Environ Law 28:19–35

Strupp K (1930) Le droit du juge international de statuer selon l'équité. R.C.A.D.I. 1930/3, vol 33

Shelton D (2009) Human rights and climate change. Buffet center for international and comparative studies. Working paper series. Working paper no. 09-002

United Nations Framework Convention on Climate Change (FCCC) Adopted in New York on 9 May 1992. Entered into force on 21 Mar 1994

United Nations General Assembly (UNGA) (1989) United Nations conference on environment and development, A/RES/44/228

Vihma A, van Asselt H (2012) Great expectations: understanding why the UN climate talks seem to fail. Finnish Institute of International Affairs, Briefing paper 109

Weyers L (2013) La chasse à la baleine dans l'Antarctique : une application du principe de l'exercice raisonnable des compétences discrétionnaires de l'Etat. RBDI (2013/2), pp 618–642

World Meteorological Organization (WMO) (1988) The changing atmosphere: implications for global security. Conference proceedings, Toronto, 27–30 June 1988. WMO-No. 710

Young M, Sullivan S (2015) Evolution through the duty to cooperate: implications of the whaling case at the international court of justice. Melb J Int Law 16:311–343

Chapter 5
Biodiversity, Human Security and Climate Change: Which Legal Framework?

Aline Treillard

Abstract The Secretariat of Convention on Biological Diversity (CBD), the Intergovernmental Panel on Climate Change (IPCC), the Food and Agriculture Organization of the United Nations (FAO), the United Nations High Commissioner for Refugees have all published alarming figures and worrying trends which call for international and local action to protect biodiversity, human security and fight against climate change. What answers can international environmental law propose with respect to these issues? This chapter attempts to alert jurists and the public about the necessity to reconsider the international foundations that are threatened by environmental and human vulnerabilities. The analysis will provide an opportunity to identify the theme of security in the international texts related to climate change and the protection of biodiversity. I will claim that despite a rather anthropocentric construction of international environmental law, human security is not explicitly taken into account in different legal regimes, thus revealing an unusual feature: the fragmentation of the law. So, after highlighting the shortcomings, some suggestions for improvements of the international system will be proposed, for example, considering intergenerational justice, reorganizing the management of institutional vulnerabilities, evolving the subjects of law, adopting a convention on environmental displaced persons, etc. Obviously, all the solutions should be devoted to the idea of interdependence.

Keywords Biodiversity · Climate change · Human vulnerabilities · Global risks · International environmental law

A. Treillard (✉)
Center for Interdisciplinary Research in Environmental Law, Land Planning and Urban law (CRIDEAU-OMIJ), Faculty of Law and Economics, University of Limoges, Limoges, France

© Springer International Publishing AG, part of Springer Nature 2019   109
M. Behnassi et al. (eds.), *Human and Environmental Security in the Era of Global Risks*, https://doi.org/10.1007/978-3-319-92828-9_5

1 Introduction

Climate change impacts biodiversity which in turn threatens human security. In fact, environmental security and human security are inseparable. How should we articulate these three elements within the international legal order which is already weakened by ethical, political and socio-economic disparities?

On the one hand, this articulation needs a legal recognition of the interrelations between the three elements. Since the 1972 Stockholm Declaration, the actors of international law proclaimed that *"man has the fundamental right to freedom, equality and adequate conditions of life, in an environment of a quality that permits a life of dignity and well-being"*. Following this awareness, the Rio Earth Summit in 1992 ended with two major conventions: Convention on Biological Diversity (CBD) and United Nations Framework Convention on Climate Change (UNFCCC). However, while the Rio Declaration recognized the integral and interdependent nature of the Earth, the conventional law separates the two objects of law, thus willfully ignoring that the current economic growth process in the North has impoverished the natural heritage of humanity. As a consequence, despite a mandatory regulation strengthened by specific protocols, the gap is widening between these two elements which are closely related to one another. So, it is time for the international environmental law to accept the interpenetration of legal regimes (Beurrier 2010).

On the other hand, the international answer to cope with current challenges calls for the introduction of an ethic of responsibility about the foundations of international law. The goal of this ethic is to create the conditions of a model based on global interest rather than states' interest (Cournil and Colard-Fabrelouge 2010:428). This transition is largely conditioned by the reintroduction of the idea of justice in inter-state relations.

The combination of these two elements (recognition of interrelations and introduction of justice) would allow for biodiversity protection and human rights to act as instruments to fight against climate change. This thinking leads to difficulties because climate change is not always the cause of the loss of biodiversity or violation of human security. It can either constitute the initial factor or just an aggravating factor. In all cases, it increases and extends risks affecting the future of humanity. For this chapter, an extensive definition of human security was chosen. Fundamental elements can be identified under article 25 of the Universal Declaration of Human Rights which stipulates that: *"Everyone has the right to a standard of living adequate for the health and well-being of himself and of his family, including food, clothing, housing and medical care and necessary social service"*. Human security corresponds to the access to safe water, food in sufficient quantity and quality, sanitation, socio-economic stability, housing, peace, respect for cultural diversity and natural heritage. In this respect, the lack of effective protection and the effects of climate change create a global injustice which aggravate the vulnerabilities of specific populations (insular populations, coastal populations, pastoral populations, indigenous populations, etc). In a global context marked by urgency, I wonder about the capacity of international laws to regulate the relations between biodiversity and human security and to tackle climate change. Firstly, I will expose the insufficient normative framework (Sect. 2) and secondly, in a forward-looking manner, I will develop an insight about the capacity of international environmental law to reduce vulnerabilities (Sect. 3).

2 The Deficiency of the Legal Framework for Biodiversity Regarding Human Security

The normative framework for biodiversity in international environmental law is guaranteed by the Convention on Biological Diversity (CBD), extended by the Cartagena Protocol on Biosafety and the Nagoya Protocol. These mandatory texts are confronted to the effects of climate change in a variable dimension. The three goals pursued by the CBD, which are the conservation, the sustainable use of the biodiversity, and the fair and equitable sharing of benefits, reveal two guidelines of the legal regime. They are equally concerned with both human (sometimes more) and the environment. Built on an anthropocentric approach, the international law of biodiversity is characterized by heavy deficiencies in such a way that principles and proceedings act as limits to protecting human security (Sect. 2.1). Moreover, when international law treats biodiversity as a challenge for the environment, the protection of human security is carried out indirectly (Sect. 2.2).

2.1 Weaknesses of the Legal Framework for Biodiversity Protection as a Challenge for Man

Two major risks threatening human security, which are linked to the loss of biodiversity and the effects of climate change, can be highlighted: looting and exhaustion of natural resources; and natural disasters caused by ecosystems' imbalances and climate impacts. Because it ignores certain populations (Sect. 2.1.1) and has failed to implement a bottom-up model (Sect. 2.1.2), the international environmental law is unsuited.

2.1.1 The Scorn of Indigenous People by the Nagoya Protocol

Before analyzing the weaknesses of the legal regime about human security, it is necessary to identify the scope of the Nagoya Protocol. Pursuant to articles 8.j, 15.2 and 15.7 of the CBD, the Nagoya Protocol defines procedures of access to genetic resources and the fair and equitable benefit-sharing. Originally, the Protocol was presented as a way to fight against looting of diverse natural resources by powerful industries. Numerous elements could lead us to believe this idea. Firstly, the preamble recalls that *"Recognizing the importance of genetic resources to food security, public health, biodiversity conservation, and the mitigation of and adaptation to climate change..."*. Secondly, writers have chosen a large definition of "the utilization of genetic resources", with which human security, particularly that of indigenous people, is better protected. Nevertheless, if the Protocol confers to them a right over genetic resources present on their territory, the legal regime frames environmental, social and economic injustices instead of tackling these kind of injustices. Indeed, the Protocol only necessitates states to respect the committment to request unanimous consent from indigenous people when an industry wants to access genetic resources (Article 6). So, we could admit that this procedure recognizes a special status for indigenous people who played a substantive role in the protection of biodiversity and in the fight against climate change. Their lifestyle, traditional knowledge on ecosystems and legal actions at different levels merit consideration. However, this regime is hypocritical because it does not force states to recognize the legal existence of indigenous people. Besides, the Protocol does not complete the notion of traditional knowledge mentioned in Article 8.j of the CBD; a lack of clarification which becomes a real problem for other legal regimes, particularly during the transposition of the Protocol at the European level (Burelli 2015). The implementation of the procedure remains *"upwind conditioned by the definition and the recognition of certain new categories of actors and certain new legal objects within signatory States"* (Burelli 2015:61). The only valid aspect is that it can confirm the specific status. In short, the Nagoya Protocol does not recognize the rights of indigenous people. Their status is either implicitly determined by the text or by the soft law.

2.1.2 The Lack of Local/Global Management as Illustrated by Regulations on Natural Disasters

Contrary to the biodiversity regulation, the natural disasters framework comes from soft law. After the tsunami in the Indian Ocean in 2005, many states adopted the Hyogo Framework for Action for 2005 to 2015. Today, this document is substituted by the Sendai Framework for Disaster Risk Reduction applied for 2015–2030. The text covers the human security dimension and stresses on the importance of international cooperation and preventive action. However, these two elements may be insufficient because they do not guarantee the articulation between the local and global level. This failure reduces the effectiveness of the international action and prevents the consideration of local particularities and local participation which are decisive elements in the fight against natural disasters. So, it is necessary to promote the local governance by promoting a negotiated networking of sectoral policies and the involvement of state and non-state actors. Thanks to researchers, some European countries are obliged to consider the information, consultation and participation rights of citizens. For example, in law planning, the population accepts more easily the interdiction of construction in flood-risk area and the destruction of wetlands or forests when they are provided with the needed information to understand the negative consequences of constructions on natural disasters, erosion of biodiversity and climate change.

2.2   Biodiversity as a Challenge for the Environment: An Indirect Protection of Human Security

Currently, the international legal regime on biodiversity is obliged to take into account human security issues. In this way, some mechanisms are used in order to link the issues, for example in the framework for genetically-modified organisms (GMO) or the regulation introducing the ecological functionality (Bonnin 2010, 2012). However, the positive effects can still be expected.

2.2.1   Cartagena Protocol on Biosafety: A Subsidiary Protection

Pursuant to the first objective of the CBD, states have adopted the Cartagena Protocol based on the precautionary principle. The Protocol concerns the transboundary movements of living modified organisms (LMOs). It determines procedures to reduce potential threats to biodiversity and, consequently to human health. So, the Protocol illustrates that biotechnology can bring benefits for human and environmental security. However, the benefits can enter in rivalry. Of course, genetic changes can improve crop yields, fight against diseases or increase nutrient inputs of food; however, they can also cause irreversible damage to biodiversity and human health. In addition, the procedures proposed by the Protocol don't reflect the necessary balance between the objects of the rivalry. The legal obligations of the Protocol are dissociated from the security objective. For LMOs to be introduced intentionally into an environment, an advance informed agreement procedure is necessary. For LMO's direct use as food or feed, it is necessary to just have a risk assessment. This hierarchy annihilates the capacity of environmental security to contribute to the protection of human security.

   If we analyze all the regimes presented, we can consider that the natural disasters regime is the most attractive because it proposes a dynamic approach and uses mechanisms which are founded on mutualisation of threatened juridical objects (Compagnon 2010:32).

2.2.2   The Incomplete Integration of Ecological Functions

Since 1992, environmental law has been characterized by a systemic approach. In this way, international law can take into account the interaction between natural elements and human beings. Based on this, I will attempt to determine the vulnerabilities linked to these relationships and propose an adequate framework as a means for an effective protection. Thanks to science, we know that the characteristics of each ecosystem play an active role in global climate regulation (Serpantié et al. 2012).

The protection of ecosystems maintains protective services which are decisive for human security (such as food security, water and climate security), (Valentin 2009). Since the Millenium Ecosystem Assessment (2005), the international community knows clearly that humans benefit from ecosystems services. Indeed, the ecosystem functioning *"contributes directly or indirectly to the population welfare"* (Doussan 2012:121). The international law has to capture this scientific discovery to minimize human vulnerabilities. In this perspective, the forest and natural disasters international frameworks constitute an important progress.

In a curative approach, some researchers agree that the legal recognition of the interaction between biodiversity and climate change can help international or internal jurisdictions to elaborate a full repair principle for environmental damages that is ecologically, morally and financially fair (Martin and Neyret 2012). It is essential because biodiversity loss and climate change aggravate poverty and non-respect for human rights. So, in a negative view, environmental and human rights legal systems are interlinked. By understanding ecological functionality, scientists and environmental lawyers will be able to come up with solutions to combat what is harming the proper functioning of interactions, and thus reduce the causes of poverty. Therefore, this interdependence must be strengthened.

3 Legal Dimensions of the Capacity of International Law to Meet the Challenges of Human Vulnerabilities

The interdependence and existing threats create new unknown risks. To deal with these issues, the international community is adopting new instruments and new positions to understand and reduce vulnerabilities. Within this perspective, human interests should be increasingly considered by governance processes (Sect. 3.1) and the intrinsic logic of human rights has to be conjured to propose a new and fair legal framework for the protection of people in a context of climate change (Sect. 3.2).

3.1 The Need to Take into Account Human Interests in Governance Processes

International environmental law and the CBD deal with human security but in a too limited way to achieve a laudable objective. The explanation is simple. The international law is built only by and for states. Today, this approach is highly criticized (Sect. 3.1.1). Researchers call for the need to extend the content of international law in order to adapt it to the interests of all stakeholders involved in international law and human security (Sect. 3.2.2).

3.1.1 Recognizing the Failures of the Nation–Centered Approach

The nation-centered approach, which is illustrated by the CBD (Articles 3, 4 and 6a), cannot be adapted to the protection of human security because it supposes a method on several scales (international, national, local). On one hand, the current approach ignores the action of non-state actors in the fight against climate change, erosion of biodiversity and threats to human security. On the other hand, this approach does not take into account the capacity of non-state actors to efficiently participate in the process. This denial causes a division between actors which is an additional difficulty when fighting against environmental and human vulnerabilities. Besides, in this way, states accept to lose power because, in fact, some NGOs for example can take decisions. So, this sovereign approach of Article 3 of the CBD runs contrary to the phenomen of globalization and global governance. Finally and implicitly, states may lose power when letting NGOs (and also multinational firms) intervene in the legal international framework.

Moreover, this approach distorts the idea of justice which was fundamental to the initial project of CBD. In Article 1 of the UNFCCC, developed countries recognized their historical responsibility in causing climate change. The justice objective, which is totally dependent on human security, seems to be included. The juridical mechanisms related to the access to genetic resources and the funding to protect developing countries could prove this objective. But in fact, CBD makes "biotechnology and access to living resources one of its major challenges as if the advances in biotechnology would necessarily solve the problem of food shortages and underdevelopment" (Arbour and Lavalle 2006:450). Areas of vulnerable countries that are rich in natural resources are viewed only as an economic issue whereas they are also an area of reconciliation between humans and nature (an essential ethical element of human security). Additionally, the legal logic of the Nagoya Protocol validates a sovereign and capitalist use of genetic resources.

The legal landscape of international law is mixed. It is true that the CBD contributes indirectly to the protection of human security but the interrelations are still not significantly promoted at the international level.

3.1.2 Adapting the Content of International Environmental Law

The content of international environmental law is still shaped in a way that is currently challenged; it is based on states' interest model instead of a model based on global interest. In the first conception, the spatial dimension dominates. The sovereign approach illustrates perfectly this conception. In the second conception, an intertemporal dimension is added, thus underlining that states have to defend new interests and assume new obligations. Sovereignty is not as important as before. The legal concerns are rather focused on interstate justice, intergenerational justice and intragenerational justice. Weiss (1993) has identified three problems caused by the lack of equity at the international level: depletion of resources; decrease of the quality of resources; and decline of access to natural resources and their benefits (that is to say, the core subject of research). The goal of future relevant multilateral conventions should be the promotion of equality between present and future generations with regard to natural and cultural heritage and human and environmental security. The introduction of the principle of 'non-retrogression' would be the main pillar of the long-awaited adaptation (Prieur et al. 2012). This adaptation has to be completed by a procedural dimension.

3.2 Strengthening Human Rights

The strengthening of human rights requires a better representation of states in international institutions (Sect. 3.2.1). This advancement could be facilitated by the expansion of subjects of rights at the international level which is an essential condition for a new ethic (Sect. 3.2.2).

3.2.1 Institutionalization of Vulnerabilities

Facing vulnerabilities from a justice perspective is currently a broad objective on international level, with a priority often granted to more vulnerable categories such as women and children. However, these vulnerable people are not subjects of international law since they are represented by states, most of which are generally powerless in international organizations. Therefore, the management of human vulnerabilities is still organized by powerful and rich countries. This setting does not respect the principles of justice and equity and the inequitable distribution of power often violate collective interests. In 2006, the World Bank once again criticized this situation and proposed a proportional participation of states according to their roles in natural processes and balance.

Today, it could be difficult to change the position of international organizations, but the establishment of a global organization on environment may be more efficient. To deal with human vulnerabilities, the power relations, which are linked to the economic international order rather than a legal international order, should be clarified. It is also very important to promote the access to justice globally. The international order may frame the needed guidelines to facilitate the achievement of this goal. The Aarhus Convention is a good example in this respect and deserves to be expanded.

3.2.2 Expansion of Subjects of Rights at the International Level

The introduction of an ethical responsibility in the international order needs more than an evolution of the power distribution because, in this context, a state-centered approach is still maintained. The first challenge is to accompany states for the establishment of a "convergent and complementary culture between public and private institutions" and between public institutions and individuals (Sedjari 2005). In fact, individuals are more affected than states and as a consequence are in a better position to tackle vulnerabilities. However, at international level, individuals are not subjects of rights. They cannot for instance take legal action in the International Court of Justice (ICJ). This paralyzing system must be reformed. Some states started to adopt new solutions by choosing a collective conception of human rights against a more classic individual conception. Such an evolution can, for instance, facilitate the protection of indigenous people and lay the foundation for a future convention on environmentally-displaced people (Cournil et al. 2015). The second challenge is to guarantee the conditions for the conservation of political, economic and social rights because, generally, the rights of environmentally-displaced people are violated. This situation illustrates the interdependence between environmental and human security. The University of Limoges which has drafted the text of a convention on the international status of displaced people has taken into account this fragile balance between human security, biodiversity and climate change.

Lastly, the recognition of the rights of humanity, which are opposable to the international community, is a claim made by developing countries and environmental researchers who have faith in human solidarity. This trend advocates for a very committed ethical perspective which has accepted that climate change and threats to biodiversity result in impoverishment and conflicts that, in turn, threaten "directly all countries who have not considered these questions as external affairs but as an internal matter" (Suy 2009). It appears difficult to consider humanity like a subject of rights, but at least, these insights empower states on the preservation of collective world interest (Suy 2009).

It is difficult to predict the future, but one can conclude that the current structure of international order is still limited by an anthropocentric rationale.

References

Arbour JM, Lavalle S (2006) *Droit international de l'environnement*, Bruylant, 2006, 450 p.

Beurrier JP (2010) *Droit international de l'environnement*, Pedone, 590 p.

Bonnin M (2010) *Connectivité écologique et gouvernance territoriale*, IRD/3ED, Guyancourt, 9

Bonnin M (2012) "L'émergence des services environnementaux dans le droit international de l'environnement: une terminologie confuse", *VertigO* - la revue électronique en sciences de l'environnement, 2012, n°3, 12 p.

Brown WE (1993) *Justice pour les générations futures*, UNU Press, Editions sang de la terre, Paris, 356 p.

Burelli T (2015) "L'Union Européenne et la mise en œuvre du protocole de Nagoya, faut-il se réjouir de l'adoption du règlement n.511/2014.", Revue Juridique de l'environnement 3/2015, 437–462

Compagnon D (2010) "Les défis politiques du changement climatique: de l'approche des régimes internationaux à la gouvernance transcalaire globale", in Changement climatique et défis du droit, Bruylant, pp. 32

Cournil C, Colard-Fabrelouge C et al (2010) Changements climatiques et défis du droit: actes de la journée d'études du 24 mars 2009, Centre d'études et de recherches administratives et politiques Bruxelles, Belgique, Bruylant, 450 p.

Cournil C et al (2015) *Mobilité humaine et environnement: du global au local*, Quae, 403 p.

Doussan I (2012) La représentation juridique de l'environnement. In: Martin G, Neyret L (eds) *Nomenclature des préjudices environnementaux*. LGDJ, Paris

Martin G, Neyret L (2012) *Nomenclature des préjudices environnementaux*, LDGJ, Paris, 456 p.

Prieur M et al (2012) La non régression en droit de l'environnement, Bruxelles, Belgique, Bruylant, 547 p.

Sedjari A (2005) Partenariat public-privé et gouvernance future. L'Harmattan, Paris, 515 p.

Serpantié G, Méral P, Bidaud C. (2012) "Des bienfaits de la nature aux services écosystémiques", *Revue électronique en sciences de l'environnement* n°3

Suy A (2009) La théorie des biens publics mondiaux, une solution à la crise, L'Harmattan, 178 p.

Valentin JM (2009) "Vers un nouveau concept stratégique: la 'sécurité climatique'", Lettre de l'Iris

Chapter 6
Adaptation and Mitigation: Relevant Governance and Risk Management Options for Pacific Island Countries

Arunesh Asis Chand

> *On a global scale, there is a growing recognition of the significant role that developing countries play in determining the success of global climate change policies, including mitigation and adaptation policy options.*
>
> IPCC Fourth Assessment Report: Climate Change 2007

Abstract Adaptation and mitigation have recently become a major concern for the governments of the Pacific Island Countries in regards to the global environmental change and human security issues. Adaptation and mitigation need to be analyzed and addressed on different scales through multilevel governance perspective, and the responses must be multilevel and multi-actor – combining simultaneously local and global scales and involving public and private actors. The linkages between national, regional, and local policies should strengthen the multilevel, regional, and urban governance to more effectively address the challenges related to adaptation and mitigation in Pacific Island Countries. This chapter elaborates on how an appropriate governance system can help in effective designing, implementing, and in particular managing adaptation and mitigation policies at different levels of governments, non-state, and non-governmental actors. This chapter further incites important discussion on the availability of a wider range of options than those embodied in the international environmental regime to effectively design and implement adaptation and mitigation policies along with networks that draw together government actors.

Keywords Adaptation · Mitigation · Climate change · Governance · Policies

A. A. Chand (✉)
Faculty of Science Technology and Environment, School of Geography, Earth Science and Environment, University of the South Pacific, Suva, Fiji

© Springer International Publishing AG, part of Springer Nature 2019
M. Behnassi et al. (eds.), *Human and Environmental Security in the Era of Global Risks*, https://doi.org/10.1007/978-3-319-92828-9_6

1 Introduction

The Islands in the Pacific region are unique and this uniqueness is attributed to the fact that they are situated in the middle of the Pacific Ocean, which is the biggest of all, and are relatively big enough to be recognized by the rest of the world. The presence of the pristine environment and abundance of nature's creativity has made some of these Islands to be known as the World Heritage Sites that should be preserved for future generations (Brooke et al. 2004). Contrary to this, Pacific Island countries are often overlooked when it comes to the global issue of climate change and its impact on them due to their geographical positioning.

Pacific Ocean has around 30,000 islands of whom around 1000 are inhabited.[1] There are three major ethnic groups: Polynesia which consists of Tonga, Cook Islands, and French Polynesian; Melanesia (Papua New Guinea (PNG), Solomon Islands, Vanuatu, and New Caledonia); and Micronesia which includes Micronesian and the Marshall Islands. The sheer vulnerability of most of these island nations to sea-level rise has once more provoked the necessity to reconsider adaptation and mitigation in light of analyzing the availability of relevant governance and risk management options. As the global warming is becoming a real threat, people in Pacific Island countries have no option but to adapt to the impacts and start a more resilient life-style towards climate change.

For this reason, adaptation to climate change is very important for Pacific Island countries. People in this region are regularly experiencing the effects of extreme events such as flooding caused by high rainfall, unusual long droughts, cyclones, and storm surges. Apart from the non-climatic forces – such as poor landuse practices, mismanagement of resources, internal migration and over-population – progress in the development sector in this region is largely hindered. Most low-lying atoll nations are expected to be severely affected in terms of economic downfall as the result of inundation of seawater, which often damages cash crops and livestock, thus putting pressure on food security as well. The disastrous effect of climate change has led to population relocation, which results in social and cultural disruption and disproportion in some of these countries.[2]

In light of the risk management options, the leaders of Pacific Island countries have voiced out to the outer world for assistance in the areas of adaptation and mitigation. According to the latest UNDP report, the Pacific Adaptation to Climate Change (PACC)[3] is the first project to act in response to this in an urgent call for action and also to support the systematic and institutional capacity in addressing adaptation in this region. The core task for the project is the capacity building in Pacific Island governments to mainstream adaptation into government policies and

---

[1] For more information, see http://fam-courtages.com/pacific-opportunities-incubator/

[2] Example, Carteret Island in PNG, refer to http://www.davisprojectsforpeace.org/media/view/1023/original/

[3] The PACC Project is the first Global Environment Facility (GEF) project in the Pacific drawing on the Special Climate Change Fund (SCCF) focusing specifically on adaptation implementation in 13 countries simultaneously.

plans, and setting off a platform to address adaptation for medium to long term at regional level.

Adaptation to crosscutting issues, such as climate change which has a huge funding factor[4] attached to it, has become a major problem to Pacific countries while mitigation is not a prior concern since the major emitters and polluters are developed – and currently emergent – nations. In addition to this, the comparative smallness and the geographical isolation from the developed world is another key factor in securing funding for adaptation works. In cases where funding was secured for some reasons, the presence of poor governance and improper planning impeded the effective delivery in adaptation works. Therefore, the immediate challenge in this area is to create a secure platform to ensure funding activities being managed in a transparent and well-organized manner.

In this context and looking at the relevant governance and risk management options for Pacific Island countries, this chapter aims to raise a number of questions such as: *what is the true meaning of successful adaptation?* This may lead to the question of: *what aspects of our life are threatened by climate change that we need to protect?* This would lead one to think about any failed adaptation strategy and its outcome for the community. For example, if the effects of climate change lead to relocation and migration, then obviously adaptation has not been successful. Finally, adaptation polices has to be in line with what people aspire about their future; in other terms what kind of life people want to live and what future they want for their future generations. It is these aspects of climate-related risk management policies that need to be addressed.

Hence, this chapter considers some important issues related to adaptation, which are often overlooked by authorities in the current development of disaster risk reduction and risk management arena, but are highly critical for poor communities in Pacific Island countries. Particular focus is on the issue of response to adaptation, the assessment of risk and associated impacts and the governance and risk management in the capacity-building process. The analysis further discusses that the development process is highly linked to the effective risk management process.

2 Responding to Adaptation and Mitigation

Conservative responses to adaptation have led to the use of General Circulation Models (GCMs)[5] by climate scientists to predict future climate events, and this has helped greatly identify and calculate potential impacts on both human and ecological systems. Using these top-down and scenario-driven approaches, adaptation options have been identified and disaster risk management plans have been drawn. However, although these kinds of climate models are useful in describing some

---

[4] For more detail refer: www.ausaid.gov.au/keyaid/adaptation.cfm

[5] GCMs, as well as Numerical Weather Prediction (NWP) models, numerically simulate the 'state' of the atmosphere, using a finite expression of the equations of motion.

general trends and dynamic interactions around the global atmosphere, they remain limited in depicting climate impacts at regional and local levels. In most cases, this scenario-based approach fails to correspond to the human interactions and their abilities to adapt, which are the most important factors in identifying relevant adaptation strategies.

The current knowledge seems preferring a new approach in addressing adaptation by simply taking a bottom-up approach or identifying the most vulnerable and assessing their past and present sensitivity to climate change. Once this process is completed, their current coping status is being identified and plans are drawn as to how these can be modified with the changing climatic conditions. According to IPCC (2001) report, vulnerability in the context of climate change means the degree of susceptibility to, or difficulty in coping with, the adverse effects of climate change, including climate variability and extremes. The use of vulnerability approaches has led to the overcoming of doubts associated with GCMs and predicted impacts as well as enhancing the capacity building of community; this in turn allowed governments to adapt to the current climate variability and future climate change.

The UNFCCC (2004) report highlights that the top-down, scenario-driven adaptation approaches are useful in providing policy-makers with critical information about the likely long-term and global climate impacts with specific details such as GHG emissions figures. But this type of information is only good for raising awareness, identifying key issues, and supporting international processes. As highlighted by Klein (2004), they do not provide policy makers with relevant information to develop efficient adaptation strategies. Hence, bottom-up/vulnerability-driven approaches are more suitable in assisting policy making at local level although the two types of approaches can be used as a mixed approach or in a harmonized manner for the formulation of adaptation policies and procedures in the future.

Climate change adaptation differs across geographical scales (local, national, regional, and global), sequential scales (assessing current impacts against preparing for long-term change) and must be addressed using multilevel and multiactor approaches – combining simultaneously local and global scales, involving public and private actors as well. Acting in response to adaptation involves interdisciplinary and multiple expertise at multilevel. This simply means that various expertise working in areas such as climatology, ecology, economics, and natural resource management – including agriculture, fisheries and forestry – will have to unite with personnel from public health, engineering, business, disaster risk management (DRM), community development, and social services (Parry et al. 2001). It is anticipated that climate change will impact other vulnerable sectors or systems that are already struggling with shocks – such as health and economic sector – hence the need for a more integrated approach for adaptation.

The IPCC Technical Guidelines for Assessing Climate Change Impacts use a set of generalized evaluation criteria and procedures for Pacific Island countries. These procedures and evaluation criteria mostly apply to the developed nations, and most cases do not reflect the true picture of the current situation of poor developing countries in the Pacific (Cutter 1996). Hence, the guidelines usually assume the availability of well-established resource management institutions and governance structures and mechanisms of robust decision-making process. In fact, these capacities are not available in most Island nations as most of them are still struggling with basic necessities such as proper housing, water, and electricity. As the outcome of the application of IPCC guidelines to the Pacific Island countries, a significant amount of uncertainties has been generated, thus creating further problems along with existing challenges. According to Barnett (2001:10), "approaching the problem of decision making about adaptation from alternative paradigms of science such as ecology, systems theory, and more generally from the social and policy sciences, acknowledges and reveals different approaches to the problem of uncertainty in adapting to climate change and accelerated sea-level rise".

3 Identifying Risks in Climate Change

The risk in climate change is differentiated generally on two categories: the frequency and intensity. It is seen whether or not it is auto-correlated (independently distributed over time) and how it is distributed among individuals and groups (idiosyncratic versus covariate) (Christoplos et al. 2009). Climate change is interacting with all the categories mentioned here. The adaptation ability of individuals, households and communities are all stressed in one way or another as well as that of the governance, social and market institutions upon which they rely.

As explained by Christoplos et al. (2009:8), "*Idiosyncratic* shocks are those that affect the individual or household (e.g., death, injury, unemployment); *covariate* shocks are those that affect localities or nations (e.g., epidemics, disasters, war)". The difficulties between these two terms were further stated in more simplified manner by Cafiero and Vakis (2006) who specified that idiosyncratic risk can be mitigated if shared within a social group or network. This simply means that at household level, the idiosyncratic risk can arise if the social network fails to eradicate it by sharing. Looking at it in this perspective indicates that a community is in fact the least required size of people needed to efficiently share the most dangerous idiosyncratic risks. In the case when risks are complex and confounding and cannot be shared at community level, then there is a need for external intervention by the government for instance (Cafiero and Vakis 2006).

Climate change has put great pressure and added extra complexity to these risk categories, since it affects at multilevel mixing the two categories. Discussions on climate change adaptation area have mostly emphasized on covariate risk, and specially focused on the occurrence of increased level of severe weather-related disasters. As highlighted by Christoplos et al. (2009), this has led to the need for a more harmonized approach in getting together the two fields of disaster risk reduction and climate change adaptation. More recently, in the Pacific Island countries it has been seen that work in this area is already underway and indeed there is an urgent need in capacity building of the Pacific's national and local governments. The expertise developed in the area of adaptation within the disaster risk reduction community and the experience developed by actors working in this area for the past years need to be an input for the policy formulation process in relation to climate change.

On the other hand, enhancing capacity development for idiosyncratic risk reduction is also important. There are two main reasons for this. Firstly, the increase in the small risk idiosyncratic as well as the idiosyncratic moving towards covariate risk has a large effect on poverty. In a nutshell, the impacts of small and medium disasters usually accumulate and surpass the impacts of large disasters. In addition, the events are periodic and the impacts are felt locally. Usually, the risks posed by small disease outbreaks, flash floods and land degradation tend to be overlooked by both media and policy makers. It has been seen that communities often rely on the informal risk-sharing mechanisms which are usually dependent on the social capital. Hence, the important capacity for addressing idiosyncratic risk is pertained to societies and local organizations.

Secondly, it was observed that households and communities which are more resilient to idiosyncratic risks are often lesser vulnerable to covariate risks. This means that different risks can compound on one another, and this in turn doubles and triples the overall impact on the household. Hence, the capacity building for community resilience and self-reliance usually strengthens the capacity to manage covariate risks. In reality, climate change will increase the complication of idiosyncratic versus covariate risks. Everything considered, there is a need to empower the local institutions and social capital with capacity development to successfully manage these risks. This in turn would strengthen the formal national and international mechanisms as well.

There is an urgent need to identify the types of changes that climate change brings in and what people need in order to adapt to these changes. For example, is it the extreme events or the sea-level rise, or both? In that case, the society needs to address the false dichotomy between humanitarian and developmental approaches to climate adaptation. As stated by Christoplos et al. (2009:9), "the gap between humanitarian perspectives and those of development actors is decreasing as development thinking comes to understand how risk is at the center of the human dimensions of poverty".

Climate change has made it obvious now that there is an urgent need to be aware of the risk out there and to incorporate it in current development processes. In fact, it is not only about mainstreaming the risk into development, but also realizing that the development itself is a risk management. This means that we need to reveal the 'content' of risk, make it 'visible and transparent', and ensure that the household and community have ample information about it to successfully make their decision as to how they are going to manage it.

It is often seen that large and sudden-onset catastrophes often attract the attention of the disaster risk reduction authorities. When it comes to developmental actors, they tend to ignore the sudden and gradual catastrophes since they see disasters as a hindrance to development rather than a signal that it is time to incorporate the effects of disasters into the development of disaster risk reduction plans. As outlined in ProVention and Active Learning Network on Accountability and Performance in Humanitarian Action (ALNAP) briefing paper (2008), most countries that experience repeated disasters often fail to coordinate relief and development efforts and fail to create corresponding structures to address emergency needs for future events of such nature.

4 Governing, Assessing, and Managing of Climate Risk

Climate change, among others, is one of the major stressors with a great influence on a system. It is vital to take into consideration the interaction of these 'multiple stressors' and their possible swelling effects when considering the vulnerability and adaptive capacity of the system in totality, including its people and institutions. It is highly important to examine the direct risks and their impacts as well as the intangible and indirect risks and impacts such as those related to the psychological health. There is a wider need for a better understanding of the changes happening in the environment, the risk impact of these changes, and the options of reducing these risks. The best way to do this is through taking an integrated approach where the wider community is trained to better understand and manage climate adaptation-related risks. This allows stronger linkages between policy and science related approaches in problem solving.

There are different types of approaches to define and measure risk and vulnerability. It is often perceived that when these approaches are combined they are likely to act in a complementary manner and give a detailed picture of the context in which these stressors perform, thus providing more options for decision-making. This is similar to a casual process of vulnerability in which different recommendations are being made to reduce risks, exposure, etc. at different points along the process. There is, in fact, an integrated approach required to understand the issues, reduce the impacts of risks, and plan for likely events in future. It can be carried out for simple risks as well as for complex processes that have greater impact on social and ecological systems. Some changes can be gradual while others can be sudden and catastrophic. Hence, this gives rise to the question about its meaning for a system and whether the current approaches to understand risk are able to capture this complexity effectively. There is a need to design effective approaches and the capacity-building process to effectively carry out these approaches.

The governance of risk management involves a set of identified rules that allow interactions in relation to elements that belong to the risk category. The role of governance actually is to decentralize the state which in turn discloses the various other agencies involved in the social ordering (O'Malley 2009). The current post-modern theory of governance sees the shift towards governance in communities, where individuals take responsibility of their own lives rather than depending on the state as the key provider (Osborne and Gaebler 1993). This shift in governance has in fact changed the societies' actual vision towards risk. As stated by O'Malley (2009:4), "to govern something as a risk is to identify a future condition as being more or less probable, and to set in train ways of responding that will affect this probability in some way". This indicates the inclusion of mitigation as an option or even broadening the scope of 'risk netting' to include some tools such as insurance.

The rising anxiety and curiosity surrounding the risks related to societal progress and modernization – such as those linked to the issue of global climate change – exerts added pressure and challenges the way risk is governed. The doubt connected to the modernization risk, in fact, obstructs the prediction of their occurrence, and therefore the governance of risk is largely under the situation of uncertainty despite the interest and commitments shown by the governments and scientists. The governance of risk is seen at different angles by various levels of governments, especially in the case of climate change as a new challenge. Researches in governance area reveal that the cross-point between science and governance has an important role in the facilitation of policy action supporting climate change adaptation (de Loë et al. 2009). It is often realized that the presence of efficient and robust governance procedures and processes allow better up take of science and support actions for adaptation.

Most Pacific Island countries comprise multi-layer governance systems which consist of three levels of government (local, national, and international), non-governmental organizations (NGOs), private sector, research entities and traditional owners; each component with a range of task influence on the policy, planning, and implementation. It is understood though that this complex network of actors has been sluggish in adopting a strategic and integrated approach to climate risk management. There are three obvious reasons for this: *firstly*, the multilevel governing body is ready to act only if it is provided with adequate convincing evidence signifying a risk that warrants management; *secondly*, the government institutions have failed to prioritize climate change, climate risk and adaptation, and as a result adaptation is yet to be incorporated in policy matters; *thirdly*, the complex network of players involved do not easily come to a common goal, and thus act as a limitation in the successful designing and implementation of vigorous adaptation strategies for climate change (Wood and Stocker 2009; Norman 2009; Wescott 2009; Lazarow et al. 2006).

The reasons discussed above, which actually account for the slow response of the governmental institution to respond to the observed climate changes, are questionable. The first two reasons seem to be predominantly weak, though this weakness does not have much impact on the dialogue part of climate adaptation and risk management. Science and policy interactions have been widely criticized by researchers in this arena for oversimplifying the manner in which science contributes to policy-making (Stokes 1997; Hansen et al. 2003; Wynne 1991, 2006; Godin 2006; Trench 2008). Decisions emerging from policies eventually reproduce effort to bring together opposing normative values within a civil society.

Therefore, the concept of 'evidence-based' inquiries itself, forcing a particular policy pathway, is contradicting in relation to public policies. In regard to this theory, the third reason, which states that the government institutions have simply been negligent in managing climate change risks, is also contradictory. The main reason is the availability of a wide range of established risk management activities in government institutions which are up and running smoothly to date. As highlighted by

Smith et al. (2008), the government institutions are usually 'duty-bound' to serve in the interest of the public safety and respond to the potential threats to civil society. Thus, because knowledge and leadership cannot be dismissed totally as constraints for adaptation, more important explanations can be found in the broader governance system, which should be analyzed to see how it works together with knowledge and leadership as well as competing values.

5   Key Risk Management Options for Pacific Island Countries

The upper most priorities for Pacific countries consist of increasing the availability of information necessary to understand the overall natural and social processes. This task belongs to the research community in both natural and social sciences. There is a wider need to produce baseline data sets and to monitor the key indicators that are necessary to understand the change and to disseminate this information to the wider community out there (in all directions) who need it the most. This would mean the development in the physical infrastructure for communication and better utilization of social practices of information exchange. The society, which is well educated, seems to be more resilient to environmental change; better focus on education policy is the key to resilience and adaptation. This does not imply that everyone need to have technical and formal qualification in order to adapt to the change but at least to have some sociological knowledge to observe elementary changes in social conditions.

The need for resilience and adaptable social-ecological system has implications for economic policy in Pacific Island countries. There is an urgent need to restore the transport infrastructure for the betterment of fast and rapid resource delivery, especially in case of extreme events; this will enable quicker movement of people to safe destinations. It is equally important to have a good supply of basic food items in reserve as a backup supply in case of emergency. As highlighted by Sen (1999), in terms of food security after extreme events, most local employment schemes may very effectively assist households with income that can be used to purchase basic necessities. Provision of basic health essentials are also seen to be effectively distributed in times of disasters as there is always danger of diseases being rampant.

There is a need for an organizational structure to be established in order to look at the climate change risk administrative system in Pacific Island countries. It should be responsible to look into the affairs of these countries at all levels: regional, national, and local. The purpose of this administrative body would be to look at the general welfare of the related climate adaptation works, including at the level of community and stakeholders. This does not mean that authorities should report to a central body and abide by its rules and regulations only, because centralization of authorities in the past have been blamed for undermining the community resilience to climate extremes. This also involves more than simply shifting all authority downward to local communities (although more of this is essential), rather it involves locating authority with appropriate regional, national, local, and non-governmental organizations in a coordinated and communication-rich system. In this scheme, regional and national organizations play the role of brokers, facilitators, and funders of local-level adaptation strategies. Such a scheme offers the possibility of re-establishing mutually supportive relationships among local communities as they liaise on the common problem of climate change and sea-level rise.

The South Pacific Regional Environment Programme (SPREP)[6] is the obvious candidate for the principal authority at the regional level. At the national level, there are two principal options: either expand on the Pacific Islands Climate Change Assistance Programme (PICCAP)[7] country teams or expands on the Disaster Management Committees established as a result of the International Decade for Natural Disaster Reduction (IDNDR).[8] At the local level, governing bodies based on the suitable administrative unit (most likely the village) could be established. Designing such a system could take account of existing organizations and linkages, and could identify key people.

As well as being a prerequisite for resilience in its own right, the proposed administrative system could be invaluable for a number of specific tasks related to adaptation: it could provide official nodes for supporting and organizing research in localities; it could provide clear lines of communication vertically; it could provide a framework for monitoring changes and communicating findings; it could provide channels and nodes for community education and awareness raising; and it can bring a wide array of people into an integrated system, thus contributing to the development of human resources. In sum, these broad policy goals aim to develop systems of purposeful exchange between informed social groups living in a social-ecological context characterized by a fair level of resource saving, a high degree of sensitivity to change, a capacity to learn, and a capacity to change. They have strong resonance with existing calls for capacity building, disaster planning, education, and human development, and it is important to stress the interdependence of these aspects as a requirement for coping with climate change.

The element of risk governance involves risk management which needs to be backed up with: efficient managing of early warning systems; mobilization of action

---

[6] The **South Pacific Regional Environment Program** (**SPREP**) is a regional organization established by the governments and administrations of Pacific region nations with the aim to promote cooperation in the Pacific islands region and provide assistance to protect and improve the environment and ensure sustainable development for present and future generations.

[7] The Pacific Islands Climate Change Program (PICCAP) is a program to help Pacific Island countries to implement the UNFCCC. It began as part of the CC-Train Program of the United Nations, but was adapted by the SPREP to be more appropriate to the Pacific countries that would carry it out.

[8] UNISDR is part of the United Nations Secretariat and its functions span the development and humanitarian fields. Its core areas of work include: the application of disaster risk reduction (DRR) to climate change adaptation; the increase of investments for DRR; the establishment of disaster-resilient cities, schools and hospitals; and the strengthening of the international system for DRR.

at multilevel with appropriate stakeholders; and capacity building through imparting of relevant knowledge and effective technologies. As there are different time frames for works, which need to be completed, policy makers must often act in an efficient manner to make progress in addressing existing challenges. Some risks may require routine-based approach especially work involving a new law or regulation. In such circumstances, it is often wise to allocate responsibilities in order to get it completed. In a more complex risk situation, it is wise to involve relevant scientific expertise and wide stakeholder consultations. In the case of an uncertain risk, it is useful to engage sustainable use and precautionary principle, especially on building resilience, to empower the capacity of a system or institutional arrangements to survive the catastrophes. There is high importance to integrate adaptation in the center of decision-making and policy formulation processes, to successfully support climate change adaptation and reduce associated risks and impacts (Huq et al. 2003).

It is also important to have a certain level of understanding about future climate projections in order to manage the associated risks and impacts. This knowledge is then used to develop new procedures or amendments to policies, programs, and projects in order to minimize the risks and enhance the capacity building. This type of structural developments in the policy work and actions to safeguard infrastructure, systems and processes against climate risks is sometimes referred to as 'climate proofing'. The inclusion of adaptation into sectoral policies and programs provides a range of opportunities to reduce the vulnerability to climate risks and embed them into the system as well as has the capacity to address changes in future (OECD 2005; Smit and Benlin 2004). Newell (2004:124) highlighted that "policy integration is perhaps the greatest contribution that governments can make towards providing climate protection and it is also potentially the least economically costly".

6 Conclusion

In order to govern properly, and considering the ever more complex risk parameters, the governments of Pacific Island countries need to adopt effective and robust governance mechanisms that can adapt to these complexities. The need to respond to climate-related risks is indeed a challenge to most of the Pacific countries if we consider issues such as infrastructure, funding, innovation, and policy-making processes which seem to be under-developed compared to other nations, especially the developed ones. Most of these issues depend largely on the capacity of global governance for production, knowledge, financial, development assistance, and security to respond in a timely and effective manner to risks. The ever-increasing risk of climate change needs a well-defined strategic approach cutting across other key sectors such as agriculture, health, and trade policy among others.

Successful adaptation to increased climate risks requires institutions to prioritize policies and engage resources and efforts to implement them. Adaptation must be addressed at all levels of decision-making process and also integrated into development planning. The risk management strategy in climate change needs to engage the multilevel governance perspective, including economic and financial entities, to effectively address risks and associated impacts at all levels. Adaptation can be streamlined with the developmental planning which would require a range of inputs and actions apart from the international assistance. It is highly essential to adopt an effective policy framework that incorporates advanced planning and climate risk management linking to the overall governance structures at national and sub-national levels. The provision of international assistance is usually ineffective in absence of proper governance policies and procedures.

References

Barnett J (2001) Adapting to climate change in Pacific Island countries: the problem of uncertainty. World Dev 29(6):977–993 http://www.sustainable.unimelb.edu.au/files/mssi/Barnett_Adapting-to-climate-change-in-Pacific-countries-2001_67062.pdf

Brooke M, Hepburn de L I and Trevelyan R J (2004) Henderson Island World Heritage Site: Management Plan 2004–2009. Foreign and Commonwealth Office, London. www.ukotcf.org/pdf/henderson.pdf. 25 October 12

Cafiero C, Vakis R (2006) Risk and vulnerability considerations in poverty analysis: recent advances and future directions. World Bank, Washington, DC

Christoplos I., Anderson S., Arnold M., Galaz V., Hedger M., Klein R.J.T., and Le Goulven K. (2009). The human dimension of climate adaptation: *The Importance of Local and Institutional Issues*. EditaSverige AB, Stockholm, Sweden. http://www.ccdcommission.org/Filer/report/HUMAN_DIMENSIONS.pdf. 20 October12

Cutter S (1996) Vulnerability to environmental hazards. Prog Hum Geogr 20(4):529–539 http://webra.cas.sc.edu/hvri/docs/Progress_Human_Geography.pdf

de Loë R C, Armitage D, Plummer R, Davidson S and Moraru L (2009). *From Government to Governance: A State-of-the-Art Review of Environmental Governance*. Final report. Prepared for Alberta environment, environmental stewardship, environmental relations. Guelph, ON: Rob de Loë Consulting Services

Godin B (2006) The linear model of innovation. Sci Technol Hum Values 31(6):639–667

Hansen J, Holm L, Frewer L, Robinson P, Sandøe P (2003) Beyond the knowledge deficit: recent research into lay and expert attitudes to food risks. Appetite 41:111–121

Huq S, Rahman A, Konate M, Sokona Y, Reid H (2003) Mainstreaming adaptation to climate change in Least Developed Countries (LDC). International Institute for Environment and Development, London

Intergovernmental Panel on Climate Change (IPCC) (2001) *Climate Change 2001*. Impacts, adaptation and vulnerability. Chapters 10, 11, 17, and 18.Contribution of Working Group III to the Third Assessment Report of the IPCC

Klein R (2004) Approaches, Methods and Tools for Climate Change Impact, Vulnerability and Adaptation Assessment. Keynote lecture to the In-Session Workshop on Impacts of, and Vulnerability and Adaptation to, Climate Change. Twenty-First Session of the UNFCCC Subsidiary Body for Scientific and Technical Advice, Buenos Aires, 8 December 2004. http://unfccc.int/files/meetings/cop_10/in_session_workshops/adaptation/application/pdf/081204_klein_adaptation_abstract.pdf. 25 October 12

Lazarow N, Fearon R, Souter R, Dovers S (2006) Coastal management in Australia: key institutional and governance issues for coastal natural resource management and planning. CRC for Coastal Zone, Estuary and Waterway Management, Indooroopilly

Newell P (2004) Climate change and development: a tale of two crises. IDS Bull 35(3):120–126

Norman B (2009) Planning for coastal climate change: an insight into international and national approaches. Victorian Government Department of Planning and Community Development, Melbourne

O'Malley P (2009) Governmentality and risk. In: Zinn J (ed) Social theories of risk and uncertainty. Blackwell Publishing, Oxford, pp 52–75

Organization for Economic Cooperation and Development (OECD) (2005) Conclusions of the Chair. Global Forum on Sustainable Development on Development and Climate Change, 11–12 November 2004, Paris, France. ENV/EPOC/GF/SD/RD (2004)15/FINAL. http://www.oecd.org/environment/climatechange/34393852.pdf. 20 October12

Osborne D, Gaebler T (1993) Reinventing government: how the entrepreneurial spirit is transforming the public sector. Plume Books, New York

Parry M, Nigel A, Tony M, Robert N, Pim M, Sari K, Matthew L, Cynthia R, Ana I and Gunther F (2001). Millions at risk: defining critical climate change threats and targets. Glob Environ Chang (11): 181–183. http://www.elsevier.com/framework_aboutus/pdfs/Millions_at_risk.pdf. 25 October 12

ProVention and ALNAP (2008), Slow-onset disasters: drought and food and livelihoods insecurity, *Briefing Paper*., www.alnap.org/pool/files/alnap-prevention_lessons_on_slow-onset_disasters.pdf

Sen A (1999) Development as freedom. Anchor Books, New York

Smit B and Benhin J (2004) Tools and methodologies for mainstreaming vulnerability and adaptation to climate change into sustainable development planning. A paper presented at the workshop integrating vulnerability and adaptation to climate change into sustainable development policy planning and implementation in Southern and Eastern Africa, 4 September 2004, Nairobi, Kenya

Smith T, Preston B, Gorddard R, Brooke C, Measham T, Withycombe G, Beveridge B. and Morrison C (2008) Regional workshops synthesis report: Sydney coastal councils' vulnerability to climate change. Prepared for the Sydney Coastal Councils Group. http://www.sydney-coastalcouncils.com.au/sites/default/files/systapproachphasetworeport.pdf

Stokes DE (1997) Pasteur's quadrant—basic science and technological innovation. Brookings Institution Press, Washington, DC

Trench B (2008) Towards and analytical framework of science communication models. In: Cheng D, Claessens M, Gascoigne T, Metcalfe J, Schiele B, Shi S (eds) Communicating science in social contexts. New models, new practices. Springer, New York, pp 119–135

United Nations Framework Convention on Climate Change (2004) Application of methods and tools for assessing impacts and vulnerability, and developing adaptation responses. Background paper to the Subsidiary Body for Scientific and Technological Advice, Buenos Aires, 6–14 December 2004. FCCC/SBSTA/2004/INF.13

Wescott G (2009) Stimulating vertical integration in coastal management in a federated nation: the case of Australian coastal policy reform. Coast Manag 37(6):501–513

Wood D, Stocker L (2009) Coastal adaptation to climate change: towards reflexive governance. Int J Sci Soc 1(3):137–146

Wynne B (1991) Knowledge in context. Sci Technol Hum Values 16(1):111–121

Wynne B (2006) Public engagement as a means of restoring public trust in science—hitting the notes but missing the music? Community Genet 9(3):211–220

Chapter 7
Progression of Policies and Laws Towards Addressing Climate Change and Sustainability Issues: Recent Initiatives from Malaysia

Maizatun Mustafa, Azlinor Sufian, and Sharifah Zubaidah Syed Abdul Kader

Abstract The United Nations Framework Convention on Climate Change (UNFCCC), of which Malaysia is a party, was adopted as a basis for a global response to the effects of climate change. Malaysia also ratified the Kyoto Protocol, but being a developing country is not subjected to any commitments towards reducing greenhouse gases (GHG) emissions at present. Nevertheless, from a domestic compliance perspective, Malaysia's commitment is evident through various national responses demonstrating its seriousness in addressing climate change issues. As a developing country however, Malaysia is cautious that any response to climate change must be balanced with its continuing need to grow, to increase its per-capita income, and to raise its living standards, in accordance with the principle of sustainable development. This chapter focuses on changes that have taken place within Malaysia's policies and laws in dealing with the compelling climate change issue, while taking into consideration its various needs. To this end, new policies have been launched, and laws passed, dedicated to climate change, and in tandem with the country's pursuance of sustainable development. This work seeks to examine how these policies and laws help Malaysia reinforce the nation's resilience to climate change implications while pursuing its priority for continued development.

Keywords Climate change · Sustainable development · Environmental policy · Environmental law · Developing countries

M. Mustafa (✉) · A. Sufian
Legal Practice Department, Ahmad Ibrahim Kulliyyah of Laws, International Islamic University Malaysia, Kuala Lumpur, Malaysia
e-mail: maizatun@iium.edu.my; sazlinor@iium.edu.my

S. Z. S. A. Kader
Civil Law Department, Ahmad Ibrahim Kulliyyah of Laws, International Islamic University Malaysia, Kuala Lumpur, Malaysia
e-mail: sharifahz@iium.edu.my

© Springer International Publishing AG, part of Springer Nature 2019   133
M. Behnassi et al. (eds.), *Human and Environmental Security in the Era of Global Risks*, https://doi.org/10.1007/978-3-319-92828-9_7

1   Introduction

Climate change is the current crisis worldwide that concerns various sectors and stakeholders. This crisis is generally caused by the increased accumulation of greenhouse gases (GHGs) in the lower atmosphere due to human activities such as, inter alia, the combustion of fossil fuels, deforestation, and growing waste dumps. A major evidence of climate change today is global warming which is perceived as an unprecedented increase in the average global temperature in the future along with other effects such as sea-level rise, oceans' acidification, melting of ice caps, spreading of climate-related diseases, higher incidence of hurricane, forest fires, and destruction of crops.

The United Nations Framework Convention on Climate Change (UNFCCC) defines climate change in its Article 1(2) as "a change of climate which is attributed directly or indirectly to human activity that alters the composition of the global atmosphere and which is in addition to natural climate variability observed over comparable time periods". Historically, the UNFCCC was established on March 21, 1994 to set an overall framework that will address issues on climate change. The Convention aims to gather information on GHGs emissions and national policies. It also intends to initiate strategies that will encourage GHGs emission reduction and to contribute to the preparation for adaptation to the adverse effects of climate change. The UNFCCC classified countries into three categories: Annex I which includes industrialized countries that are members of Organization for Economic Co-operation and Development (OECD) and countries with economies in transition (EIT); Annex II which includes member countries of OECD in Annex I only, but not EIT Parties; and Non-Annex I which mostly includes developing countries.

Developed countries were considered the major contributors to the high level of GHGs in the atmosphere due to excessive industrial activities as a result of more than 150 years of industrial activity (Pattberg and Zelli 2015). Thus, a number of nations approved the addition of a legally binding measure called the Kyoto Protocol. The Protocol – which was put in force on February 16, 2005 – is an international agreement that sets mitigation targets for 37 industrialized countries and European communities starting from 2008 to 2012. Thus, in contrast to the UNFCCC, the Kyoto Protocol sets quantified and binding commitments for limiting or reducing GHGs emissions of anthropogenic origin for countries that are developed or in the transition process towards a market economy. These countries are also referred to as Annex I parties. The developing countries, including Malaysia,[1] did not have any legally binding targets under the Protocol.

In the face of new climate challenges, Malaysia, similar to other countries, has already started to take measures to protect its population and productive capacities from a range of possible climate-induced effects on ecosystems and economic

---

[1] Malaysia signed the UNFCCC on 9 June 1993 and ratified it about one year later on 17 July 1994. It also signed Kyoto Protocol on 12 March 1999 and ratified it on 4 September 2002. The country is a part of the non-Annex Parties of the UNFCCC.

development. At the same time, the country prefers to share the global responsibility to reduce GHG emissions. For Malaysia, the particular needs in adapting to climate change are of critical importance given the interdependence of climate change and development. As a developing country, economic growth is essential to improve public health, economic livelihood, and the quality of life (Jabatan, Perdana, Menteri 2016). At the same time, economic growth is also essential to increase the adaptive capacity of developing countries. But historically, increased economic development and the corresponding increase in energy use led to increased emissions of GHGs. The challenge of addressing climate change is therefore to break the link between economic development and carbon emissions.

While Malaysia has ratified the Kyoto Protocol in 2002, as a developing country, the country has different obligations from that of developed countries. For this reason, Malaysia is not subjected to any commitments towards reducing GHGs emissions at present. Nevertheless, from a domestic compliance perspective, Malaysia continues to deploy efforts in addressing climate change as evident through various national responses it has undertaken thus far in terms of adaptation and mitigation. As a developing country, however, Malaysia is cautious that these responses must be balanced with its needs to continue to grow, to increase per-capita income, and to raise living standards, in accordance with the principle of sustainable development. This chapter seeks to examine how these policies and laws help Malaysia reinforce its resilience to climate change implications. To do so, the analysis assesses Malaysia's climate initiatives and the progress achieved within the context of the overall framework of national development policies and laws.

2   Malaysia in General

Malaysia, a relatively small country, is located in Southeast Asia and consists of Peninsular Malaysia, and the Borneo states of Sabah and Sarawak. The two regions are separated by the South China Sea. The topography of Malaysia is variable, ranging from mountainous regions, flat coastal plains, to coastal areas. The country is considered a maritime nation and is bounded by over 4800 km of coastline (Ong and Gong 2001). Located near the equator, Malaysia's climate is categorized as equatorial, being hot and humid throughout the year with average rainfall of 250 centimeters a year.

In addition, Malaysia is a tropical country and its forests are very rich in species with extremely complex ecosystems. The country is considered as one of the world's mega-diverse countries according to the National Biodiversity Index. The terrestrial biodiversity is concentrated in the tropical rainforests whereas marine biodiversity is found among island, marine and coastal ecosystems such as coral reefs and sea grasses. Biodiversity resources offer economic benefits, food security, environmental stability as well as scientific, educational and recreational values. Biodiversity also enables the provision of myriad ecological services which ensure environmental stability (Ong and Gong 2001).

Malaysia is a multi-racial country with a population of about 30 million people of which 91.8 per cent are Malaysian citizens and 8.2 per cent are non-citizens. Malaysian citizens consist of the ethnic groups Bumiputera (67.4 percent), Chinese (24.6 percent), Indians (7.3 percent), and others (0.7 percent). Among the Malaysian citizens, the Malays are the predominant ethnic group in Peninsular Malaysia, which constitutes 63.1 percent. Average annual population growth rate for Malaysia was 2.0 percent for the period 2000–2010. The average life expectancy at birth shows an upward trend. In 2015, life expectancy of females increased to 77 while male life expectancy increased to 72 years (Ministry of Statistics Malaysia 2016). In Malaysia, the provision of Article 12(1) of the Federal Constitution, which guarantees the right to education for all Malaysians regardless of gender, has enabled all Malaysians to gain equal access to education and training. Overall, the trend shows an increase in the literacy levels to 96.4 percent for male, and 93.9 percent for female in 2015 (Ministry of Health 2015).

Since the 1980s, Malaysia has progressed rapidly in economic development and social transformation. The gross domestic product (GDP) of Malaysia has shown an upward trend from 2000–2007 (Ministry of Statistics Malaysia 2016). The average growth rate for GDP at constant prices from 2000–2007 is approximately 5.6 percent. In terms of per capita growth, the GDP per capita grew from RM 15,169 in year 2000 to RM 18,633 in year 2007. For 2015, the GDP was 5.0 percent as compared to a 6.0 percent growth rate for 2014. Data from the Department of Statistics (2016) show that the main contributions to GDP for the year 2015 were from the services sector for around 54 percent of GDP whereas the manufacturing sector accounted for 25 percent of GDP, followed by mining and quarries (9 percent), and agriculture (9 percent).

3   The Impacts of Climate Change on Malaysia

As one of the Southeast Asia's most vibrant developing economies, and one of the top biodiversity-rich countries in the world, Malaysia is very vulnerable to climate change impacts. The country has also seen an increase in the number of extreme weather episodes over the past few years, some on a scale not experienced before. It experienced devastating monsoon floods affecting some states of the Peninsula Malaysia in December 2005, and in Sabah and Sarawak in December 2006 and January 2007 (Meteorological Department 2009).

The Intergovernmental Panel for Climate Change (IPCC) 2007 Report on the 'Science of Climate Change' noted small increases in temperature and rainfall through the Southeast Asia Region in the last decade. Increase in average temperatures as a result of global warming is of concern for Malaysia. Studies presented in the Malaysia's Second National Communication (NC2) to the UNFCCC claim that there is a moderate change in average temperature from 1 °C to 2 °C. In addition, changing behavior patterns of ocean circulation systems due to events such as El Nino and monsoons are triggering weather extremes, flood and drought intensities in the region, and causing haze pollution, slope failures, and the emergence of certain diseases (Adaptation Knowledge Platform 2011).

One area of concern for Malaysia is climate-induced decline in food production capacities and other environmentally-driven economic systems (Ministry of Agriculture 2016); given that fact that climate change can reduce crop yield. Areas prone to drought can become marginal or unsuitable for the cultivation of some crops – such as rubber, oil palm, cocoa, and rice – thus posing a potential threat to national food security and export earnings (Siwar et al. 2009). As much as six percent of land planted with oil palm and four percent of land under rubber may be flooded and abandoned as a result of sea-level rise (Husaini 2007). Climate change can also aggravate disease infestation on forest plantation species. The impact on Malaysia's rich biodiversity is thus of great concern, where, with the intricate inter-relationships between plant and animal species, the impact on any species can have consequences for other species as well.

Climate change is also expected to cause adverse health consequences. A direct impact could be deaths due to heat stress or respiratory diseases due to air pollution, while indirect effects could include increased food and water-borne diseases, resulting from changes in rainfall pattern. There could also be an increase in mosquito-borne diseases such as malaria and dengue fever, as changes in temperature will increase the availability of suitable breeding habitats for the vectors (Ministry of Health 2014).

Other areas of concern include climate change ethical-justice issues such as: environmentally-induced displacements and migration; the deprivation and sustenance of certain livelihood activities; and the safety and well-being of the marginalized sectors of society (Adaptation Knowledge Platform 2011).

4 Climate Change Responses

It is a concern that while Malaysia has given a commitment to reduce its greenhouse gas (GHG) emissions intensity of GDP by 45% by 2030, relative to the emissions intensity of GDP in 2005 (Malaysia 2015), the figures in Malaysia are above the global average in the energy sector (Adham and Siwar 2012; Zaid et al. 2015). While Malaysia is a fast developing country with high growth rates in the last decade, this growth rate has been dictated mainly by the economic sector that is still dependent on the characteristics of its climate regime. Inevitably, climate change would have a significant impact on Malaysia's future economy if this sector does not undertake measures to increase its resilience to climate risks.

Malaysia has been able to absorb climate change impacts to date, given its strong environmental management programs, backed by stringent economic policies including effective poverty eradication and food production programs (Adaptation Knowledge Platform 2011). However, it must be understood that these efforts address only the 'environmental change threat' and not specifically the 'climate change threat'. For over a decade since the ratification of the UNFCCC and the

Kyoto Protocol, Malaysia has designated the Department of Environment, presently under the Ministry of Natural Resources and Environment, for the implementation of climate policies. As a result, at the initial stage, climate change concerns were mainly addressed as environmental issues. In 1974, the Environmental Quality Act was declared as the main national legislation on environmental protection, and the Department of Environment, which was officially formed in 1975, became the country's leading environmental agency until today (Mustafa 2009).

Malaysia's five-year Development Plan for the first time incorporated environmental protection in 1976 under the Third Malaysia Plan (1976–1980) (Jabatan Perana Menteri 1976). Subsequent, Malaysia Plans have been built-up on environmental commitments applying sustainable development as their guiding principle which requires the balance between economic development and environmental protection. For example, the Sixth Malaysia Plan (1991–1995) provides for a specific discussion concerning the issue of economic development and the environment, and applies the sustainable development principle in the context of balancing the needs of environmental protection and economic development (Jabatan Perdana Menteri 1991). In 2001, the 8th Malaysia Plan (2001–2005) (Jabatan Perdana Menteri 2001) proposed a new development concept centering on sustainable development and harmony between human beings and nature (Mustafa 2011). However, it was not until the era of the 9th Malaysia Plan (2006–2010) that Malaysia began to directly address climate change, or indirectly contributed to the management of climate change-related issues (Jabatan Perdana Menteri 2006). Generally, the 9th Malaysia Plan put an emphasis on preventive measures to mitigate and minimize pollution, and to address other adverse environmental impacts arising from development activities. In addition, the Plan suggests steps to be taken to identify and adopt action to promote sustainable natural resource management practices in relation to land, water, forest, energy, and marine resources. The strategic thrusts for the Plan in addressing environmental and natural resources issues are as follows:

- Promoting a healthy living environment.
- Utilizing resources sustainably and conserving critical habitats.
- Strengthening the institutional and regulatory framework as well as intensifying enforcement.
- Expanding the use of market-based instruments.
- Developing suitable sustainable development indicators.
- Inculcating an environment-friendly culture and practice at all levels of society.

To address climate change-related issues, the 9th Malaysia Plan initiates and promotes the various mitigation programs including: the increase of the supply and utilization of alternative fuel – such as renewable energy; encouraging energy efficiency in industrial, building and transport sectors; and protecting forest areas via sustainable forest management in order to ensure that the forest areas are maintained as carbon sinks (Jabatan Perdana Menteri 2006).

While the 9th Malaysia Plan and other previous Development Plans were focusing on mitigation, the 10th Malaysia Plan (2010–2015) was designed to specifically and directly address climate change (Jabatan Perdana Menteri 2011). For this purpose, the 10th Malaysia Plan adopted a dual strategy in addressing climate change impacts. First, adaptation strategies to protect economic growth and development factors from adverse impacts. The following measures were mentioned: developing a robust risk assessment framework to assess and quantify the various climate risks that may impact the economy and prioritize measures to address those risks; implementing policy decision frameworks to ensure that future infrastructure investments are climate resilient; and enhancing capacity in the field of climate prediction and modeling to develop stronger malaysia-specific and sector-specific knowledge. Second, mitigation strategies to reduce emission of GHGs. Efforts introduced under the 10th Malaysia Plan focused on five areas, namely: creating stronger incentives for investments in renewable energy; promoting energy efficiency to encourage productive use of energy; improving solid waste management; conserving forests; and reducing emissions to improve air quality.

Specifically, for the purpose of adaptation and mitigation, the 10th Malaysia Plan sought to intensify energy efficiency measures to harness energy savings potential and reduce Malaysia's carbon emissions and dependence on fossil fuels (Jabatan Perdana Menteri 2011). In terms of mitigation, the Plan also sought to enhance the efficiency and effectiveness of solid waste management to help recover the methane produced from the waste and use it to generate energy. Forest conservation is another part of mitigation efforts; especially that deforestation is considered as the second most important human-induced source of GHGs and is responsible for approximately 20 percent of total global emissions.

The current Plan, the 11th Malaysia Plan (2016–2020) continues to uphold Malaysia's target of becoming an advanced economy by 2020 using a greener trajectory of 'Green Growth'. The pursuit of green growth is based on mitigation and adaptation initiatives (Jabatan Perdana Menteri 2016), namely:

- strengthening the enabling environment for green growth through fundamental transformations in regulations, enhancing awareness of all stakeholders, and establishing sustainable financing mechanisms;
- adopting the sustainable consumption and production concept in prominent sectors such as industry, power generation, infrastructure and transportation based on strategies for creating green markets, increasing share of renewable energy mix, enhancing demand side management, encouraging low carbon mobility and managing waste holistically;
- conserving natural resources by laying the regulatory framework on access to biological resources and benefits sharing in order to ensure natural resources security, and to enhance alternative livelihood for indigenous and local communities through their involvement in biodiversity conservation and alternative economic opportunities; and
- strengthening the resilience of growth against climate change and natural disasters, including through the adoption of a comprehensive disaster risk management framework. These will be achieved via strategies to strengthen disaster risk management, improving flood mitigation, and enhancing climate change adaptation by developing a national adaptation plan and strengthening resilience of infrastructure.

According to the 11th Malaysia Plan, the green growth strategy is meant to ensure sustainability of natural resources, minimize pollution and strengthen energy, food and water security. By conserving biodiversity, the continuity of their function as a natural buffer against climate change and natural disaster can be strengthened. This buffer, complemented by structural approaches such as innovative flood mitigation and green infrastructure, as well as non-structural approaches like hazard risk maps and warning system, will strengthen disaster risk management and ultimately improve the well-being and quality of life.

At present, as a result of long practices of sustainable forest management, more than 55 percent of land area in Malaysia is still covered with forests. Malaysia recognizes the importance of forest ecosystem in the global carbon cycle. Malaysia's forests are very rich in species and extremely complex ecosystems. Studies in Malaysia have shown that a significant amount of carbon is sequestered by existing forested areas and managed landuse areas (Ministry of Natural Resources and Environment 2016). In order to ensure that forests continue to function as carbon sinks, an initiative to reduce the rate of forestry-related GHGs emissions has been taken through converting less forested land to other land uses by declaring forests as protected areas (Jabatan Perdana Menteri 2016). Another mitigation effort taken was the reduction of emissions to improve air quality. Among the main focus areas include: reducing emissions from motor vehicles through stricter enforcement on emission standards that should be implemented and harmonized with global standards; preventing haze pollution from land and forest fires through partnership with neighboring countries; and reducing emissions from industries through the review of emissions regulations with new emission standards for specific industries (Mustafa et al. 2012).

Overall review of the five-year Development Plans reveals that, prior to the era of the 10th and 11th Malaysia Plans, Malaysia's approach in addressing climate-related issues had been one which treats climate change as an environmental issue rather than an issue within the broader context of development. As a consequence, climate change issues had mainly been confined to an environmental policy context. This approach, in addition to the previous institutional set up, made the task for a cross-sectoral integration of climate change adaptation in the wider policy context as a daunting approach.

Fundamentally, climate change issues extend far beyond the scope that can be handled through environmental perspectives alone, and it is insufficient to approach climate change initiatives in the context of environmental protection *per se*. This is because important development issues – such as food security, human health, water supply and natural resources protection – are also impacted by climate change. Taking these issues into consideration, it is pertinent to consider climate change initiatives within a much broader context of multisectoral planning and management. Subsequently, with an increasing understanding of the importance of climate change for the overall development of Malaysia, expertise on other related aspects is needed to cope with this alarming issue. As evident, currently, Malaysia has formulated new climate change policies and laws taking into consideration the priorities of other important key sectors/actors, which are cross-sectoral in nature and targeting both climate adaptation and mitigation. These new policies and laws are elaborated below.

5 Recent Changes in Policies

5.1 The National Climate Change Policy 2009

For nearly five decades, Malaysia's overall development has been shaped by the five-year Development Plans, starting with the 1st Malaysia Plan (1966–1970) (Jabatan Perdana Menteri 1966) until the 10th Plan (2011–2015). These Development Plans have established national priorities and the policy framework for the country's development. As already mentioned, the Development Plans, starting from the 3rd one (1976–1980), are the focal point for: reflecting on the approaches Malaysia's environmental policies have taken; and explaining Malaysia's policy moves in reaction to climate change impacts as what they are today. Just as Malaysia's environmental policies have always been identified with Development Plans (Mustafa 2011), so have climate change policies in the past years. Thus, during the past decade, when the need for climate policies has been most prevalent, Malaysia has also undergone notable changes in approaches towards climate change. Malaysia has now envisaged that even with the most stringent implementation of mitigation measures, the impacts of climate change in the next few decades cannot be totally avoided. On the contrary, without mitigation, climate change impacts are likely to be more severe and adverse, thus making adaptation an uncertain challenge and would involve very high social and economic costs.

Realizing the importance of having a new policy directive that specifically focuses on both adaptation and mitigation, while at the same time taking into consideration economic, social and other needs, Malaysia finally adopted a new policy in 2009, known as the National Climate Change Policy (Ministry of Natural Resources and Environment 2010). Through this Policy, Malaysia aimed at implementing strategies to move towards a low-carbon economy and sustainable development. The Policy promotes the implementation of both adaptation and mitigation in an integrated and balanced manner, including: reviewing and harmonizing existing laws and policies; incorporating climate change in national development plans; and establishing inter-ministerial bodies to implement climate change measures. The Policy also served as the framework to mobilize and guide government agencies, industry, community as well as other stakeholders and major groups in addressing climate change challenges in a concerted and holistic manner.

It is acknowledged that climate change policy is closely related to energy policy. Malaysia's energy development strategy continues to give priority to energy conservation and energy restructuring to diversify energy supply, all of which contribute towards the mitigation of climate change. At the same time, focus is also given to the development of renewable energy. Malaysia's move to embark on this new alternative energy started way back in the era of the 8th Plan (2001–2005) (Jabatan Perdana Menteri 2001) when the Five Fuel Policy was introduced in 2001. The Policy's main objective was to diversify Malaysia's fuel sources which were dependent on conventional sources such as fossil fuels and hydro. This 2001's Policy indicates Malaysia's great urgency to seek alternative forms of energy sources because the cost of fossil fuel is on the rise and the resources are finite. Coupled with the fact that the energy sector contributes the highest amount of carbon dioxide in the atmosphere and is deemed to be the major contributor to the global warming phenomenon, the prospect of looking for other alternative sources of energy prompted Malaysia to introduce a new policy relating to energy known as the 2009 National Renewable Policy.

6   The 2009 National Renewable Energy Policy

In recent years, renewable energy has been recognized as an option to reduce dependence on fossil fuels. Malaysia introduced the National Renewable Energy Policy in 2009 as a policy for change that emphasizes on the importance and necessity of sustainable development (Ministry of Energy, Green Technology and Water 2009). Malaysia's ratification of the Kyoto Protocol in 2002 was a move to show the government's seriousness and commitment to reduce carbon footprints through escalating green technology applications in all sectors of the economy. Thus, effective mitigation measures to reduce GHGs need to be taken with special focus on main carbon-emitting sectors such as electricity generation and transport. Malaysia has made the voluntary commitment at the COP15, 2009 to reduce 40 percent of its emission intensity of GDP by the year 2020 compared to 2005 levels.

For Malaysia, the 2009 National Renewable Energy Policy is intended to qualify the nation as a leader in renewable energy and also in green technology applications. This will ensure the transformation of the nation from a fuel-importing country to a country capable to satisfy its own energy needs through domestic renewable energy sources and a leader in green technology development, while being able to conserve its natural environment so that it can also be enjoyed by future generations. According to this Policy, a sustainable local renewable energy industry could ensure the mitigation of GHG emissions due to power generation from fossil fuel while creating a new source of economic activity. At that moment, several new initiatives anchored upon the National Renewable Energy Policy 2009 have been undertaken to achieve a renewable energy target of 985 MW by 2015, contributing 5.5 percent to Malaysia's total electricity generation mix (Jabatan Perdana Menteri 2010).

6.1   The 2009 National Green Technology Policy

Considering the important role of green technology as a measure to adapt to climate change, Malaysia launched in 2009 another new policy, the National Green Technology Policy in order to spearhead green technology sector development in the country. The Policy defined 'green technology' as 'the development and application of products, equipment, and systems used to conserve the natural environment and resources, which minimizes and reduces the negative impact of human activities' (Ministry of Energy, Green Technology and Water 2009). 'Green technology', for the purpose of this Policy, also refers to products, equipment, or systems which satisfy the following criteria: minimizes environmental degradation; has zero or low GHG emissions; safe for use and promotes healthy and improved environment for all forms of life; conserves the use of energy and natural resources; and promotes the use of renewable resources.

The 2009 National Green Technology Policy has a strategic role which spans beyond achieving energy autonomy and mitigating climate change, which are among the emerging drivers of economic growth for Malaysia. The four pillars of this Policy – energy, environment, economy, and social – seek to achieve the following: achievement of energy independence and promotion of efficient utilization; conservation and enhancement of the natural environment and promotion of mitigation measures; enhancement of the national economic development through the use of technology; and improvement of the quality of life for all.

Both the 2009 National Renewable Energy Policy and the 2009 National Green Technology Policy come under the purview of a newly-established ministry, namely the Ministry of Energy, Green Technology and Water. This Ministry was set up in 2009 following the executive's reshuffling and restructuring of the ministry. Prior to that, the Ministry was known as the Ministry of Energy, Water and Communications which was established in 2004 through the restructuring of the Ministry of Energy, Communications and Multimedia. Following the 2009 reshuffle, the Ministry now holds one new function on 'green technology'. Specifically, the Ministry is responsible for the planning and formulation of relevant policies and programs in order to lead a new initiative addressing global issues such as environmental pollution, ozone depletion, and global warming (Ministry of Energy, Green Technology and Water 2014).

Arguably, Malaysia's new policies related to climate change reflect a national strategic transformation. Through these policies, Malaysia seeks to discard its protective stance to initiate more proactive strategies in an effort towards climate adaptation and gaining new market opportunities in green industries. These new policies are a reformulation of key policies that concentrate, among others, on reducing GHG emissions, lessening dependency on fossil fuels, and developing alternative energies. In order to transform these policies into action, several new laws have been passed as mechanisms to achieve policy targets. These laws are examined below.

7   Recent Changes in Law

Previously, since climate change used to be considered as an environmental issue, Malaysia's strategies targeting mitigation were incorporated mainly within environmental-related legislations. These legislations – such as the 2010 Wildlife Conservation Act, the 1960 Land Conservation Act (Revised 1989), and the 1984

National Forestry Act – are examples of laws meant to conserve natural resources and biological diversity, and are indirectly relevant to climate change. However, with the increasing concern towards climate change both at the global and domestic levels, new laws were passed to deal with climate change issues which are beyond the scope of environmental protection.

In 2011, one specific law enacted by the Malaysian Parliament for the purpose of climate change adaptation is the Renewable Energy Act. Considering the importance of energy efficiency improvement as a key element of Malaysia's energy security, this law was passed aiming at promoting renewable energy and supporting the utilization of new and renewable energy. The adoption of this law was also critical for Malaysia in providing clarity and certainty to the regulatory framework which has the effect of encouraging firms to enter the renewable energy power generation business. It is argued that the introduction of the new Act is significant considering that the lack of appropriate environmental regulations has the exact opposite effect as firms and businesses in Malaysia adopt least-cost options, and continue to pollute (Mustafa 2009). Once the 2011 Renewable Energy Act introduced, the aim was to provide a support measure to improve the environmental standards (Alias 2012).

Another significant renewable energy law newly passed by Parliament is the 2011 Sustainable Energy Development Authority Act. This Act gave birth to the new regulator of renewable energy, namely the Sustainable Energy Development Authority (SEDA). SEDA is a statutory body whose functions, as provided in section 15 of the Act, include: advising the government on all matters relating to sustainable energy, including recommendations on policies and laws to promote sustainable energy; promoting and implementing the national policy objectives for renewable energy; promoting and developing sustainable energy; implementing the feed-in tariff system; implementing sustainable energy laws; and acting as a focal point to assist the Minister on climate change and sustainable energy-related issues.

Under the 2011 Renewable Energy Act, only the following renewable resources are recognized: biogas, biomass, small hydropower, and solar photovoltaic. As such, other renewable resources – such as those from wind or wave farms – are not covered by the current feed-in tariff system. Electricity generated from these energy sources will enjoy a significantly higher tariff than conventional sources. By implementing this scheme under the new Act, the objective is to provide a financial incentive to the public to adopt renewable energy technology. Under the same Act, any person who wants to generate renewable energy must first obtain approval from SEDA. The maximum permitted generating capacity for each renewable energy installation is no more than 30 MW unless the government approves otherwise. Indeed, the adoption of these energy laws are significant for Malaysia since they pave the way for greater development in the field of renewable energy as well as in the context of climate change mitigation.

In 2014, a new law on air pollution control was enacted, known as the Environmental Quality (Clean Air) Regulations. Its approach to control air pollution departs substantially from that of the previous ones in term of scope, approaches and targets specifically through the application of 'Best Available Techniques Economically Achievable' (BAT) concept. This concept generally imposes self-regulation on the part of industries or activities that cause air pollution. The BAT's overall purpose is to reduce industrial discharges and ensure a high level of protection for the environment as a whole, taking into account different types of environmental impacts. The BAT also encourages the premises to develop and introduce new and innovative technologies and techniques which meet air quality criteria, and look for continuous improvement in their overall environmental performance as part of the sustainable development process and climate change mitigation (Mustafa 2016). It might be too premature to gauge the success of these laws. However, having new energy and clean air laws, and a new statutory body enforcing some of the new laws may transform Malaysia's legal framework, especially that this kind of laws is considered as a catalyst for renewable energy generation and climate change mitigation. These laws are also aligned with Malaysia's aims to reduce carbon emissions and achieve 5.5 percent renewable energy in total energy mix by 2015.

8 Conclusion

To this date, climate change impacts on Malaysia can still be tackled by the strong foundations of the country's environmental management programs (as charted in the five-year Development Plans and other policy documents), which are also backed by Malaysia's stringent economic policies. However, this scenario can change if the gradual increase in global warming is left unchecked and unabated. Malaysia realizes that it is counterproductive to create stand-alone institutions charged with responsibility for climate risk management. Climate change cannot be the sole responsibility of any single institution or professional practice. Additionally, having isolated adaptation goals, without considering cross sector effects and linkages with other goals, could also lead to missed opportunities for Malaysia. Government agencies responsible for the provision and management of public goods, food production, and water management need to be fully accountable for maximizing the efficiency of public goods and services, while minimizing the fiscal burden from climatic losses. Promoting increased resilience to climate change impacts is closely intertwined with development choices and actions that cover a variety of sectors such as energy, agriculture, health, water, and infrastructure.

Malaysia realizes that climate change must be addressed through two approaches: adaptation which refers to actions taken to help communities and ecosystems cope with actual or expected impacts of climate change; and mitigation which refers to actions taken to reduce carbon emissions and enhance carbon sinks to lessen the impacts of climate change. From these perspectives, Malaysia begins to reconsider its climate change initiatives through the adoption and implementation of several policies and laws, dedicated to climate change, and in tandem with the country's pursuance of sustainable development. For Malaysia, strategies within these new policies and laws show the shifting stance from mitigation to proactive strategies that take advantage of new market opportunities. They also reflect the industrial competitiveness to survive the era of GHG reduction. The climate challenge is immense for Malaysia. While recent adaptation efforts are more notable, the adaptation approach is still very much a new landscape with various challenges for the country. Given the long-term, uncertain, and cross-sectoral nature of climate risks, Malaysia needs to continue adjusting with flexibility and foresight and across traditional boundaries.

References

Adaptation Knowledge Platform (2011) Scoping assessment on climate change adaptation in malaysia. AIT-UNEP RRC.AP, Bangkok

Adham KN, Siwar C (2012) Empirical investigation of Government Green Procurement (GGP) practices in Malaysia. OIDA Int J Sustain Dev 4(4):77–88

Alias LS (2012) Malaysia. In: Schwartz DL (ed) The energy regulation and markets review. Law Business Research Ltd, London

Husaini A (2007) Flood and drought management in Malaysia. In: National seminar on socio-economic impact of extreme weather and climate change. Ministry of Science Technology and Innovation, Putrajaya 21–22 June 2007

Jabatan Perdana Menteri (1966) First Malaysia plan 1966–1970. Jabatan Percetakan Negara, Kuala Lumpur

Jabatan Perdana Menteri (1976) Third Malaysia plan 1976–1980. Jabatan Percetakan Negara, Kuala, Lumpur

Jabatan Perdana Menteri (1991) Sixth Malaysia plan 1991–1995. Jabatan Percetakan Negara, Kuala, Lumpur

Jabatan Perdana Menteri (2001) Eighth Malaysia plan 2001–2005. Jabatan Percetakan Negara, Kuala, Lumpur

Jabatan Perdana Menteri (2006) Ninth Malaysia plan 2006–2010. Jabatan Percetakan Negara, Kuala Lumpur

Jabatan Perdana Menteri (2011) Tenth Malaysia plan 2011–2015. Jabatan Percetakan Negara, Kuala Lumpur

Jabatan Perdana Menteri (2016) Eleventh Malaysia plan 2016–2020. Jabatan Percetakan Negara, Kuala Lumpur

Malaysia (2015) Intended Nationally Determined Contribution of the Government of Malaysia

Meteorological Department Malaysia (2009) Climate change scenarios for Malaysia 2001–2009. Ministry of Science Technology and Innovation, Putrajaya

Ministry of Agriculture Malaysia (2016). http://www.moa.gov.my. Accessed on 20 Dec 2016

Ministry of Energy, Green Technology and Water Malaysia (2009) National green technology policy. Ministry of Energy, Green Technology and Water Malaysia, Putrajaya

Ministry of Energy, Green Technology and Water Malaysia (2014). http://www.kettha.gov.my/. Accessed on 20 Dec 2016

Ministry of Health Malaysia (2014). http://www.moh.gov.my. Accessed on 20 Dec 2016

Ministry of Health Malaysia (2015) Health indicators 2015. Ministry of Health, Putrajaya

Ministry of Natural Resources and Environment Malaysia (2010) National Policy on Climate Change. Ministry of Natural Resources and Environment Malaysia, Putrajaya

Ministry of Natural Resources and Environment Malaysia (2016.) http://www.nre.gov.my/. Accessed on 20 Dec 2016

Mustafa M (2009) Environmental quality Act 1974: development and reform. MLJ Article Supplement, Vol 2 March–April 2009

Mustafa M (2011) Environmental quality Act 1974: a tool towards the implementation and achievement of Malaysia's environmental policy. IIUM Law J 19(1):1–34

Mustafa M (2016) Environmental law in Malaysia. Wolters Kluwer, The Netherlands

Mustafa M, Syed A, Kader SZ, Sufian A (2012) Coping with climate change through air pollution control: Some legal initiatives from Malaysia. Int Conf Environ Energy Biotechnol IPCBEE 33(2012):101–105

Ong JE, Gong WK (2001) The encyclopedia of malaysia, vol 6. The Seas. Archipelago Press, Kuala Lumpur

Pattberg PH, Zelli H (2015) Encyclopedia of global environmental governance and politics. Edward Elgar, Glos

Siwar C, Alam MM, Murad MW, Al Amin AQ (2009) A review of the linkages between climate change, agricultural sustainability and poverty in Malaysia. Interv Rev Bus Res Pap 5(6):309–321

Zaid SM, Myeda NE, Mahyuddin N, Sulaiman R (2015) Malaysia's rising GHG emissions and carbon 'lock-in' risk: A review of Malaysian building sector legislation and policy. JSCP 6(1):1–13

Part II
Environmental and Climate Change: Case Studies on Conflicts and Human Security

Chapter 8
Corporate Interests vs Grassroots Environment Movements in India: A Losing Battle for the People

Himangana Gupta and Raj Kumar Gupta

Abstract The problem of environmental security in the Indian context can be linked to the deficit of democracy in recent times. On the face of it, India is the largest democracy in the world, but the issues directly affecting the lives and livelihoods of poor people in the resource-rich regions are often pushed to the background. With the climate change agenda moving into the hands of corporate interests, what was earlier done in the name of development is now done in the name of environment. Although India is among the countries most vulnerable to climate change and loss of biodiversity, citizens have little say in environmental policy-making. India has a strong tradition of living in harmony with nature since the Vedic times beginning 1700 BC. The 1970's and 80's saw the emergence of major successful grassroots resistance movements such as Chipko, Silent Valley, and Save Narmada movements. More such spontaneous movements are failing now in the absence of political support. There is no 'green' party in India which can highlight the issues facing the poor and vulnerable people. Even the mainstream media controlled by the corporate sector does not reach the darkness in which the poor people live. Only once did the Indian Parliament take notice of climate change negotiations when the country decided to take voluntary emission reduction commitments at the Copenhagen Summit in 2009. Biodiversity has never been discussed in Parliament though India stands to lose its natural treasure due to the policies pursued by the Government. This chapter traces the history of popular and successful environment movements and discusses the reasons behind the failure of such movements in present times.

Keywords Environmental security · Equity · Climate change · India · Democracy

H. Gupta (✉)
National Communication Cell, Ministry of Environment, Forest and Climate Change, New Delhi, India

R. K. Gupta
Journalist and Independent Analyst on Environmental and Social Policy, New Delhi, India

© Springer International Publishing AG, part of Springer Nature 2019
M. Behnassi et al. (eds.), *Human and Environmental Security in the Era of Global Risks*, https://doi.org/10.1007/978-3-319-92828-9_8

1 Introduction

Ancient Indians in the Vedic period, beginning of 1700 BC, perceived God's presence through Nature (Desai 2009) and interweaved environmental protection into their cultural ethos, which is preserved to this day (Singh 2011). Every Hindu religious ritual begins with seeking forgiveness of Mother Earth for the insolence of placing foot on Her and of the plants for plucking flowers to offer prayers to God. In western Rajasthan, there is a unique community of nearly one million people called Bishnois (twenty + nine) who live by 29 commandments or rules and protect plants and animals with their lives (Chandla 1998). There have been strong messages from Indian religious and other leaders in favor of conserving resources and their sustainable use. Jenkins and Chapple (2011) and Nugteren (2005) catalogue Indian religious traditions and rituals that show respect for nature and vegetation.

In contrast, the modern society, entrenched in consumer culture, perceives Nature as a resource to be exploited for profit. When this exploitation resulted in imbalances from unsustainable emissions caused by the burning of fossil fuels, a global crisis emerged, threatening the existence of life itself on the planet. This unsustainable exploitation of resources was acknowledged by the two major legally binding conventions negotiated at the Rio Earth Summit in 1992 – the United Nations Framework Convention on Climate Change (UNFCCC) and the Convention on Biological Diversity (CBD) – to prevent climate change and loss of diversity of biological species (Drexhage and Murphy 2010; UNCED 1997).

The Kyoto Protocol to the UNFCCC was negotiated in 1997, under which 37 developed countries took legally-binding commitments to reduce six greenhouse gases (GHGs) by 5.2 per cent from their 1990 levels (Kyoto Protocol 1997). However, the Protocol offered these countries additional means of meeting their targets by way of three market mechanisms – Emissions trading, Clean development mechanism (CDM), and Joint Implementation (JI).

The foremost critic of the market mechanisms is no other than British economist Nicholas Stern who characterized climate change as "the greatest example of market failure" because of the negative externality when the actions of firms or individuals cause potential harms to others (Stern 2007). Ironically, the Kyoto Protocol's solution to the consequence of market failure was more market instruments. One instrument, more relevant to India and other developing countries, is the CDM, which allows a country with a binding commitment to implement an emission-reduction project in developing countries and count the saleable carbon credits earned towards meeting their Kyoto obligations. Complex rules have turned the governing structures of these mechanisms into environmental imperialism.

Of course, the blame does not rest entirely on the international agreements. They only provide a regulatory framework and financing options. The socio-environmental and ecological aspects are left entirely to the host governments. Indian businesses, like businesses around the world, are primarily interested in finance part of the projects but not in sharing the burden of cleaning part of the mess. For instance, Indian business sector wants money-earning CDM projects, but do not want the Government to take emission commitments (FICCI 2007; Gupta et al. 2015).

2 Forestry in Clean Development Mechanism

CDM projects have raised concerns for their low contribution to emission reductions as well as sustainable development and uneven distribution of projects (Bakker et al. 2011). Afforestation/reforestation projects designed to sequester carbon dioxide from the atmosphere by planting fast-growing monocultures are causing conflicts between local populations and companies promoting industrial plantations. It is more for money than for environmental commitment because the companies get a double benefit: they can claim that they are green; and can earn money through carbon credits. They also ensure their raw material supply through these plantations. Such projects cause irreparable ecological damage and negatively impact the livelihoods of local people (Gupta and Kohli 2014).

The main cause of resistance to such plantations is the corporate control over land and resulting displacements of local people who depend on natural ecosystems and agriculture. Gerber (2011) analyzed 58 conflict cases round the globe and found that authorities often responded by repression. While these movements can be regarded as classical land conflicts, they usually also have an ecological content. The interest of local communities living in and around forests may be compromised by treating the forest as carbon sinks when these areas play a complex role in their lives (Parikh et al. 1997).

At least 50 million people have been displaced by big dams, mines, factories, and firing ranges since India's Independence in 1947. While the majority of displacement in the name of 'development' has been in rural areas, the number of poor displaced in cities in the name of 'property development' is also huge (Padel 2015). The question is, how is it that in the land of Gandhi, where farmers in Champaran stood up against the British for forced indigo plantations (Brown 2007), farmers in the villages of Andhra Pradesh, Odisha, and Chhattisgarh cannot stand up against forced plantation of eucalyptus? In a pure technical sense, it is not forced plantation. The farmers were lured into growing eucalyptus by promising higher profits without farm labor. But now they feel trapped as the money will flow only once in five to six years after the first harvest. Many of them want to opt out, but they cannot as they are bound by agreements (Gupta 2015; Mate and Sahu 2011; RCDC 2012). This state of affairs makes more sense when looked at from a wider perspective of successful green movements in India and the tools the Government has acquired to suppress the emergence of more such uprisings.

3 Grassroots Green Movements

In the 1970s and 80s, India witnessed three major grassroots environmental movements in the Gandhian non-violent tradition (Karan 1994). These movements differed from those in the West in as much as they are concerned with economic equity and social justice apart from environmental preservation. The first one, the *Chipko* Movement led by Chandi Prasad Bhatt and Sunderlal Bahuguna, started spontaneously in 1973. The villagers of the Garhwal Himalayas, now part of a separate Uttarakhand state, did it by simply hugging the trees when the workers of timber contractors came to axe them. Forests are the main source of livelihoods in the Garhwal Himalayas and over a dozen conflicts were reported between the villagers and contractors in the 70s, but villagers prevailed in every case.

The movement was sparked by the decision of the state government to allot a plot of forest area to a sport goods company. This angered the villagers as their earlier demand to use the wood for making agricultural tools had been refused. The success of the movement led to similar protests in other parts of the country. The involvement of Sunderlal Bahuguna, a Gandhian activist and philosopher, gave a direction to the wider movement, and his appeal to the then Prime Minister of India, Indira Gandhi, resulted in a 15-year ban on tree felling in the Himalayan forests. The government finally cancelled the timber contracts.

Similar methods were adopted in the 1980s by *Appiko* (translation of *chipko* in Kannada language) Movement in the Western Ghats of the southern Indian state of Karnataka to stop the commercial exploitation of forests. In the 1950s, forest covered over 81 percent of Uttara Kannada district in northern Karnataka. The government declared the district as 'backward' and initiated the process of 'development'. Industries such as pulp and paper and plywood factories came up in the region. Hydroelectric dams were constructed to harness the rivers, submerging forest and agricultural areas and displacing local people. The forest shrunk to nearly 25 percent of the district's area by 1980. The local people were displaced by the dams and the conversion of the natural forests into teak and eucalyptus plantations dried up the water sources. The movement, led by Panduranga Hegde, was focused on saving the remaining tropical forests of the region.

The original inventors of the *chipko*-like movement were the Bishnois, when 363 people in Khejadli village of Jodhpur district were massacred by the king's men who came to clear the forest for building a palace while men, women and children hugged the trees to protect them in 1730 A.D. When the king heard of this, he called off the operation and took the village in his own protection (Bishnoi 1992; Bishnoi and Bishnoi 2000). The Bishnoi movement was launched in 1485 A.D. when a visionary leader, Jambho Ji, used a powerful medium of religion to purvey his message of complex bond between man and nature. Even today, the 29 commandments, which formed the basis and nomenclature of the newly-founded religion in 1485, act as a reference guide for the adherents to live in harmony with their environment and establish a symbiotic relationship with nature.

The second major green movement in modern India started in 1978 against the Silent Valley hydroelectric dam project over the Kunthipuzha River in the southern part of the Western Ghats. The project would have submerged the unique and the only remaining undisturbed rainforests in peninsular India holding 50-million-year-old genetic heritage (Swaminathan 1979). The movement was led by Kerala People's Science Movement, a network of rural school teachers and citizens promoting environmental scientific projects in the villages. The father of the Indian Green Revolution, M.S. Swaminathan, who was then Secretary to the Ministry of Agriculture and Irrigation, recommended the scrapping of the project, arguing that electricity can be generated in other ways but the genetic heritage will be lost for ever (Swaminathan 1979). The project was scrapped in 1983 by the Government, again headed by Indira Gandhi, after another review committee headed by M. G. K. Menon recommended that the project be scrapped.

The third major grassroots environmental movement was the *Narmada Bachao Andolan* (Save Narmada Movement) against one of the world's largest multipurpose water projects, the Narmada River Development Project, funded by the World Bank (Karan 1994). This turned out to be the world's largest and most successful environment campaigns against 'destructive development'. The movement, led by Medha Patkar, Anna Hazare, and Arundhati Roy, brought into focus major environmental issues, including the very rationale of big dams. The campaign could not stop the project from going ahead but it helped in extracting better resettlement compensation for the displaced people.

The common thread that runs through all the successful grassroots movements is that the governments were responsive to the prevailing mood of the people. Instead of remaining peaceful, the movements are now showing a trend towards turning violent. In April 2010 in Chhattisgarh, Maoist rebels killed 77 men of the Central Reserve Police Force (CRPF) to protest against the award of mines to companies which are destroying local water resources (Majumder 2011).

4 Judiciary to the Rescue

When the government began to fail in its duty, the Supreme Court of India came to the rescue of environment and people affected by political corruption and unaccountability. The last decade of the twentieth century saw several high-profile environmental cases being brought before the Supreme Court when judicial activism emerged as a powerful tool of social change. Several judges re-interpreted court procedures and evolved Public Interest Litigation (PIL) to empower ordinary citizens to draw the attention of India's apex court through a simple letter (Verma 2001). The Supreme Court has evoked a great deal of interest for its activism and the role which it has begun to play in Indian governance (Khosla 2009).

The Indian Constitution is one of the first constitutions of the world containing specific provisions for environmental protection and applying the human rights approach through various constitutional mandates (Jaswal 2003). The 42nd

Amendment to the Constitution of India in 1976 added Article 48-A under the Directive Principles of State Policy to protect and improve the environment and to safeguard the forests and wildlife of the country.

The first recognition of the right to live in a healthy environment as a part of Article 21 of the Constitution came in *R.L. and E. Kendra, Dehradun v. State of U.P*[1] case. The PIL started with a letter written by the Dehradun-based Rural Litigation and Entitlement Kendra which complained that mining has denuded the Mussourie hills of forest cover and accelerated soil erosion resulting in landslides and blockage of underground water channels which fed many rivers and springs in the valley. The letter was treated as a writ petition under Article 32 (right to constitutional remedies), and on the advice of the expert committee appointed by the Court, a number of limestone quarries were closed. The Court also took care of lessees and those rendered unemployed by the closure of quarries by directing the State of U.P. to give priority to the claims of displaced lessees in other parts of the state and rehabilitate workmen in afforestation and soil conservation programmes to be undertaken in the reclamation of the area by the Eco-Task Force of the Department of Environment.

In the *Banwasi Seva Ashram v. State of U.P.*[2] case, the Supreme Court protected the interests of forest dwellers who were being evicted from their homes to acquire the land and make way for a super thermal plant of the National Thermal Power Corporation (NTPC). Without making any reference to Article 21, the Supreme Court implicitly treated the rights of Adivasis (tribal people) under the Article by observing that they used the forest areas as habitat for generations and used forest produce for their livelihoods (Jaswal 2003). The Court gave detailed directions for safeguarding the interests of the *Adivasis* and permitted the acquisition of land only after NTPC agreed to provide certain facilities to the ousted forest dwellers.

In the case of *Tehri Bandh Virodhi Sangarsh Samiti and Ors. Vs. State of U.P. and Ors,*[3] the Supreme Court considered the safety aspects of the Tehri Dam and dismissed the case after the opinion of the a high-level committee that the design of the dam had incorporated the adequate defensive measures on the recommendations of the International Congress of Large Dams. The judgment appreciated the petitioners' concern for the safety of the project but did not find any good reason to restrain the respondents from proceeding with the project.

In *M.C. Mehta v. Union of India*[4] case, the Supreme Court ordered the relocation of 168 industries to other towns of the National Capital Region to protect Delhi from the environmental pollution caused by hazardous/noxious/heavy/large industries. To mitigate the hardship of the employees of such industries, the Court specified the rights and benefits to which these employees were entitled on relocation/

---

[1] The Supreme of Court of India: A.I.R. 1985 S.C. 652, Doon Valley Case.

[2] The Supreme Court of India: A.I.R. 1987 S.C. 374.

[3] The Supreme Court of India: 1992 Supp (1) SCC44, [1990] Supp2 SCR 606, 1991(1) UJ121 (SC).

[4] The Supreme Court of India: (1996) 4 SCC 750.

shifting of these industries. The Court also issued the package of compensation for employees of industries which decided to close down instead of relocating.

5   Dilution of Laws

When the highest court of the land begins to stand for the oppressed people, the only way left for the governments is to dilute the laws to bring about faster environment clearance of projects and further reduce the power of other stakeholders apart from the corporate sector. In order to address the shortcomings in Indian laws, the outgoing Congress-led United Progressive Alliance (UPA) government replaced the antiquated 1894 Land Acquisition Act of the colonial era with the 2013 Right to Fair Compensation and Transparency in Land Acquisition, Rehabilitation and Resettlement Act, which required social impact assessment and consent of 80 percent of land owners. But the new government of the National Democratic Alliance (NDA), led by the Bharatiya Janata Party (BJP), brought a Bill to amend the law for exempting certain projects from its provisions of social impact assessment and 80 percent consent. Under corporate pressure, the government was in such a hurry that it promulgated an ordinance during Parliament recess pending the passage of the Bill. During Parliament session, the Bill was passed by the Lok Sabha (Lower House), but the Government could not get it passed in the Rajya Sabha (Upper House of Parliament) due to lack of majority. Despite the widespread criticism of the move as undemocratic, the ordinance was re-promulgated.

This is just one example of how the first government to win absolute majority in the Lok Sabha in 30 years undermines democracy. Within three months of coming to power, the NDA government, through ministerial notifications, diluted several environmental laws. Through a series of notifications, the Ministry of Environment, Forest and Climate Change eased rules for mining, roads, power, industrial, and irrigation projects (Sethi and Jha 2014). The first casualty was the 2006 Forest Rights Act, which requires 'prior informed consent' of *gram sabhas* (village councils) before forests are used for prospecting or industrial activity. Without amending the law, the powers of local communities have been transferred to the district administration.

The Government has reconstituted the standing committee of the National Board for Wildlife under the 1972 Wildlife Protection Act to reduce the authority of independent experts, exempted small coal mining projects from public hearings, and allowed irrigation projects without clearance. In doing so, it has flouted the Wildlife Protection Act's requirement for independent experts. It allowed industrial projects to come within 5 km of a national park without the approval of the National Board for Wildlife and without Environmental Impact Assessment. It also curtailed the powers of the National Green Tribunal which first hears all challenges to environmental clearances before they can be passed on to the Supreme Court.

6 No Green Politics

In principle, governments have to protect the interests of their subjects in perpetuity. However, governments are not elected forever. They only have a limited time period in which to balance the call on resources before the next election. Business lobbies have more money and power to influence the decisions than the masses who are kept poor in perpetuity due to the policies that are framed as a result of diverse, and often conflicting, pressures. In the case of environment, it is more complex. When intragenerational equity is missing from considerations, it is impractical to even expect the politicians to give a thought to inter-generational equity.

One of the reasons for the government having a free run on the national laws is that the green parties have not taken root in India. Medha Patkar blames 'the reality of neo-liberalism' for the weak political will on the environment, but warns that this will not pass muster with India's increasing, and increasingly vocal, green voters (Majumder 2011).

For millions of poor tribal people living off forests, environmentalism is not a lifestyle, it is livelihood. On the other hand, for politicians seeking young urban votes, green politics means pandering by saying things like: "I use CFL bulbs in my house. I shut down my SUV engine at the red light" (Majumder 2011).

Grassroots democracy has been the hallmark of Green parties ever since they grew out of the new social movements and established themselves as political parties (Poguntke 2002). Green Politics came into existence for the first time in early 70s when campaigns around the world started on a predominantly environment platform. One of the first political Green parties began campaigns in Australia, New Zealand, and Britain. The first Green party to achieve national prominence was German Green party, which gained popularity opposing the nuclear power.

Earlier, green parties emphasized only on environmental causes. With time, the Green parties usually follow a coherent ideology, which includes environment as well as social justice, consensus decision making, grassroots democracy, non-violence, and in general those issues which are inherently related to one another as a foundation for world peace. Today Green parties exist in nearly 90 countries (Shrivastava 2014).

The mainline political parties in India do not consider environment as an important issue. A comparison of the election manifestos of the two main political parties during the last election in May 2014 reveals that both the BJP and the Congress treat environment with industry focus. The BJP's guide for environmental management finds mention under the subject of 'industry' rather than 'flora, fauna and environment'. The emphasis was on framing environmental laws in a manner that encourages speedy clearances, removal of red tape and bottlenecks, which was borne out by the dilution of the laws after the party came to power. Like the BJP, the Congress manifesto also focuses on a business-friendly governance structure, and clearly mentions that the party intends to streamline the regulatory structures and create a business-friendly environment.

7   Conclusion

International environment agreements have given the Indian national government tools to propagate neo-liberal policies and still claim that its policies are environment-friendly. Through the process, the Government has acquired tools, and justification, to suppress the common people, and their movements, fighting for their rights and livelihoods.

India has seen some of the world's best-known grassroots non-violent green movements but their leaders have failed to capitalize on their popularity to turn these movements into a political force. The mainstream political parties do not realize that for millions of poor people, green politics is not about lifestyle but about livelihoods. Where the politicians failed, the judiciary filled the space with its activism. But there is a limit to judicial activism when the Government is bent upon diluting the laws that favored the common people. With such policies, the Government also limits democracy.

Disclaimer The views expressed in this chapter are those of the authors and do not reflect the official policy and position of any ministry or department of the Government of India.

References

Bakker S, Haug C, Van Asselt H et al (2011) The future of the CDM: same same, but differentiated? Clim Pol 11:752–767. https://doi.org/10.3763/cpol.2009.0035

Bishnoi RS (1992) A blueprint for environment: conservation as creed. Surya, Dehra Dun. ISBN: 978-81-85276-27-4

Bishnoi V, Bishnoi NR (2000) Bishnoism: An eco-friendly religion. Religion and Environment 1

Brown JM (2007) Gandhi's rise to power: Indian politics 1915–1922. Cambridge University Press, Cambridge. ISBN: 978-0-521-09873-1

Chandla MS (1998) Jambhoji: Messiah of the Thar Desert. Aurva Publications, New Delhi

Desai FP (2009) Ecological ethics in Vedic metaphysics an effectual method to indoctrinate environmental awareness. J Environ Res Dev 4:2

Drexhage J, Murphy D (2010) Sustainable development: From Brundtland to Rio 2012. United Nations, New York

FICCI (2007) FICCI climate change task force report. Federation of Indian Chambers of Commerce and Industry

Gerber J-F (2011) Conflicts over industrial tree plantations in the south: who, how and why? Glob Environ Chang 21(1):165–176. https://doi.org/10.1016/j.gloenvcha.2010.09.005

Gupta H (2015) Gaps and linkages between climate change and biodiversity conventions: Science, politics and policy. Doctoral Thesis, Panjab University, Chandigarh, India

Gupta H, Kohli RK (2014) Policy conflict between forest land use and climate change. Presented in The Third Xishuangbanna International Symposium: Botanical Gardens and Climate Change, XTBG, Chinese Academy of Sciences. http://xtbgsymposium2014.csp.escience.cn/dct/attach/Y2xiOmNsYjpwZGY6NzQ0Ng==

Gupta H, Kohli RK, Ahluwalia AS (2015) Mapping "consistency" in India's climate change position: dynamics and dilemmas of science diplomacy. Ambio 44:592–599. https://doi.org/10.1007/s13280-014-0609-5

Jaswal PS (2003) Environmental law, Second Edition. Pioneer Publications, Allahabad

Jenkins W, Chapple CK (2011) Religion and environment. Annu Rev Environ Resour 36:441–463. https://doi.org/10.1146/annurev-environ-042610-103728

Karan PP (1994) Environmental movements in India. Geogr Rev 84:32–41. https://doi.org/10.2307/215779

Khosla M (2009) Addressing judicial activism in the Indian supreme court: towards an evolved debate. Hast Int Comp Law Rev 32:55

Kyoto Protocol (1997) Kyoto protocol to the United Nations framework convention on climate change. United Nations

Majumder K (2011) Stop! Or I'll vote green. Tehelka 8:23

Mate N, Sahu SK, Yasmin H (2011) Forestry CDM case study: ITC social forestry project in Khammam. Andhra Pradesh Mausam 3(1):25–31

Nugteren A (2005) Belief, bounty, and beauty: rituals around sacred trees in India. Brill

Padel F (2015) What is the main strength of the 2015 Land Bill? Implicit rejection of 1894 Land Acquisition Act. The Economic Times, New Delhi. http://articles.economictimes.indiatimes.com/2015-04-26/news/61542745_1_land-bill-infrastructure-projects-land-acquisition-act

Parikh JK, Culpeper R, Runnalls D, Painuly JP (1997) Climate change and north-south cooperation: indo-Canadian cooperation in joint implementation. North South Inst, New Delhi

Poguntke T (2002) Green parties in national governments: from protest to acquiescence? Environmental Politics 11(1):133–145

RCDC (2012) From cash crops to carbon sinks: a new identity to virtual land alienation in Odisha. Regional Centre for Development Cooperation. Bhubaneswar

Sethi N, Jha S (2014) Govt eases environment rules to attract investments. Business Standard, New Delhi

Shrivastava S (2014) AAP – India's own green party. The World Reporter. http://www.theworldreporter.com/2014/01/aap-indias-green-party.html

Singh RP (2011) Environmental concerns (the Vedas) – a lesson in ancient Indian history. Shaikshik Parisamvad Int J Educ 1(1)

Stern N (2007) The economics of climate change: the stern review. Cambridge University Press. ISBN: 978-0-521-70080-1

Swaminathan MS (1979) Report on the visit to the Silent Valley area of Kerala. Government of India. http://krishikosh.egranth.ac.in/handle/1/2049186

UNCED (1997) UN conference on environment and development (1992). http://www.un.org/geninfo/bp/envirp2.html. Accessed 11 Jan 2015

Verma A (2001) Taking justice outside the courts: judicial activism in India. The Howard J Crim Just 40:148–165. https://doi.org/10.1111/1468-2311.00198

Chapter 9
Real or Hyped? Linkages Between Environmental / Climate Change and Conflicts – The Case of Farmers and Fulani Pastoralists in Ghana

Kaderi Noagah Bukari, Papa Sow, and Jürgen Scheffran

Abstract Violent conflicts between farmers and pastoralists are reported in national media and public discourse to have increased in Ghana and are intricately linked to human security related issues. Some media and research reports attribute farmer-pastoralist conflicts to resource scarcity, environmental and climate change, thus reiterating the environmental scarcity/security and neo-Malthusian postulations. These conflicts accordingly arc induced by pastoralists' migrations southwards mainly due to poor climatic conditions from Northern Ghana and the Sahel. This work examines the perceptions of farmers and Fulani pastoralists in Ghana on the role of environmental/climate change in triggering conflicts between them. To do this, the analysis compares data sets (of climate change indicators, especially rainfall data and temperature) and primary data taken from field studies in the Agogo (southern Ghana) and Gushiegu (northern Ghana) districts of Ghana to assess if environmental/climate change has contributed to conflicts between farmers and Fulani pastoralists. Based on the analysis of interview outcomes, farmer communities and pastoralists perceive environmental and climate factors as indirectly influencing conflicts between them, especially through increased pastoralists' migrations and competition for pasture lands. The data sets (of rainfall), however, reveal that despite climate variability, there were basically no major changes in rainfall figures. We also found that the abundance of resources and increases in the value of land in Agogo were major drivers of conflicts between farmers and pastoralists.

Keywords Climate and environmental change · Conflicts · Fulani pastoralists · Farmers · Ghana

---

K. N. Bukari (✉) · P. Sow
Center for Development Research (ZEF), University of Bonn, Bonn, Germany
e-mail: papasow@uni-bonn.de

J. Scheffran
Researcher Professor, Head of the Research Group Climate Change and Security (CLISEC), Institute of Geography, University of Hamburg, Hamburg, Germany
e-mail: juergen.scheffran@uni-hamburg.de

© Springer International Publishing AG, part of Springer Nature 2019
M. Behnassi et al. (eds.), *Human and Environmental Security in the Era of Global Risks*, https://doi.org/10.1007/978-3-319-92828-9_9

1 Introduction

West Africa has experienced different types of conflicts that range from ethnic, religious, resource, communal to political and electoral conflicts (Adetula 2006). One of such conflicts is pastoralist-farmer conflicts related to natural resources. These types of conflicts, though very subtle, have always been a prominent feature of the economic livelihood and agriculture productivity in West Africa; thus it is not a recent phenomenon (Hussein et al. 1999; Hagberg 1998). Local farmers have often competed and clashed with pastoralists over varied issues such as crop destruction, natural resource uses (i.e. water, pasture, farming and grazing land), castle rustling, problems of land tenure and ownership, marginalization of pastoralists and ethnic differences (Diallo 2001; Tonah 2006; Shettima and Tar 2008).

At the same time, there has been cooperation between farmers and pastoralists over resource uses in a bid to increase their economic productive activities – crop production and livestock rearing. This cooperation has often been mutual, particularly in the exchange of crop production and livestock rearing. The Fulani (also called *Fulbe*[1]), who are the focus of this research, are nomadic and semi-sedentary pastoralists. They have taken care of farmers' livestock, settled side by side with farmers and traded with farmer communities. Widespread regional distribution of Fulani has added to their interaction with a large array of groups, including many subsistence farmers, community leaders, chiefs and landowners (Davidheiser and Luna 2008).

Although Fulani herders have had conflicts with the local farmers for a long time (Harshbarger 1995), their recent conflicts with local farmers have been a result of depleting natural resources – the search for water, forage and grazing pastures of land for their livestock. A violent trend in conflicts has been observed, with increased deaths, destructions, security threats to communities and less co-existence between them and sedentary farmers. There have been numerous reports of violent clashes between Fulani pastoralists and farmers in Ghana (Dosu 2011), southwest and southeast Burkina Faso (Brockhaus 2005; Brockhaus et al. 2003), Nigeria (Blench and Dendo 2003), and Cote d'Ivoire (Diallo 2001). Several studies from other West African countries have already suggested that environmental and climate changes are indirectly leading to competitions for scarce resources, thereby resulting in conflicts between farmers and Fulani pastoralists (see Brown and Crawford 2008; Turner et al. 2011; Benjaminsen et al. 2012; Obioha 2008). However, in the case of Ghana, except few policy and media reports on the subject, no scientific study has been carried out on the environmental/climate-conflict nexus. This work contributes empirically to the literature on the environment-conflict debate on pastoralism in Ghana.

---

[1] Fulani, which is derived from Hausa, is the English term for the Fulani. Fula is derived from the Manding languages and is also used in English. Fulβe is the original term for the people. This has been adopted into English, often spelt Fulbe. In French, Fulani are called Peuls. See Davidheiser and Luna (2008).

2   Environmental Change and Farmer-Pastoralists Conflicts

Theoretically, environmental security/scarcity links the environment to security. Environmental security is opposed to environmental insecurity. The latter implies: "the vulnerability of individuals and groups to critical adverse effects caused directly or indirectly by environmental change" (Barnett 2001:17). Environmental change is either the result of human anthropogenic actions or natural factors which reduce the capacity to mitigate the change, thereby making the world environmentally insecure (Barnett 2007a, b). These changes are seen through the effects of climate change, depletion of natural resources (water, pasture, forest, etc.), and environmental degradation. Environmental change/resource scarcity is said to have the potential of causing violent conflicts and insecurity. This argument has had a long history spanning from the time of Thomas Malthus through to Robert Kaplan (*The Coming Anarchy,* 1994) to the present. It is believed that environmental conflicts are driven by environmental stress and the competition for scarce resources (Page 2000). The causal link between environmental change and conflict is represented in Fig. 9.1 (Scheffran 2011).

In Fig. 9.1, Scheffran (2011) shows that stunted rainfall, droughts, land degradation, desertification etc. put pressure on natural resource use – i.e. land, pasture, water – and these turn to negatively impact human needs for survival. Consequently, when humans are unable to fulfill their needs, their response may contribute to conflict, migration, violence, crime, etc., thus negatively impacting security. Similarly, Homer-Dixon (1999) observes that the stress on humans to fulfill their needs of natural resource use leads to intense competition and impingements (conflicts). According to Homer-Dixon (1999), resource scarcity puts human needs under stress which may result in violent conflict when combined with social effects such as population growth, ethnicity, marginalization and social inequality. Homer-

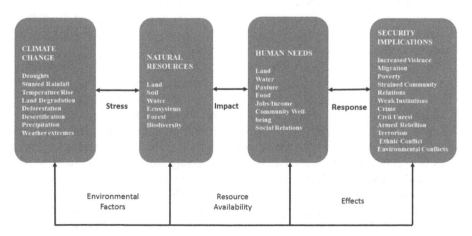

Fig. 9.1 Linkages of environmental change, natural resource scarcity and security
Source: Adapted from Scheffran (2011)

Dixon (1999) sees resource (environmental) scarcity as the degradation and depletion of natural resources such as cropland, water, forest and fish stocks. This scarcity is perceived in resource capture and ecological marginalization. The conflict in Darfur, Sudan has often been cited as a classic case where environmental/resource scarcity has led to violent conflict. The United Nations Environment Programme (UNEP) (2007:8) claims that: "there is a very strong link between land degradation, desertification and conflict in Darfur". The UNEP (2007) further argues that the environmental constraints acted as a trigger of conflict which was sustained by other political, ethnic and tribal factors while conflicts have in turn caused environmental degradation.

As a result, many interests have arisen over the interaction between the environment and conflicts among farmers and Fulani herders, especially within the ongoing discussions about the impacts of climate change and variability in many parts of the world (Hummel et al. 2012). The West African region is expected to be most severely affected by climate change in the near future due to the low adaptive capacity of the region (Christensen et al. 2007; IPCC 2007). Environmental changes such as stunted rainfall, droughts, deforestation, desertification and land degradation are increasing in Africa. With the region's vulnerability to climate change, some authors argue about the likelihood of resource scarcity, especially in the dry lands of the region (Alden 2010; Flintan 2012). Anderson et al. (2004) state that as Africa is covered by 40% of dry lands, variable and erratic climatic conditions in many dry lands continue to push people away from their areas, threatening their livelihoods. Within this perspective, farmer-pastoralist conflicts are seen as a result of environmental changes and resource shortages.

Davidheiser and Luna (2008) and Bevan (2007) observe further that climatic conditions and patterns in interaction with changing demographics, modes of production and land tenure are escalating farmer-pastoralist conflicts. Environmental conditions such as droughts and predicted climate change seem to be occurring more frequently, and the ability of farmers and pastoralists to overcome them have been reduced (Mwangi and Dohrn 2006). This is increasing aridity in the Sahel, decreasing rangeland vegetation, making natural resources scarce, and leading to likely contestations (Davidheiser and Luna 2008).

The Savannah-Sahel region where many of the Fulani pastoralists come from is one of the most unstable and harsh environments in the world (Oppong 2002). The climate in the Sahel is mainly characterized by long droughts, and much of the land is degraded. Resource constraints in the Sahel worsened following the Sahelian droughts of the 1960s, 1970s and 80s (Adebayo 1997; Toulmin 1986). This increased the fragility of pastoralists since the constraints reduced their adaptive capacity and livestock production, and forced many of them to migrate to the southern savannah zones and fringes of the forest zones of West Africa in search of water, pasture and grazing land for their cattle. Many of these pastoralists migrated southwards into Benin, Ghana, Cote d'Ivoire, Nigeria and Togo where environmental conditions were better (de Bruijn and Han van Dijk 2003). The conditions in the Sahel region are not better today as the rainfall period has become shorter and much of the vegetation is degraded inducing more migrations (Turner et al. 2011). According to

Trémolières (2010) for example, tensions between farmers and Fulani herders in Burkina Faso's Comoé province have been driven mainly by environmental factors (settlement and sedentarization of pastoralists as a result of the drought and soil degradation) in combination with expanded agricultural production and population pressures. Thus, farmer-pastoralist conflicts are interpreted by some scholars from the neo-Malthusian and environmental scarcity school.

Climate change predictions for Ghana, according to the National Climate Change Adaptation Strategy (UNEP and UNDP 2012), are expected to see major changes in temperature, precipitation, humidity, and annual rainfall. This presupposes decrease in resources for both farmers and herders. However, large parts of Ghana are relatively better than the Sahel since rainfall has remained higher, though erratic and variable (interview with Ghana Meteorological Agency, 2014). Dosu (2011) states that unpredictable climate patterns and unsustainable practices have intensified conflicts between farmers and Fulani herders in the northern and central part, as well as the Middle Volta Basin of Ghana. According to Dosu (2011), Fulani herdsmen from neighboring Burkina Faso, Mali and Niger, who depend on the wet season of the north to sustain their herds, are losing their cattle to starvation and dehydration due to droughts and degraded resources. When faced with such negative environmental conditions, the herders migrate to Ghana where better environmental conditions exist. The Volta Basin of Ghana has also experienced erratic and stunted rainfalls which increased the competition for the diminishing resources and resulted in conflicts (Dosu 2011).

Similarly, the abundance of resources could trigger conflicts between resource users (Welsch 2008; Brunnschweiler and Bulte 2009). Benjaminsen et al. (2012) note that conflicts increase between farmers and pastoralists during the rainy and wet season where resources – such as water and pasture – are available. These increases in resources can trigger conflicts where farmers and Fulani pastoralists compete for the use and ownership of the same resources. Basset (1988) states that farmer herder-conflicts in Cote d'Ivoire are intense in places where grazing areas are relatively abundant. Also, Abdulai and Tonah (2009) claim that a major reason for the increasing conflicts between farmers and herders in the middle part of Ghana is due to the fertile and abundant resources in the area, which breed intense competition between resource users.

The argument that environmental change is increasing conflicts between farmers and pastoralists is debatable. Multiple, varied as well as complex factors account for farmer-pastoralist conflicts in West Africa (Adano et al. 2012; Shettima and Tar 2008). Political ecologists will, therefore, argue that multidimensional and complex factors – such as social, political and resource-related issues – account for farmer-herder conflicts rather than being solely resource scarcity-driven (Turner 2004). Thus, both environment and politics play important roles in farmer-herder conflicts (Robbins 2004). However, rising interests in the interactions of environmental factors (including climate change and variability) and increasing conflicts between farmers and Fulani pastoralists in Ghana need to be scientifically examined to see how environmental change influences such conflicts.

3   Study Area and Methodology

3.1   Study Area

This study was carried out in the Asante Akim North District (AANDA, located in southern Ghana) and Gushiegu District (located in northern Ghana). The Asante Akim North District Assembly (AANDA, Fig. 9.2) was created in 2012, out of the then Asante Akim North Municipal Assembly (now Asante Akim Municipal Assembly). The capital of the AANDA is Agogo, which is located in the eastern part of the Ashanti Region, geographically classified as southern Ghana. The district covers a total land area of 1125.69 sq. km. Based on the population of the 2010 Ghana Housing and Population Census, the population of the AANDA is 68,186, representing 1.4% of the region's total population. Males constitute 48.8% and females 51.2%. The AANDA has a wet semi-equatorial climate and experiences a double-maxima rainfall, ranging between 1250 mm and 1750 mm per annum from May to July and from September to November. With a double-maxima rainfall, climatic conditions and resources in the area – such as fertile land, pasture and water – are available the entire year. Due to this fact, migrant farmers, mainly from northern Ghana, cattle owners from both within and out of Ghana, and Fulani herders are regularly attracted to the area, especially in the dry season in northern Ghana and the Sahelian countries.

The Gushegu District (Fig. 9.2) is one of the 26 administrative districts of the northern region of Ghana that was carved out of the then Eastern Dagomba District in 1988, following the introduction of Ghana's decentralization programme in 1988. It is located in the north-eastern corridor of the northern region, geographically called northern Ghana. The District has a land surface area of approximately 5796 Km². The administrative capital of the district is Gushiegu, which is about 105 Km north-east of Tamale, the regional capital. The Gushiegu District Assembly (GDA), according to the 2010 GDA Medium Term Development Plan, has the largest cattle market in the region; hence the presence of many Fulani herders in the area. According to the Ghana 2010 Population and Housing Census, the Gushiegu District has a population of 111,259 inhabitants, distributed in 395 communities. The district lies in the tropical climate marked by the alternation of two seasons, a dry season from November to April and a rainy season occurring between May to October. Much of the vegetation and grasses dry up in most parts of the year. Despite these dry conditions and resource constraints (pasture and water sources), many Fulani herders and cattle owners still move to the area. The number of Fulani herders in Ghana is statistically unknown since they are often not consciously counted in national census or immigration census.

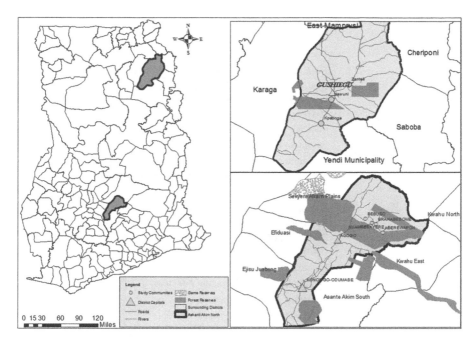

Fig. 9.2 Map showing Gushiegu and Agogo districts
Source: Authors (2015)

3.2 Methodology

The data presented in this work is part of 9 months of intensive fieldwork carried out by one of the authors from June 2013 to February 2014 as part of a PhD study. The approach used is qualitative in nature based on ethnographic interviews, narratives, participant observations and focus group discussions involving farmers, Fulani migrants (comprising both herders and cattle owners), sedentary herders and cattle owners, traditional chiefs, community leaders, government officials, and community members purposefully sampled to obtain in-depth information of the topic. Among herders, the sampling criteria were both purposeful for sedentary herders and snowball sampling for the seasonal migrant herders and their cattle owners. The breakdown of the interviews is shown in Table 9.1. In all, 295 interviews were conducted.

In all, twelve communities were identified and selected in Agogo[2] and 16 in Gushiegu[3] (Fig. 9.2). The choice of these communities was informed by the pres-

---

[2] Study communities in Agogo are Ahomaporawa Beposo, Nyamebekyere, Bebome, Abrewapong, Kowereso, Bebuso, Kwame Addo, Onyemso, Mankala, Matuka, Kansanso and Kowereja.

[3] Study communities in Gushiegu are Bulugu/Nawuni, Kpasinga/Kpatinga, Damdaboli, Offini, Sugu, Zamanshagu, Zamashiegu, Lamalim, Zanteli, Toti, Jingbani, Timya, Nnagmaya, Makpedanya, Kpakpaba and Kpug/Yawungu.

Table 9.1 Information on interviews conducted

Respondents	Specific Individuals	Numbers	
		Agogo	Gushiegu
Farmers	Both indigenous farmers and migrant farmers	43	44
Fulani herders	Resident/permanent Fulani and non-resident/seasonal Fulani	36	38
Cattle owners	Fulani cattle owners local/indigene cattle owners	4	7
		4	9
Traditional chiefs	Paramount chiefs	1	1
	Community chiefs,	10	13
	Kontehene (assistant to the paramount chief),	1	–
	Queen mother of Agogo	1	1
	Fulani chiefs/spokespersons, secretaries of the traditional councils	2	1
		1	1
Government officials	District chief executive, member of parliament (MP)	–	1
	Elected assembly members	1	–
	Deputy District coordinating directors	3	5
	Secretary of district assembly	1	1
	Veterinary officials	–	1
		1	1
Opinion leaders	Fulani elders,	3	2
	Community chairmen (in case of Gushiegu)	–	4
Community residents / households		25	28
Total		137	158

Source: Fieldwork (2013/2014)

ence of Fulani in the area who could give information on the topic. Data were also taken at Sekyere Kumawu and Karaga because a number of Fulani migrated from Agogo to Sekyere Kumawu and Gushiegu to Karaga following violent attacks. A total of four FGDs were carried out – two with Fulani herders and two with cattle owners (two in each of the study areas). At some points, the Fulani herders were followed deep into the fringes of the forest and bush to interview them and observe their daily activities. We spent several hours following and trying to locate them, especially the seasonal migrants. Those interviewed were mainly from Burkina Faso, Nigeria, and Mali. Meteorological data were also obtained from the Ghana Meteorological Agency, analyzed and graphs were generated from them to understand the climatic conditions to enable us to compare them with respondents' perceptions.

The interviews were transcribed and a content analysis of the data was done. From the transcribed data, we deduced themes and concepts for the analysis. Respondents in many cases have been anonymously cited. Some responses have been mentioned verbatim to highlight relevant expressions on the topic under study. Besides, the literature involved documents, reports as well as book chapters and articles on the topic, and were used in the study to help understand the issues. A challenge in the data collection stage was the thorough digestion and understanding of the terms 'environmental change' and 'climate change' to the Fulani herders and other respondents in their languages as they understood this in relation to mainly rainfall. We, therefore, needed to explain both terms using them in relation to rainfall and how the terms affected their livelihoods. Besides, statistics on Fulani conflicts and migrations were obviously difficult to obtain. Several attempts to get this data from government offices were unsuccessful.

4 Findings

4.1 Farmer and Pastoralist Knowledge and Perceptions of Environmental Change

Knowledge of farmers and pastoralists of environmental change is important for our understanding of how these changes affect their relations and activities. Schareika (2001) has already observed that pastoralists' ecological knowledge of the changes in the seasons and resource quality remains high. He maintains that they know when and where to find nutritional pasture and water for their cattle in dry seasons and the decisions to do so hinges on their high knowledge of their environment. Farmers equally have high knowledge of the changes that occur within their environment. Their knowledge about the weather and climate – such as changes in rainfall, temperature and winds – is also high (Mertz et al. 2009; Ifejika et al. 2009; Thomas et al. 2007). Farmers' knowledge of the weather, especially rain, helps them understand the planting seasons, the type of crop to cultivate, and generally what adaptation strategies to take in cases of rainfall fluctuations – stunted or abundant rain. For herders, their understanding of the climate over long periods determines their migration in search of quality pasture for their cattle. Table 9.2 below shows how farmers and herders each perceive environmental change. Although their perceptions of environmental change are similar and conform to general categories of environmental change independent of who perceives them, their interpretation of these changes are different. Herders, for instance, interpret poor rainfall by reference to its impacts on pasture while farmers consider it by reference to its amount (whether it is too much or small).

Just like Table 9.2 displays, Yaro (2013), Roncoli et al. (2011), West et al. (2008), and Tschakert (2007) report that farmers in the Sahelian region of West Africa identify shrinking water bodies, disappearing plants and crops, and changing settlement patterns as evidence of reduced rainfall over the last three decades of the twentieth century. The weather, especially rainfall, has largely remained the most visible environmental change perceived by farmers. Since much of their agricultural activities in the study sites are rain-dependent, they listed rainfall as the most important environmental change they perceive. Farmers in Gushiegu, especially, observed that the weather has largely remained unpredictable with rainfall seeing the most changes.

Table 9.2 Perceptions of environmental change

Farmer perceptions	Herder perceptions
Poor /stunted rainfall	Poor /stunted rainfall
Change in timing of rains/ late start of rain	Change in timing of rains/ late start of rain
Dry spells	Dry spells
Reduced crop yield due to soil infertility	Lack of pasture land/less pasture
Lack of good harvest	Less water
High temperature – the weather is hotter than previously	High temperature – hot weather (in fact pastoralists claim this dries up water bodies and fresh pasture)
Desertification and environmental degradation – bush burning, tree falling, over-grazing	Destruction/burning of fresh pasture by farmers
Floods[a]	Floods[a]

Source: Authors' extraction based on field interviews and FGDs (2013)
[a]Both farmers and herders think that floods are no longer common

> "The weather has changed significantly. We all know that rainfall normally starts in April but these days it will be July and yet we will have no rain. When the rain eventually comes, we are unable to sow our traditional crops like early millet and groundnuts. By the end of July, we have continuous rain until the mid of October when it stops all of a sudden [...]. And the poor rainfall has been consistent for the past years – especially for 2009-2013 [...]. Harvests for even the late millet have been very poor [...]. The changes are happening rapidly" (Farmer at Zantile-Gushiegu, 16/12/2013).

While herders equally see rainfall as an important indicator of environmental change, their observations of environmental change are looked at from the availability of fresh quality pasture for their cattle. The herders, thus, deduce that reductions in green pasture, scarcity of forage resources and the quality of pasture clearly show changes in the environment. This is important since Zampaligré et al. (2013) had found similar results in their study of pastoralists and farmers in Burkina Faso.

> "[…] Decline in crop yields, decreased soil fertility and increased erosion and land degradation were the major impacts of climate change as perceived by the crop farmers. For pastoralists and agro-pastoralists, major impacts of climate change on their livestock husbandry systems were the shrinkage of grazing areas and the decline of forage resources with consequently lowered animal productivity (offspring numbers, milk and meat yields). Since pasture areas and livestock corridors are increasingly cut-off by crop fields that cannot be trespassed during the rainy season until crop harvest, livestock mobility is restricted […]" (Zampaligré et al. 2013:7).

Farmers' perceptions of rainfall are mainly hinged on its onset and cessation. Farmers and herders in Gushiegu were mainly those who stated to have observed considerable environmental changes. To them, the changes are so rapid that the whole rainy and farming season has changed. However, the question we posed to them was '*what environmental changes are observed in the area (Gushiegu) that lie in the Savannah woodland, generally characterized by semi-arid climatic features such as dry spells and erratic rainfall*'. In response, farmers noted that although the area is semi-arid, there is an intensity of dry spells and stunted rainfall. They contended that the dry season (Harmattan[4]), instead of lasting only three to 5 months, has extended and thus impacting the rainy season. This, according to them, has changed the weather significantly. The elderly among the farmers believed that the changes have occurred over long periods of more than 10 years and have affected their farming. In Agogo, farmers, especially, see environmental change mainly in the migration response of herders from their countries to Agogo because of the resource availability in the area. They agreed that herders' migrations are the result of bad climatic conditions. Farmers in Agogo also see environmental change as being caused by herders' activities through over-grazing and bush burning which they believe is turning the land into savannah.[5]

Environmental change in the language of Fulani herders is often described as *'Yori kanye Toroka din wula yoomudu din tobi she koŋ no wuodi'*,[6] which is literally translated as *'climatic conditions (the weather[7]) have changed'*. Herders also see migration as a response to environmental change. Using historical narrations, herders maintained that their migrations were propelled by droughts, an important indicator of environmental change. Sedentary Fulani (*Bouboji*), for instance, traced their migration to Ghana to the early 1970s when the Sahelian droughts were severe in their countries whereas the seasonal migrants (*Yaligonji/Jerigoji*) indicated that migration was a seasonal affair and that environmental change has worsened, hence their continuous migrations. Thus, Fulani indicated that their migration and eventual presence in Ghana was occasioned by environmental change – long dry spells and stunted rainfall. A Fulani cattle owner in Karaga stated that when they were expelled in 1988, they returned to Ghana in 2001 from Burkina Faso because of lack of pasture for their cattle.

"[…] our migration back to Ghana was because there is pasture for our cattle. When you settle in a place and it is good for you, you would wish to remain there […]. In some years in Burkina, it was very difficult to find enough pasture for our cattle […]. Indeed, it is because of climatic conditions we are here" (Fulani Herder at Karaga, 17/11/2013).

4.2   Farmers and Pastoralists' Knowledge and Perceptions of Resource Scarcity

Farmers and herders look at natural resources from their livelihoods' perspective. Farmers, for instance, list fertile and arable land, sources of water for crops, and seed crops as important resources needed by them, whereas herders name arable

---

[4] The period of the dry season is called the Harmattan. This is where the north-east monsoon winds blow dust from the Sahara Desert into much of West Africa from the end of November to the middle of March. See more at http://www.britannica.com/EBchecked/topic/255457/harmattan

[5] Some parts of the land in Agogo have had features of savannah land after the 1983 bush fires that destroyed much part of it. See Sect. 3 for more details.

[6] Based on FGD with herders in Bulugu - Gushiegu, 22/11/2013.

[7] Herders take the weather and climate to be identical. These terms are of no difference to them mainly because it difficult to distinguish them in their language.

pasture land, green forage/pasture, and sources of water as important resources required by them. The two, therefore, consider resource scarcity from the perspective of resource absence or reduction. Farmers perceive scarcity as 'less/lack of water for crops, lack of good harvest, lack of land for agricultural expansion and infertile land'. Herders on the other hand perceive scarcity as including 'lack of fresh pasture land, less pasture, less water and less forage resources'. To migrant herders, one of the reasons that account for scarcity is the large herds of cattle in their countries (Burkina Faso, Nigeria, Niger, and Mali). They explained that due to many cattle, herders compete for the little available pasture which propels them to migrate. Their coping strategy to this scarcity and competition for limited resources is to migrate to regions where these resources are available and cattle fewer; hence their migration into Ghana. This is why Jónsson (2010) observes that Fulani use migration as the most visible coping strategy in response to the vagaries of the Sahelian climate.

Due to the scarcity of fresh pasture for cattle, especially during the dry season, herders are resorting to tree species to feed their cattle. The Shea (*Vitellaria paradoxa*) and Dawadawa (*Parkia biglobosa*) have become the most commonly used trees by them. Both the *Shea* and *Dawadawa*, as drought-resistant trees, are available all-year-round, and therefore become readily available as resources for the cattle. Herders cut the leaves, branches and fruits of the Shea and Dawadawa to feed their cattle. This has invariably put them into conflict with women who see the Shea and Dawadawa[8] as economically valuable to them. Other forage resources used by herders include thatching grasses (*Hyparrhenia involucrate Stapf*), which are also used by the local people for roofing houses. The use of the thatching grasses by herders to feed their cattle remains another controversy between them and the farmers. Their usual grasses used by the cattle, such as the elephant grass (*Pennisetum purpureum*), cat tail grass (*Sporobolus pyramidalis*), torpedo grass (*Panicum repens*), and others accordingly, are becoming unavailable for their use.

Theoretically, environmental/climate change affects important resources needed by herders and farmers to increase productivity and maintain their livelihoods. The argument of scholars is that environmental/climate change further constrains resource availability in areas that already suffer from resource shortage such as those of Africa (IPPC 2013; Scheffran 2011). In this vain, Scheffran et al. (2014) observe that agricultural societies with low levels of economic income are more dependent on natural resources and ecosystem services than industrial societies with high income. These agricultural societies are, therefore, likely to suffer from resource constraints due to environmental and climate change. Livestock production is likely to be affected in relation to grassland and rangeland productivity, lack of water and droughts. This partly accounts for Fulani herders' resort to tree species such as the Shea, Dawadawa and thatch grasses to feed their cattle.

---

[8] Shea and Dawadawa are valuable to farmers especially women since their nuts and seeds respectively are sources of income to them because they are sold out for export whilst locally used for making cooking oil, pomade and local meat.

Scheffran et al. (2014:357) note that "since human beings need resources to live, produce and consume, resource scarcity can increase the motivation to acquire or defend resources by the use of violence, individually or collectively. This can in turn contribute to resource destruction. When people are forced to migrate, this can cause conflicts due to scarce resources or cultural differences in receiving areas". Thus, to the authors, the consequences of resource scarcity are the motivation to use violence for resource acquisition through competition or collective motivation. On the other hand, resource scarcity could also lead to cooperative opportunities for the use of resources by two users as has been the case of farmers and herders in some situations. Therefore, relations between resources and environmental/climate change are such that their scarcity or abundance could be catalyst for either violence or opportunity for cooperative interactions.

4.3 What Data Sets Say About Environmental Changes in Ghana?

There has been a 1% rise in temperature in Ghana over the last decades and a 20% reduction in annual rainfall since 1960 (EPA 2000). According to the *National Climate Change Adaptation Strategy* (NCCAS) (UNEP and UNDP 2012), temperatures in Ghana in all the ecological zones are rising whereas rainfall levels and patterns have been generally reducing and increasingly becoming erratic.

> "Historical data for Ghana from the year 1961 to 2000 clearly shows a progressive rise in temperature and decrease in mean annual rainfall in all the six agro-ecological zones in the country. Climate change is manifested in Ghana through: i) rising temperatures, ii) declining rainfall totals and increased variability, iii) rising sea levels, and iv) high incidence of weather extremes and disasters. The average annual temperature has increased 1°C in the last 30 years [...]. The major challenges in all zones are weather extremes such as flooding, droughts and high temperatures. In the Transitional zone, the projected trends that are most likely to pose the major problem are the early termination of rainfall which is likely to convert the current bi-modal regime to a uni-modal one [...]. National Climate Change Adaptation Strategy (NCCAS) Ghana (UNEP and UNDP 2012)"

The report further projects rainfall variability in the near future and increases in temperatures between 0.8 °C and 5.4 °C for the years 2020 and 2080 respectively with estimated decline in total annual rainfall of between 1.1%, and 20.5% for this period. The UNEP and UNDP (2012) notes that the agricultural sector would be the worst affected by these changes. This presupposes decrease in resources for both farmers and herders alike.

Interestingly, both data sets and interviews with officials of the Ghana Meteorological Agency show a different picture of changes in the weather and for that matter the environment. Rainfall data (Figs. 9.3 and 9.4) reveal that whilst there is rain variability, there were basically no significant changes in average rainfall numbers. Officials of the Ghana Meteorological Agency rather believe that what is happening is mainly variability, especially with regard to rainfall.

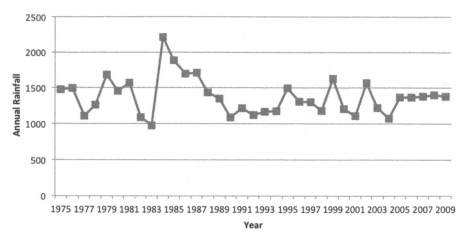

Fig. 9.3 Annual rainfall figures for Agogo (1975–2009)
Source: Based on figures provided by the Ghana Meteorological Agency

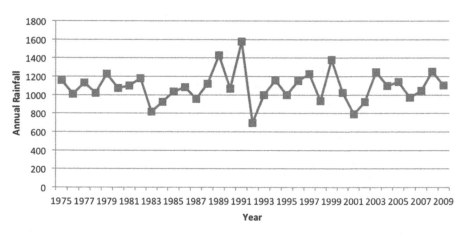

Fig. 9.4 Annual Rainfall figures for Tamale (Figures for Tamale were used for Gushiegu because rainfall figures for Gushiegu were unavailable. Besides, Gushiegu and Tamale lie in the same climatic region and actually experienced the same amount of rainfall.) (Gushiegu) (1975–2009)
Source: Based on figures provided by the Ghana Meteorological Agency

The annual rainfall for Agogo, as projected by the Ghana Meteorological Agency, has always been between 1000 mm and 1700 mm, although the year 1984 saw rainfall above 2212.3 mm (Fig. 9.3). As seen in Fig. 9.3, there is no relevant change in the average rainfall over time. The mean rainfall in Agogo is 1376.1 mm. For Gushiegu (Fig. 9.4), there are higher variations in the annual rainfall. Significant fluctuations of decrease in the rainfall can be seen in 1983 (which cut across many parts of Ghana including Agogo) and again in 2001. The mean rainfall for Gushiegu is 1087.8 mm. Both figures actually do show variations in the annual rainfall, but not significant changes.

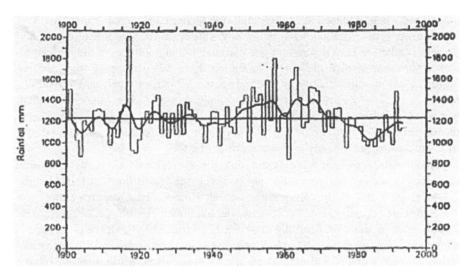

Fig. 9.5 Rainfall trends in northern Ghana (1900–1993)
Source: Dietz et al. (2004:156)

Dietz et al. (2004) had shown earlier that rainfall from 1900–1993 has seen recurrent pattern of variations and fluctuations in northern Ghana (Fig. 9.5), which are generally a feature of the Soudano-Guinean Ecological Zone (encompassing the whole area of northern Ghana). Rainfall patterns from 1960 to 1991 have generally been characterised by highly variable and erratic patterns (Dietz et al. 2004) which are not different from those experienced in recent times. Nicholson (2000) also notes that rains have generally recovered in the West African Sahel and the region has increasingly become wetter since 1998. Similarly, Yaro (2013) states that periods of poor rainfall in Ghana have been followed by years of good rains. This appears to contradict the views of farmers and herders who have identified depreciating rainfall in Ghana.

4.4   Farmer–Herder Conflicts: Are they the Result of Resource Scarcity and Environmental Change?

From the data sets, as well as farmers' and herders' understanding of environmental/climate change, we look at the linkages between conflicts and environmental change. Whilst the climate data presented above clearly shows climate variability and not significant changes in rainfall and temperature, the views of herders and farmers are that there is a link between environmental changes/scarce resources they are experiencing – as identified above – and the conflicts that have 'increased' between them. The linkages between climate change, resource scarcity and conflict have always remained difficult to establish. This is why Hussein et al. (1999)

questioned whether indeed all studies claiming increased resource scarcity and violent farmer-herder conflicts have really any figures on the frequency of past and present conflicts. Many theoretical postulations, ethnographic as well as quantitative studies that involve the use of modeling, have failed to prove that resource scarcity/environmental change are the only responsible for farmer-herder conflicts. Frerks et al. (2014:17) note that present debates recognize that "the environment and associated factors like environmental degradation, resource scarcity and more recently climate change, do or may play a role in the rise and continuation of conflict, but are seldom the only or most important factor".

Other studies as well have found that the correlation/link between natural resources (scarcity) and conflict is not a direct one (Le Billon 2001; Ballentine 2004). Quite a few studies have argued along Homer-Dixon's scarcity theory in which resource scarcity is said to increase the likelihood of violence. Many of these studies, such as those of Schilling et al. (2012), Njiru (2012), Opiyo et al. (2012) and others have found an indirect link between resource scarcity and violence escalation between farmers and herders or among pastoralists. More directly, Hsiang et al. (2013:1–12) proclaim, after a study of a variety of data, that "past climatic events have exerted considerable influence on human conflict".

In lieu of this study, a key question posed to farmers and herders was: '*Is resource scarcity/environmental change a cause of conflict between you and how?*'. Out of a total of 87 farmers, 53 of them responded in the affirmative whilst the rest either said they did not know or there was no link. For the total of 98 herders and cattle owners interviewed, 72 believed that changes in the climate (rainfall and dry spells) were mainly responsible for the increased conflicts. However, they admitted that resource scarcity or environmental change was not the only reason; it was a major factor. The basis for this argument is that resource scarcity induces herders' migrations to semi-arid areas in search of resources for their cattle which leads to competition for land in particular, thereby leading to violent confrontations. There are no figures about the number of Fulani migrants into Ghana as their migrations are irregular since they pass through unapproved borders into the country. Herders argue that due to the lack of pasture in their countries, they regularly migrate into Ghana. These competitions tend to affect their relations.

> "As pastoralists, we turn to be sensitive to changes in the environment because our activities and livestock depend on the availability of resources (water and pasture) which are negatively affected by climate change. When these resources are not available, we need to migrate to places where they are available […]. I have been here (Gushiegu) for over ten years now but there are new Fulani frequently coming in here with cattle just in search of pasture for them. Even those of us here also move further to places like Yeji, Buipi and Atebubu to look for pasture in hard years […]. Due to the many farms these days, we have to careful or else we would destroy crops which is a recipe for confrontations" (Interview with 42 Year Fulani Cattle owner, Gushiegu, 29/11/2013).

Thus, for herders, their migrations are in search of natural resources due to poor environmental conditions – such as poor rainfall, dry spells, and lack of pasture and water – indirectly causing conflicts between them and herders. The cattle owner in the interview above admits that herders' migrations are frequent which indirectly indicates competitions for land/pasture and water. Farmers equally stated that the continuous migration of Fulani from their countries to Ghana has put pressure on resources, particularly land, and hence their conflicts with farmers. Farmers at Agogo particularly argued that the influx of many cattle in the area is leading to a competition over land, thereby destroying crops, which aggravate conflicts. Owusu, a farmer at Bebome, maintained that: "the Fulani migrate to Agogo with their cattle because of harsh climatic conditions in other parts of the sub-region. Many herders do not like other places, but rather prefer Agogo due to the availability of pasture, easy access to water and the absence of tsetse flies". Farmers, thus, admit that the area has endowed resources (fertile lands and fresh pasture as well as sources of water) and that scarcity is not the problem, but competition for the rich resources. This is why de Bruijn and van Dijk (2005) maintain – after carrying out case studies on some farmer-herder conflicts in West Africa – that violent conflicts between farmers and herders are taking place in resource-endowed areas rather than in resource-scarce areas. This is the case of Agogo where fertile land, pasture and sources of water are readily available.

According to farmers (in Agogo), what rather cause scarcity are the activities of migrant Fulani herders to use the seemingly abundant resources in the area. Famers believed that the Fulani herders themselves are drivers of resource scarcity/environmental change. They argued that the migration of the Fulani herders to Agogo tends to change the environment rapidly causing environmental degradation and poor land use through bush burning, deforestation, and over-grazing by the cattle. Officials of the Ghana Forestry Commission equally agreed that lands, which have been designated as forest reserves, are being destroyed by many of the herders making much of Agogo grassland.

These observations of farmers and herders resonate with the findings of Henku (2011) that scarcity of grazing areas in the Great Lakes Region by cattle herders has forced them to constantly move in search of pasture and water on the lands of farmers or national parks, which have always resulted in constant clashes among the two. The perceptions that migrations of Fulani pastoralists are responsible for conflicts are also emphasized by NGOs. The national coordinator of West African Network for Peacebuilding (WANEP), Justin Bayo[9] is reported saying: "They (Fulani herders) cross over into the country fully armed and with no regard for our laws. If unchecked they will escalate the many dormant conflicts especially in the north over land".[10] Even the media discourse on the conflicts between farmers and herders tends to suggest a link between climate change and conflicts.[11] Writings in the media have sort to blame the harsh climatic conditions in the Sahel and the northern part

[9] He was the coordinator at the time of the statement. He has since left WANEP.

[10] See IRIN (29 April 2010). Ghana: Police crackdown on migrant Fulani herdsmen. http://www.irinnews.org/report/88957/ghana-police-crackdown-on-migrant-fulani-herdsmen   (Accessed 24/10/2014).

[11] See, for instance, Dakurah C. (August 17, 2012), The FULANI menace: a product of climate change, education and misconception. http://opinion.myjoyonline.com/pages/articles/201201/79940.php. (Accessed 24/10/2014).

of Ghana which propel herders to migrate to other parts of Ghana and have invariably resulted in clashes between the two. For instance, a journalist stated categorically in an article that:

> "[...] Climate change is affecting the availability of forage land for nomadic herdsmen in Ghana, as a result they have slowly begun to migrate towards the middle savannah areas and in lower proportions to some southern parts of Ghana [...]".[12]

Many of the statements, like the one above, have no scientific evidence to prove that it is environmental/climate change that actually propels herder migrations, and hence farmer-herder conflicts; yet they categorically provoke them. The data sets presented above have shown that there are no significant changes of rainfall and temperature over time.

Also, herders and farmers (in Gushiegu) observed that due to the change in weather events (i.e. stunted rainfall, dry spells and general reduction in rainfall), resources (such as pasture and water) are becoming scarce. They stated that rivers and streams easily dry up making it difficult to find water in the dry season. They pointed out that the lack of water and pasture has put pressure on the Bulugu Dam (the biggest dam in the area) for domestic use, dry season farming and cattle use. Farmers and community members in Bulugu, Yawungu, Kpug, Jingboni, Offini, Sugu, Toti, and Makpedanya claimed that dug-out bore holes always dry up during the dry season making them resort to dams and streams as sources for drinking water and domestic uses. This puts them into competition with herders for the use of water resources. The argument is that due to scarcity, mainly caused by weather events, competition for resource use has intensified. A farmer in Bulugu observed:

> "You know these days it doesn't rain as it used to be and so our harvests have been very poor. The Fulani are also struggling to find grass and water for their cattle. So, you have many of them moving from many parts of the district to Bulugu here. And as the largest dam and forest in the district are here, this puts pressure on the dam and resources here. That makes Fulani destroy our crops and this causes the conflicts we have with them. (Researcher: *Are you then saying that it is scarce resources that bring conflict between you and the Fulani?*). Well, if you look at it is all because the resources are not available. If not why do they have to leave and come to this community – it is because it is not raining which leads to lack of resources. In fact, this community I think hosts the largest number of Fulani and if they all come here like this they destroy more crops which can lead to conflicts" (Interview with 32 male farmers at Bulugu - Gushiegu, 16/10/2013).

Thus, the farmer's argument is based on the resource-scarcity postulations. He, just like herders (seen in the interview with the herder below), assumes that there is a direct link between droughts and conflicts between them.

> "[...] Indeed, dry spells are longer than expected. The grasses and water bodies have dried up in the north. Pasture availability is therefore a big problem. When we first arrived in Yendi, there was abundance of pasture and water, but that has changed so much. The time you expect the rain, it never comes. This is why we moved to Agogo and now we have to contend with many farms here which is leading to crop destructions and conflicts. Because we have to move through farms to search for pasture, it often leads to crop destruction and violent confrontations with farmers" (Interview with Fulani herder at Agogo, 2013).

---

[12] Ibid.

Schilling et al. (2012) found similar results in north-east Kenya among Turkana pastoralists that reduction in pasture, water and livestock has made raiding the only way of survival, which consequently leads to violent conflicts.

Historically, the Sahel has had many long droughts and famine spanning from the 1960s to the 1990s (Dai et al. 2004) with devastating consequences. Hardest hit by these were Niger, Mali, Burkina Faso, and Chad. Dietz et al. (2004) had shown that rainfall from 1900–1950s has generally been low in the Soudano-Guinean Ecological Zone. Similarly, rainfall patterns from 1960 to 1991 were characterised by highly variable and erratic patterns (Dietz et al. 2004). Mean average rainfall for the region ranges from 600–800 mm (Cour 2001). Aside rainfall, land degradation and hot temperatures are higher in the region. Much part of the Sahel is degraded and occupied by desertification. Livestock extensive movements over long distances and transhumance are particularly common in the Sahel as established patterns of migration of millions of Sahelians are usually southwards to better watered coastal countries, and northwards to the oases of the Sahara and onto the Maghreb coast (Hesse et al. 2013). Their migrations were said to be higher during the 1960s, 70s, and 80s droughts.

Migration in Sahel is often a coping mechanism to political, economic or environmental stress. In the midst of climate change, a particular human adaptive mechanism is migration (Scheffran 2011; Podesta and Ogden 2000; IPCC 2007). It is estimated that 60 million people will eventually move from the deserted areas of sub-Saharan Africa towards North Africa and Europe by the year 2020.[13] This is partly due to droughts, changing patterns of rainfall and scarcity of resources in the Sahel. Herder migration out of the Sahel is very common and is often seasonal. Interestingly, de Bruijn and van Dijk (2005) note that despite high levels of resource scarcity and poor rainfall in Hayre in Mali, which is located in the Sahel, violent conflicts between Fulani pastoralists and farmers are less and hardly occur. The two cooperate on the use of resources more than they conflict. The study by de Bruijn and van Dijk (2005), therefore, raises the question of whether scarcity indeed is the result of increased farmer-herder conflicts. Despite these arguments that migration is a response to climate change and the subsequent consequence of migration on resources (breeding competitions for resource use and conflicts), de Haas (2007) maintained that migration is not a phenomenon that is particular to the Sahel and that there are various migration patterns across the entire continent, which are not resulting in violent conflicts.

Importantly, the Fifth Assessment Report of the IPCC, especially of the Working Group II (2014b), raises doubt about the evidence that climate change does actually lead to long-term war or conflicts. Similarly, Gleditsch and Nordas (2014:85) note that:

> "[…] There is 'some agreement' that climate variability is associated with non-state conflicts (generally smaller and localized conflicts), but that the risk is mediated by the presence of conflict-management institutions. The chapter finds that 'Many of the factors that

increase the risk of civil war and other armed conflicts are sensitive to climate change', citing poverty, slow economic growth, economic shocks, and inconsistent political institutions [...] This is not controversial. However, there are (at least) three problems here: first, none of the studies on climate and conflict, with the possible exception of literature on heat and individual aggression, assume that climate has a direct influence on violence. The assumption, usually if not always made explicit, is that climate change (be it increasing heat or changes in precipitation) influences other factors, which in turn lead to conflict. Without these intervening factors (or mechanisms), the relationship between climate change and conflict simply cannot be understood [...]".

Gleditsch and Nordas (2014:86) particularly note that the links between climate change and conflict via migration are presented in vague terms, with reference to "interaction of climate change, disaster, conflict, displacement, and migration" since not much details and evidence are given regarding the nature of this interaction. The authors stated that the evidence on the linkages, as noted by the Fifth Assessment Report of the IPCC, is rather too minute and empirically limited.

Following arguments advanced by farmers and herders that resource scarcity and environmental change play a role in conflicts between them, we refer to an important observation by Frerks et al. (2014:17): "African environments display much more variability than commonly assumed and so environmental change are much more complex and varied than usually portrayed". Yaro (2013:1260–1261) therefore notes that:

> "Even though there is a noticeable decline in rainfall, the emerging pattern shows that periods of generally low rainfall running for about 20 years are followed by cycles of improved rainfall lasting almost same time periods…Both annual rainfall and mean temperatures are not very useful in Africa. What is really important is variability, not only from year to year, but within each growing season [...]."

Frerks et al. (2014) and Yaro (2013) raise the important question of whether the climatic conditions that are experienced are indeed new or whether climate variability has always characterized the West African zone. Yaro (2013) especially notes that variability has been common in history in the African environments and is thus not new.

Political ecologists have already established multi-causal reasons for farmer-pastoralist conflicts following a processual approach and mode of explanation that evaluates the influence of a number of variables. Turner (2004) has emphasized that environmental conflicts are bigger than usually assumed. He maintained that conflicts – such as farmer-herder struggles in society – are multi-faceted and go beyond just issues of resource scarcity or historical ethnic cleavages. He notes particularly that "struggles over resources are often only superficially so – they in fact reflect not only broader tensions between social groups, but also within these groups" (Turner 2004:866). Abundance of pasture, fertile lands, and increases in the economic and productive value of land, the role of mobilization and politics rather tend to trigger conflicts between farmers and herders in Agogo (a resource-endowed area) and that of Gushiegu cannot be said to be resource-driven since the area is a semi-arid zone and challenges with rainfall and dry spells are not a recent phenomenon. The data sets and the Ghana Meteorological Agency (Figs. 9.3 and 9.4) have already revealed that the erratic nature of rainfall in terms of its onset and cessation is a sign of variability and not a significant change in rainfall amounts.

Besides, since there are no data on the resource endowment of the area prior to the so-called scarcity and environmental change, it is difficult to determine the extent of scarcity of resources, and how this has affected herders and farmers in terms of conflict. Homer-Dixon (1999) and Spillmann and Bächler (1995) state that to describe a conflict as occurring due to resource scarcity/environmental change, one must be able to establish and have knowledge of the origins of this resource scarcity/environmental change and its consequences before. It is therefore difficult to state categorically that conflicts between farmers are mainly or partly caused by resource scarcity/environmental change. Also, de Bruijn and van Dijk (2005:57) mention that:

> "[…] To establish the precise relationship between environmental scarcity and violent conflict, one needs, firstly, to ascertain whether there is indeed a growing scarcity of natural resources and that cases of violent conflict are in fact on the increase. Secondly, this scarcity needs to be perceived as such by the parties involved in violent conflict. Thirdly, is scarcity one of the reasons why the various stakeholders are engaging in conflict and resorting to violence? [...]".

It is important to be careful to link scarcity of resources to just environmental and climate change. This scarcity could emanate from several causes, including temporary factors such as periods of short dry spells, variations in the climate, type of farming practices, human environmental activities, and even the natural resource capability of the area prior to the so-called changes. Therefore, direct impacts of climate and environmental change on conflicts between farmers and Fulani herders in the study areas are not known. Local discourses of environmental change do not mean that indeed farmer-herder conflicts are caused by environmental/climate change and resource scarcity. One must not also lose sight of the fact that in the light of the so-called environmental change/resource scarcity, the adaptive capacity of farmers and pastoralists alike are higher. This resonates with the Boserupian thinking in which scarcity could be an opportunity for technological innovation.

5   Conclusion

The purpose of this work was to investigate the linkages between environmental change and resource scarcity on one hand, and farmer-herder conflicts on the other hand. Theories linking resource scarcity and conflicts are hinged on the basic assumption that scarce resources – such as water, land and pasture – will eventually force people to migrate to resource-endowed places, thus leading to competitions and subsequently to violent confrontations. This also resonates with the perceptions of farmers and herders in our study areas. Conceptualization of farmers and herders perceptions of the environment-conflict nexus can be summed up systematically as:

1. Environmental/climate change and resource scarcity has worsened in the Sahelen region and some parts of northern Ghana.
2. These have propelled herders to migrate to resource-endowed areas. At the same time, these areas have also seen expansion in farming activities.
3. As a result, competitions are higher for land for both farming and pasture.
4. These competitions are leading to crop destructions and killing of cattle that are found grazing on farms. Farmers and herders are therefore defending their livelihoods by 'fighting' each other through violent means by acquiring weapons.

Whilst it is worth stating that media reports from 2009 to 2016 reveal that there were increases of gory clashes between farmers and Fulani herders in many parts of Ghana, and there has been some penetration of Fulani herders and cattle owners into Ghana over the last decades, it does not mean conflicts between them were the result of their migrations (which fit into the broader issue of environmental change/resource scarcity). In many cases, herders' migrations were not just in search of scarce pasture and water, but also cattle herding jobs and quality pasture. Data sets (of rainfall) reveal that whilst there is climate variability, there were basically no major changes in rainfall figures/patterns. Thus, whilst environmental change is perceived to be indirectly causing farmer-herder conflicts, climatic evidence suggests otherwise. Overall, evidence of the climate-conflict link in our case-study region remains limited.

References

Abdulai A, Tonah S (2009) Contemporary social problems in Ghana. In: Tonah S (ed) Contemporary social problems in Ghana. Department of Sociology, University of Ghana, Research and Publication Unit, Accra

Adano WR, Dietz T, Witsenburg K, Zaal F (2012) Climate change, violent conflict and local institutions in Kenya's drylands. J Peace Res 49(1):65–80

Adebayo A (1997) Contemporary dimensions of migration among historically migrant Nigerians. J Asian Afr Stud 32(1–2):93–109

Adetula HV (2006) Development, conflict and peace building in Africa. In: Best GS (ed) Introduction to peace and conflict studies in West Africa: a reader. Spectrum Books, Ibadan, pp 383–406

Alden WL (2010) Fodder for war: getting to the crux of the natural resources crisis. Rights and Resources Initiative, Washington

Anderson J, Bryceson D, Campbell B, Chitundu D, Clarke J, Drinkwater M, Fakir S et al (2004) Chance, change and choice in African drylands: a new perspective on policy priorities

Ballentine K (2004) Program on economic agendas in civil war: principal research findings and policy recommendations. International Peace Academy, New York

Barnett J (2001) The meaning of environmental security: ecological politics and policy in the new security era. Zed Books, London

Barnett J (2007a) Climate change, human security and violent conflict. Polit Geogr 26(6):639–655

Barnett J (2007b) Environmental security and peace. J Hum Secur 3(1):4–16

Bassett TJ (1988) The political ecology of peasant-herder conflicts in the northern Ivory Coast. Ann Assoc Am Geogr 78(3):453–472

Benjaminsen TA, Alinon K, Buhaug H, Buseth JT (2012) Does climate change drive land-use conflicts in the Sahel? J Peace Res 49(1):97–111

Bevan J (2007) Between a rock and hard place: armed violence in African pastoral communities. United Nations Development Programme

Blench R, Dendo M (2003) The transformation of conflict between pastoralists and cultivators in Nigeria. J Africa (special issue), Mark M (ed) Cambridge, UK

Brockhaus M (2005) Potentials and obstacles in the arena of conflict and natural resource management: a case study on conflicts, institutions and policy networks in Burkina Faso, 1st edn. Cuvillier Verlag, Göttingen

Brockhaus M, Pickardt T, Rischkowsky B (2003) Mediation in a changing landscape : success and failure in managing conflicts over natural resources in Southwest Burkina Faso. IIED, London

Brown O, Crawford A (2008) Assessing the security implications of climate change for west Africa country case studies of Ghana and Burkina Faso. International Institute for Sustainable Development, Winnipeg/Manitoba

Brunnschweiler C, Bulte EH (2009) Natural resources and violent conflict: resource abundance, dependence, and the onset of civil wars. Oxf Econ Pap 61(4):651–674

Christiansen JH et al (2007) Regional climate projections. In: Solomon S, Qin D, Manning M, Chen Z, Marquis M, Averyt KB, Tignor, Miller HL (eds) Climate change 2007: the physical science basis. Contribution of Working Group I to the fourth assessment report of the Intergovernmental Panel on Climate Change. Cambridge University Press, Cambridge, pp 847–940

Cour JM (2001) The Sahel in West Africa: countries in transition to a full market economy. Glob Environ Chang 11:31–47

Dai A, Lamb PJ, Trenberth KE, Hulme M, Jones PD, Xie P (2004) The recent Sahel drought is real. Int J Climatol 24(11):1323–1331. https://doi.org/10.1002/joc.1083

Davidheiser M, Luna AM (2008) From complementarity to conflict: a historical analysis of farmer-Fulbe relations in West Africa. Afr J Confl Resolut 8(1):77–104

de Bruijn M, Han van Dijk R (2003) Changing population mobility in West Africa: Fulbe pastoralists in central and South Mali. Afr Aff 102(407):28–307

de Bruijn M, van Dijk H (2005) Natural resources, scarcity and conflict: a perspective from below. In: Chabal P, Engel U, Gentilli A-M (eds) Is violence inevitable in Africa? Theories of Conflict and Approaches to Conflict Prevention. Brill, Leiden, pp 55–74

de Haas H (2007) The myth of invasion: irregular migration from West Africa to the Maghreb and the European Union. University of Oxford International Migration Institute, Oxford

Diallo Y (2001) Conflict, cooperation and integration: a west African example (Cote d' Ivoire). Max-Plank Institute, Halle

Dietz AJ, Millar D, Dittoh S, Obeng F, Ofori-Sarpong E (2004) Climate and livelihood change in north east Ghana. In: Dietz AJ, Ruben R, Verhagen A (eds) The impact of climate change on drylands with a focus on West Africa. Kluwer Academic Publishers, Dordrecht, pp 149–172

Dosu A (2011) Fulani-Farmer conflict and climate change in Ghana: Migrant Fulani herdsmen clashing with Ghanaian farmers in struggles for diminishing land and water resources. http://www1.american.edu/ted/ICE/fulani.html. (Accessed 20 Apr 2013)

EPA (2000) Climate change vulnerability and adaptation assessment of the agricultural sector of Ghana. Final report. Accra: Ministry of Environment Science and Technology

Flintan F (2012) Making rangelands secure: past experience and future options. Rome, International Land Coalition (ILC)

Frerks G, Diet T, van der Zaag P (2014) Conflict and cooperation on natural resources: justifying the CoCooN programme. In: Bavinck M, Pellegrini L, Mostert E (eds) Conflicts over Natural Resources in the Global South – Conceptual Approaches. London, Taylor and Francis Group

Gleditsch NP, Nordas R (2014) Conflicting messages? The IPCC on conflict and human security. Polit Geogr 43:82–90

Hagberg S (1998) Between peace and justice: dispute settlement between Karaboro agriculturalists and Fulbe agro-pastoralists in Burkina Faso. Acta Universitatis Upsaliensis, Upsula

Harshbarger CL (1995) Farmer-herder conflict and state legitimacy in Cameroon. Dissertation presented to the graduate school of the University of Florida

Henku AI (2011) Environmental security and peace in the Great Lakes region of Africa. University of Peace, San José

Hesse C, Anderson S, Cotula L, Skinner J, Toulmin C (2013) Building climate resilience in the Sahel. International Institute for Environment and Development. Available at www.pubs.iied. org/pdfs/G03650.pdf. (Accessed 11 Nov 2014)

Homer-Dixon T (1999) Environment, scarcity, and violence. Princeton University Press, Princeton

Hsiang S, Burke M, Miguel E (2013) Quantifying the influence of climate on human conflict. Science, 341(6151):1235367-1 -1235367-12

Hummel D, Doevenspeck M, Samimi C (2012) Climate change, environment and migration in the Sahel - selected issues with a focus on Senegal and Mali. MICLE working paper No. 1. Frankfurt/Main: MICLE

Hussein K, Sumberg J, Seddon D (1999) Increasing violent conflict between herders and farmers in Africa: claims and evidence. Dev Policy Rev 17(4):397–418

Ifejika SC, Kiteme B, Ambenje P, Wiesmann U, Makali S (2009) Indigenous knowledge related to climate variability and change: insights from droughts in semi-arid areas of former Makueni District, Kenya. Clim Chang 100:295–315. https://doi.org/10.1007/s10584-009-9713-0

IPCC (2007) Fourth assessment report of the intergovernmental panel on climate change. IPCC, Geneva

IPCC (2013) Climate change 2013: the physical science basis. IPCC, Geneva

IPCC (2014) Climate change 2014: mitigation of climate change. IPCC, Geneva

Jónsson G (2010) The environmental factor in migration dynamics – a review of African case studies. International Migration Institute working paper no. 21, University of Oxford

Kaplan RD (1994) The coming anarchy: how scarcity, crime, over population, tribalism and disease are rapidly destroying the social fabric of our planet. Atl Mon 273:44–65

Le Billon P (2001) The political ecology of war: natural resources and armed conflicts. Polit Geogr 20(5):561–584

Mertz O, Mbow C, Reenberg A, Diouf A (2009) Farmers' perceptions of climate change and agricultural adaptation strategies in rural Sahel. Environ Manag 43:804–816

Mwangi E, Dohrn S (2006) Biting the bullet: how to secure access to resources for multiple users. CAPRI working paper No. 47. Washington: IFPRI

Nicholson SE (2000) The nature of rainfall variability over Africa on time scales of decades to millennia. Glob Planet Chang 26:137–158

Njiru BN (2012) Climate change, resource competition, and conflict amongst pastoral communities in Kenya. In: Scheffran J, Brzoska M, Brauch HG et al (eds) Climate change, human security and violent conflict. Springer, Berlin, pp 513–527

Obioha EE (2008) Climate change, population drift and violent conflict over land resources in northeastern Nigeria. Hum Ecol 23(4):311–324

Opiyo EOF, Wasonga VO, Schilling J, Mureithi MS (2012) Resource-based conflicts in drought-prone North-Western Kenya: the drivers and mitigation mechanisms. Wudpecker J Agric Res 1(11):442–453

Oppong YPA (2002) Moving through and passing on: Fulani mobility, survival, and identity in Ghana. Transaction Publishers, Accra

Page E (2000) Theorizing the link between environmental change and security. Rev Eur Commun Int Environ Law 9(1):33–43

Podesta J, Ogden P (2000) The security implications of climate change. Wash Q 31(1):115–138

Robbins P (2004) Political ecology: a critical introduction. Wiley-Blackwell, Oxford

Roncoli C, Orlove B, Kabugo M, Waiswa M (2011) Cultural styles of participation in farmers' discussions of seasonal climate forecasts in Uganda. Agric Hum Values 28:123–138

Schareika N (2001) Environmental knowledge and pastoral migration among the Woɗaaɓe of south-eastern Niger. Nomadic Peoples 5:65–88

Scheffran J (2011) Security risks of climate change: vulnerabilities, threats, conflicts and strategies. In: Brauch HG, Oswald Spring Ú, Mesjasz C, Grin J, Kameri-Mbote P, Chourou B, Dunay P et al (eds) Coping with global environmental change, disasters and security. Springer-Verlag, Berlin, pp 735–756

Scheffran J, Ide T, Schilling J (2014) Violent climate or climate of violence? Concepts and relations with focus on Kenya and Sudan. Int J Hum Rights 18(3):369–390. https://doi.org/10.108 0/13642987.2014.914722

Schilling J, Frleier KP, Hertig E, Scheffran J (2012) Climate change, vulnerability and adaptation in North Africa with focus on Morocco. Agric Ecosyst Environ 8(156):12–26

Shettima AB, Tar UA (2008) Farmer-pastoralists conflicts in West Africa: exploring the causes and consequences. Inf Soc Just 1(2):163–184

Spillmann KR, Bächler G (1995) International project on violence and conflicts caused by environmental degradation and peaceful conflict resolution. Center for Security Studies and Conflict Research, Swiss Federal Institute of Technology, Zurich

Thomas D, Twyman C, Osbahr H, Hewitson B (2007) Adaptation to climate change and variability: farmer responses to intraseasonal precipitation trends in South Africa. Clim Chang 83:301–322

Tonah S (2006) Migration and farmer-herder conflicts in Ghana's Volta Basin. Can J Afr Stud 40(1):162–178

Toulmin C (1986) Pertes de bétail et reconstitution du cheptel aprés la sécheresse en Afrique subsaharienne. Addis-Abeba, Ethiopia

Trémolières M (2010) Security and environmental variables: the debate and an analysis of links in the Sahel. SWAC/OECD. Available at https://www.oecd.org/swac/publications/47092980.pdf

Tschakert P (2007) Views from the vulnerable: understanding climatic and other stressors in the Sahel. Glob Environ Chang 7:381–339

Turner MD (2004) Political ecology and the moral dimensions of "resource conflicts": the case of farmer–herder conflicts in the Sahel. Polit Geogr 23(7):863–889

Turner MD, Ayantunde AA, Patterson KP, Patterson ED (2011) Livelihood transitions and the changing nature of farmer-herder conflict in Sahelian West Africa. J Dev Stud 47(2):183–206

UNEP (2007) Sudan: post-conflict environmental assessment. United Nations environment programme (UNEP)

UNEP and UNDP (2012) National Climate Change Adaptation Strategy (NCCAS). Accra: UNDP. Retrieved from https://s3.amazonaws.com/ndpc-static/CACHES/PUBLICATIONS/2016/04/16/Ghana_national_climate_change_adaptation_strategy_nccas.pdf

Welsch H (2008) Resource abundance and internal armed conflict: types of natural resources and the incidence of 'new wars'. Ecol Econ 67(3):503–513

West C, Roncoli C, Ouattare F (2008) Local perceptions and regional climate trends on the central plateau of Burkina Faso. Land Degrad Dev 19:289–304

Yaro JA (2013) The perception of and adaptation to climate variability/change in Ghana by small-scale and commercial farmers. Reg Environ Chang 13(6):1259–1272

Zampaligre N, Dossa LH, Schlecht E (2013) Climate change and variability: perception and adaptation strategies of pastoralists and agro-pastoralists across different zones of Burkina Faso. Reg Environ Chang 14(2):769–783

Chapter 10
Environmental Change Impacts and Human Security in Semi-Arid Region, the Case of Nuba Mountains of Sudan

Taisser H. H. Deafalla, Elmar Csaplovics, and Mustafa M. El-Abbas

Abstract The links between climate change, environmental degradation, and conflict are highly complex and poorly understood as yet. Resources always constitute risks of conflict in developing countries, particularly if poverty is widespread, the resource is scarce, and property rights are insecure and/or unclear. Some of these causal factors are manifested in and driving forces behind armed conflicts in Nuba Mountains of Sudan, often with broader geographical, socioeconomic, and political repercussions, displacement of people, and unsustainable use of natural resources. The aim of this study is to raise understanding of the links between patterns of local-level economic and demographic changes of Nuba Mountains region, specifically of those systems in poor condition under environmental change. The methodology applied in the current research is three-pronged: a formal literature review; semi-structured interviews of household heads; and mulit temporal optical satellite data (i.e. LANDSAT TM and ASTER) to study the land use/land cover changes during the period from 1984 to 2014. The qualitative and quantitative techniques were used to analyze the socio-economic data. Post Classification Analysis (PCA) as well as Post Change Detection (PCD) techniques were applied. The study showed that there are increasingly rapid changes in land cover, social structure, institutional and livelihood transformation across broad areas of the state. More information exchange is needed to inform actors and decision makers regarding specific experiences, capacity gaps and knowledge to address the environmental change.

Keywords Environmental change · Conflict · Semi-arid region · Remote sensing · Control of resources

T. H. H. Deafalla (✉)
Institute of Photogrammetry and Remote Sensing, University of Dresden, Dresden, Germany

E. Csaplovics
Researcher Professor, Faculty of Environmental Sciences, TU Dresden, Dresden, Germany
e-mail: elmar.csaplovics@tu-dresden.de

M. M. El-Abbas
Researcher Professor, Faculty of Environmental Sciences, TU Dresden, Dresden, Germany

Faculty of Forestry, University of Khartoum, Khartoum, Sudan

© Springer International Publishing AG, part of Springer Nature 2019   187
M. Behnassi et al. (eds.), *Human and Environmental Security in the Era of Global Risks*, https://doi.org/10.1007/978-3-319-92828-9_10

1 Introduction

Environmental Change (EC) is defined as a change or disturbance of the environment caused by human influences or natural ecological processes; EC is a complex dynamic system (Johnson et al. 1997). Accordingly, global environmental change is the one that has the potential to undermine human security, i.e. the needs, rights, and values of people and communities. 'Global' in this sense does not mean that responsibility for EC is shared equally among all people, or that the impacts of these changes are uniformly distributed among all places. Instead, global refers to the linkages between EC and social consequences across distant places, groups, and time horizons (UNEP 2007; Barnett et al. 2008, 2010).

The characteristics and importance of each driving force may vary from one area to another. However, the long-range interactions of pressures may lead to the environmental change. In addition to that, the ability to create and transfer environmental pressures or stress onto the environment of other societies varies from one area to another. Developed countries are significantly contributing in global and transboundary environmental pressures than the less affluent societies which have minimal interaction with the environment in which they live (UNEP 2007).

EC is only one of the many security, climatic and developmental challenges facing Africa. EC is, in effect, a 'threat multiplier' that makes existing concerns, such as drought, desertification, water scarcity, food insecurity and land degradation, more complex and intractable. However, the non-climate factors such as poverty, conflict management, governance, regional diplomacy etc. largely determine whether and how climate change moves from being a development challenge to presenting a security threat (Brown and Crawford 2009).

There is a gap in the studies that focus on the intersections between global environmental change and human security (Behnassi and McGlade 2017). Indeed, over the past two decades the concept of human security has been used to frame and analyze problems of social change. Recently, and since the late 1990s, the concept of human security has been used for these purposes with respect to global environmental change (O'Brien and Barnett 2013).

The changes caused by human-induced impacts on atmospheric composition, climate, land, water, and biodiversity are occurring so rapidly that the natural systems are increasingly losing their adaptive capacity. The resulting degradation of the planet's resources and life-support systems may be irreversible at scales relevant to present human society. Moreover, the inertia in many natural processes means that the damage may take only years or decades later to become fully apparent (IPEC 2003).

The natural resources play a key role in increasing the likelihood of onset, the duration and the return to conflict; particularly in African developing countries, e.g. Sudan where poverty is widespread (Bannon and Collier 2003; Collier 2003). The links between environmental resources and conflicts are, however, highly complex, nonlinear and influenced by a combination of factors, including political, social, economic, environmental, historical, and the different aspects of vulnerability (Notaras 2009). Actually, the interaction of these factors plays a role in either preventing or stimulating conflict. Nevertheless, large open access natural resources can undermine the quality of governance, aggravate corruption, weaken economic performance and, thereby, increase the vulnerability of countries to conflicts. Moreover, conflicts can also occur over the control and exploitation of resources (Deafalla 2012). On other hand, the forced movement of people in those areas undermines, sometimes for decades, the economic development, sustainable livelihoods as well as the capacities of societies and nations. The shortage or degradation of natural resources contributes directly to lower levels of well-being and higher levels of vulnerability. However, in many developed societies, they clearly recognized the contribution of natural resources to poverty reduction efforts (UNEP 2007; Deafalla 2012).

Environmental factors are rarely the sole source of conflict. However, the related environmental stresses have a determining influence on peace and security. Furthermore, resource abundance or possession of specific high-value resources in socio-political contexts of weak institutions and poor governance are not only associated with low economic growth, but can contribute in increasing the likelihood and incidence of civil or armed conflicts, stimulates violence between rival groups as well as the stakeholders of the resource (UNEP 2007; Ekbom 2009; Deafalla et al. 2014). The changes in and depletion of natural resources linked to climate change have been considered as a causal factor in the current crisis in many regions of Sudan.

Low education levels, high poverty, population characteristics, high dependence upon natural resources, ethnic and religious fractionalization, geography of the area, as well as previous conflicts, have facilitated the armed conflict in Nuba Mountains. This study is an attempt to raise understanding of the links between patterns of local-level economic and demographic changes of Nuba Mountains region, specifically of those systems in poor condition under EC. It also outlines some of the actions being taken to help a country adapting to the changing environment, and makes recommendations for how such actions could become more effective.

2 Study Site

The study area is located in south Kordofan States. It lies between latitudes 10° and 13° N and longitudes 29° and 33° E (Fig. 10.1). It is composed of five provinces (Kadugli, Rashad, Abu Gubeiha, Talodi and El Dilling). It covers an area of approximately 141,096 km² (Fig. 10.1). The region has a varying climate, ranging from semi-desert in the north to rich savanna in the south. Annual rainfall ranges from less than 50 mm on the northern border to more than 800 mm on the southern border. The rainy season varies from about 5 months or less, with rains occurring between May and October. The average daily temperature ranges from 10 to 35 °C with an annual variation of 15 °C. April to June is the hottest period and December to February is the coldest. Wind direction differs according to seasons: northeast in winter and southwest in summer (El Tahir et al. 2010). Total population of Nuba Mountains in 2008 was 1.3 million distributed into 120,986 households (CBS 2009). The livelihood activities found in the area are agro-pastoralism, nomadic

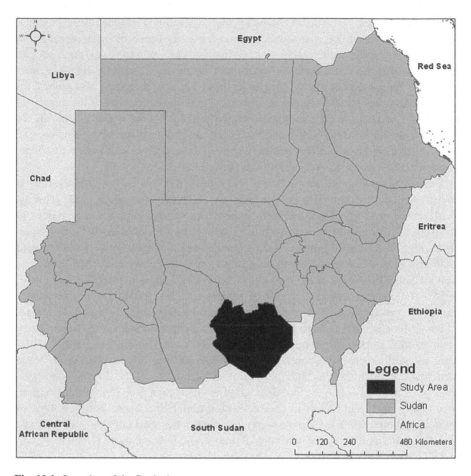

Fig. 10.1 Location of the Study Area
Source: DIVA-GIS, developed by the authors

pastoralism and rain-fed agriculture; both traditional farming for subsistence and mechanized farming for commercial operations. In addition to that, a third source of livelihood is related to natural forests in form of woody and non-woody production derived from various tree species (UNDP 2006).

3   Research Methods

3.1   Socioeconomic Data

Firstly, human, social and financial data were collected through social survey of households; stratified sampling was used to represent different geographical areas and different income groups. This helped ensure better precision and reduce time, effort, and monetary costs. This assessment was conducted in June–July 2014. The total sample size was 255 questionnaires, 200 for heads of households (covered displaced and non-displaced respondents) distributed among different units of Nuba Mountains according to the Principle of population Proportional to Size (PPS) in the selected sites. Rapid rural appraisal (RRA) with a focus on group discussion, interviews, free listing and key informants techniques were applied in two refugee's camps (Abkorshola in Rashad town and Jabarona Omdurman, Khartoum), with total samples size of 55 contributions. Focus group discussions were facilitated for 5 different groups from randomly selected locations (separated by sex and age). Participants for the discussions were selected purposively amongst representatives in the communities, as well as various sub-groups within it. The questionnaire was designed to identify main war impacts such as: agricultural situation; livestock conditions; macroeconomic situation; availability of food; types of food consumed; and number of meals per day.

Secondly, a formal literature review was used to cover the areas which are directly related to difficulties of data collection due to security reasons.

Regarding data analyses, the Statistical Package of Social Sciences (SPSS) was used. More precisely, descriptive statistical methods were applied to data related to household social characteristics and respondents' perspectives about different aspects of the change. Moreover, the Food Consumption Score (FCS) was used to measure both diversity and frequency of food consumption in surveyed areas. The following equation was used according to WFP (2008):

$$FCS = {}^a staple \times {}^{xi} staple + {}^a pulse \times {}^{xi} pulse + {}^a vegetable \times {}^{xi} vegetable + {}^a fruit \times {}^{xi} fruit + {}^a animal \times {}^{xi} animal + {}^a sugar \times {}^{xi} sugar + {}^a dairy \times {}^{xi} dairy + {}^a oil \times {}^{xi} oil$$

Where:

FCS= Food Consumption Score

^{ai}= Weight of each food group.

^{xi}= Frequencies of food consumption = number of days for which each food group was consumed during the past 7 days.

*(7 days was designated as the maximum value of the sum of frequencies of the different food items belonging to the same food group).

The FCS used was borrowed directly from the World Food Programme (WFP). The information was collected from a country specific list of food items and food groups, using Emergency Nutrition Assessment software.

Finally, correlation analysis was applied to describe and test the relationship between these factors.

3.2   Remotely Sensed Data

Two types of temporal optical satellite data from Advanced Spaceborne Thermal Emission and Reflection Radiometer (ASTER) and Landsat Thematic Mapper (TM) were utilized. Environmental Visualization (ENVI) software was used for the image pre-processing and enhancement. The image processing, analysis and classification have been carried out using Earth Resources Data Analysis System (ERDAS) Imagine and ArcGIS softwares. These processes include radiometric, spatial and spectral enhancements, maximum likelihood classification, and post-change detection. The main approach includes supervised classification that attempts to identify spectrally homogeneous groups within the image that are later assigned to information categories of LU/LC classes (Richards 1993; Chuvieco 1996; Lillesand et al. 2008).

3.2.1 Image Registration

Image registration technique was applied based on ground control points using the polynomial geometric model, so that the geometry of the images (1984, 1994, 2004, 2011, and 2014) has been normalized accordingly.

3.2.2 Supervised Classification

The study applies supervised classifier as a per pixel approach based on spectral properties. Maximum Likelihood (ML) classifier was selected as one of the common used method, which it quantitatively evaluates both the variance and covariance of the category spectral response patterns when classifying an unknown pixel (Lillesand et al. 2008).

3.2.3 Change Detection

After classifying the imagery of the selected dates, multi-temporal classified maps (1984, 1994, 2004, 2011, and 2014) were introduced to the PCD to determine changes in LU/LC during the study period using mean-shift and outlier-distance metrics. In order to perform an appropriate multi-temporal analysis, PCD might reduce the possible effects of atmosphere, sun angle, seasonal variation of acquired date and multi-sensor variability (Singh 1989). However, rather than using multi-spectral imagery, the classified images were used. In addition, the post classification comparison provides class changes from-to and a change matrix of classes.

4   Results and Discussion

A tradition of intensive interactions between the rural and urban areas has long been acknowledged, but recent changes in the global political economy and environmental systems, as well as local dynamics of the study area such as war, drought, and deforestation have led both to a new rapidity and depth in rural transformation, as well as a significant impact on urban area. Like most environmental problems, the effects of these drivers are complex and stressed differentially across varied geographies by the socio-political processes that underlie recent economic and cultural globalization. These interactions and processes have increasingly brought rapid changes in land cover, social, institutional and livelihood transformation across broad areas of the state. Furthermore, the study unveiled new dynamics such as: high rates of migration and mobility for the indigenous population; and the increasing domination of market-centric livelihoods in many villages that were once dominated by rural agricultural and natural resource based on socio-economic systems.

4.1   Migration and Mobility

As mentioned earlier, the links between EC and conflict are very complex. Given the history of ethnic and political conflicts in Kordofan region, EC could aggravate territorial and border disputes and complicate conflict resolution and mediation processes in the future. Now, Nuba Mountains is a critical flashpoint zone in Sudan, and have communities living in fragile and unstable conditions, which make them more vulnerable to the risk of violent conflict and environmental change's effects. The degraded resources and conflicts forced migration to other regions such as Khartoum and North Kordofan states (Fig. 10.2), in addition to republic of south Sudan with negative consequences for political stability in the area. The current study supports Thomas and Blitt (1998) who noted that the absence of a socioeconomic adaptation strategy to EC and resource scarcity often exacerbates ethnic conflicts, migrations and insurgencies; and in directly impacts the international community.

According the definition of 'migration' by the National Geographic Society (NGS) in 2005, the study identified five types of migration in the region: external migration; internal migration; impelled migration; return migration; and seasonal migration.

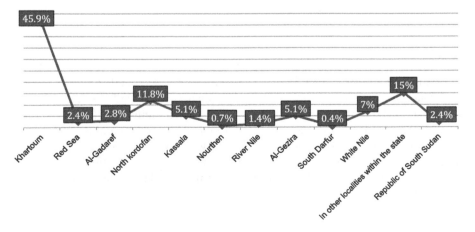

Fig. 10.2 Rate of migration and mobility

4.2 Impelled Migration

The conflict between the Sudanese Armed Forces (SAF) and the Sudan People's Liberation Movement-North (SPLM-N) has continued in parts of south Kordofan since June 2011, causing major destruction of assets, reduced access to farms for cultivation, damage to harvests, diminished agricultural labor opportunities, disruption to livelihoods, civilian displacement, and an uncertain number of deaths according to the Famine Early Warning Systems Network (FEWS NET) 2011.

The displacement and migration rates have increased. The Office for the Coordination of Humanitarian Affairs (UNOCHA) reported in 2013, more than 45,077 individuals were estimated to be severely affected and displaced to neighboring states (IFRC 2013). Unfortunately, this number was increased to be 116,000 in 2014 according to aid agencies and the SPLM-N (UNOCHA 2014). Meanwhile, 25,900 people in the same period have sought shelter in Government controlled areas, according to the Government of Sudan's Humanitarian Aid Commission (HAC), the International Organization for Migration (IOM) and other aid agencies (UNOCHA 2014). To date, an estimated 600,000–630,000 people in south Kordofan have been directly affected and/or displaced by the conflict since the start of the war in 2011(FEWS NET 2016). Disappointingly, the situation in the Nuba Mountains is worse since 42.0% of household's state they are currently displaced, 95% of displaced households fleeing due to fighting. The other causes of migration were 5.2% for lack of livelihood opportunities, poverty (2%) and drought (0.6%). These figures in 2004 were different since the causes of migration were mainly attributed to the lack of livelihood opportunities (54%), followed by drought (12.7%), poverty (10.4%), tribal conflict (8.9%) and search for modern life (14.4%). 31% of respondents said they had families living in refugee camps compared with 37% in 2013. Furthermore, 80% of households stated that they don't feel safe at home, compared to findings in 2013 (an increase of 13%).

4.3 Education

The education level in the study area is very low. Illiteracy between respondents is relatively high (34%), while the basic education (38%) and religious school (Khalwa) (13%) are main education levels attained by respondents. The number of educated peoples with intermediate, secondary, and university levels are low. This could directly be related to the non-existence or limited numbers of schools (55% of respondents), followed by the inability to pay for school fees which was given as the predominant reason (35%), insecurity (30%), in addition to the transportation problem for students living far from the villages where schools are located (10%).

4.4 Agricultural Situation

The land tenure system has been considered as a major underlying factor behind the use of natural resources. Rain-fed agriculture has mainly depended on the natural base of available land and natural water sources from rainfall. Generally, agriculture in the study area is labor-intensive, where 88.5% of the respondents (for both displaced and non-displaced respondents) are farmers, and provides the main livelihood source. Its seasonality (June to October) pushes households to seek other types of employment during the slack period.

Severe and prolonged droughts and loss of arable land due to war, desertification, and soil erosion are reducing agricultural yields and causing crop failure and loss of livestock, which endangers both pastoralists and rural peoples. The study indicated that, for the current season, fewer households were able to plant crops (12%), and the acreage they were able to plant was also smaller compared to 2 years ago (median acreage is one third of the size of 2 years ago). The household land under cultivation for the current season has dropped by 75% that significantly correlated with bombardment (0.044) with Correlation Coefficient (0.175*). Some of these areas, extremely damaged, and they are unable to cultivate and produce goods. In the Abkarshola unit, farm equipments and agricultural crops had been nearly destroyed. Actually, in 2014 most farmers on large-scale semi-mechanized farms (located in Sudanese government-controlled areas) were likely to avoid risks associated with farming in Nuba Mountains, due to the potential damage of crops and looting of cultivation equipment and/or the harvested crop itself by SPLM-N. In 2015, the late rains causing poor harvests in all districts of South Kordofan.

According to FEWS Net (2015), grain production of agricultural season 2015–2016 was estimated to be about 40 percent less than the average of 5 years. The poor crop production was aggravating in 2016 with an average household 2–3 bags of maize for example, compared with 15–25 bags in 2014. Crop production is predicted to be declined substantially for the season of 2017. This was expected also for horticultural products namely: Mango (*Mangiferaindica*), Limon (*Citrus aurantifolia*), Guava (*Psidiumguajava*), and Bananas (*Musa acuminata*), in addition to forest loss due to soil infertility.

4.5 Livestock Conditions

16.8% out of total respondents declared that they rely on livestock to improve their diet and food security. The current condition of livestock is very critical since 65.3% of respondents are poor and 30.5% are very poor. The study noted an increase in livestock deaths and prices: 72.1% of households are experiencing animal disease outbreaks and livestock deaths, with 56.8% having lost at least half their livestock. This critical period for livestock is due to high insecurity around the main seasonal grazing routes in border areas (85%), in addition to the severe pasture and water shortages (15%).

4.6 Extreme Poverty

Despite the potentiality of the study area, the poverty level of households is extremely high as shown in Fig. 10.3. Faki et al. (2011) have already mentioned that Kordofan is believed to be among leading states in poverty levels, compared with other states of Sudan. This situation is due to weak infrastructure, poor natural resource management, and high dependency on rain-fed agriculture for their livelihoods. Skoufias et al. (2011) and Deafalla et al. (2014) mentioned that the change in the environment might affect the path of poverty. The majority of responders are subsistence poor (89%) and highly dependent on natural resources for daily life and income generation. Additionally, there is a lack of sufficient financial and technical capacities to manage environmental risk, ability to adapt, as well as to access credit and safety nets.

On the other hand, the current research agrees on the significant overall negative impact of global environment change on livelihood, which may even exacerbate other real or potential situations such as discrimination or terrorism and poverty (Thomas and Blitt 1998). EC impedes poverty reduction; hence it is a key factor for poverty-reduction strategies such as the Sustainable Development Goals (SDGs) and UN's Millennium Development Goals (Lisk 2009; Deafalla 2012). In the study area, the wealth measure was done according to the number of assets owned (such as large food supply, animals, farms), as clearly shown in Fig. 10.3 below.

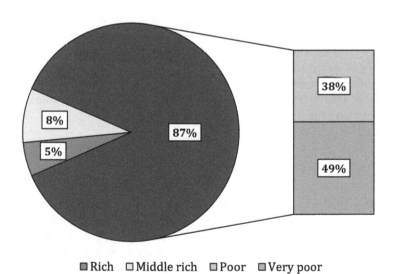

■Rich   □Middle rich   □Poor   ■Very poor

Fig. 10.3 Wealth status of households in 2014

4.7   Water

There is scarcity in freshwater sources, due to the unsuitability of the area for drilling boreholes and poor development of other water sources, which is causing tension between the local communities. On other hand, sanitation and hygiene indicators are very low in the area. 27% of households are using an improved water source as their primary source. Only 18% of the residents have sanitation facilities in their homes.

4.8   Health Statistics

The health situation of the population in the study area is very critical due to low access of medical service and medicines. The withdrawal of some NGOs such as Médecins Sans Frontières (MSF) from the area due to insecurity has seriously affected the access to primary health care for the local population. 49% of households reported at least one child had diarrhea in the preceding two weeks and 69% of them at least have one child with Malaria in the preceding four weeks.

4.9   Food Insecurity

Food insecurity is of greatest concern in Nuba Mountains of Sudan, where more than 250,000 people now face crisis to emergency levels of food insecurity (FEWS NET 2012). The South Kordofan and Blue Nile States Food Security Monitoring Unit (FSMU) reported about 242 people, including 24 children, had died in Kau-Nyaro and Werni areas of Nuba Mountains from lack of food and hunger-related illness in eight villages in the second half of 2015 (UNOCHA 2016). The food security situation has dramatically deteriorated in study site. 84% of households in 2014 surviving on one meal a day, compared to only 61, 5% in 2013, 57 and 5% in 2011. Furthermore, the food consumption score test was used to gauge both diversity and frequency of food consumption over a week. The results reveal that 79.2% of non-displaced household and 86.5% displaced household have unacceptable consumption levels (Fig. 10.4).

Additionally, households are consuming food in low feeding frequency with poor diversification. Food insecurity is expected to deteriorate through the scenario period of the upcoming months, as households exhaust coping strategies due to current restrictions on humanitarian access, no access to markets, trade flows, poor rainfall, limited food production and population movements. The net result of all these factors are extreme hunger during the "lean season" and perhaps even worse to come in 2017 if the situation continues under these circumstances.

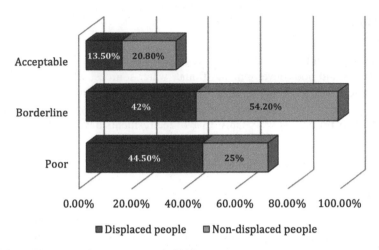

Fig. 10.4 Food consumption score status in 2014

4.10 Macroeconomic Situation

Industry and economy sectors are very sensitive to disturbances of EC, therefore there is an increasingly needs to mitigate those effects. The current research was designed to study the macroeconomic situation of Nuba Mountains in three axes as follow.

4.10.1 Decreased Levels of Income

95% of households report having less income than normal and 56% of responders have no income at all (Fig. 10.5). This effectively precludes the majority of households from the capacity to buy food.

4.10.2 Increase of Inflation

The continuous increase of inflation led to devaluation of the local currency. During the last years till now, and due to the impacts of conflict on oil production, the Sudanese government faces shortfalls in foreign currency to fund imports of essential commodities. In July 2014, the official inflation rate was 46.8%, and the unofficial black market exchange rate was over SDG 11 per EURO, more than double of the official Bank of Sudan exchange rate. Actually, the fall in the official exchange rate masks the true size of the fall in purchasing power as prices for locally produced, imported food and the non-food items have increased in local currency terms.

Concerning the high prices of these products, they are driven by the reduced supply to markets due to the poor production, as well as restricted market access in areas affected by conflict in south Kordofan. The study indicates that there is an

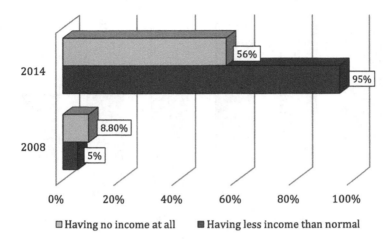

Fig. 10.5 Decreased levels of income in 2014

increase in all the commodity prices: i.e. nominal sorghum prices reportedly between 1–1, 5 EURO per malwa, an increase of over 100% higher than last year and 130% above the five-year average and about 75% above the reference year (2009/2010); and millet prices in June were 96% higher than last year and 150% above the five-year average. The availability in local markets of some commodities such as maize, sorghum, and wheat are very limited (Fig. 10.6).

4.11   Assessment of Land Use/Land Change (LU/LC)

Environmental degradation, which is pervasive in the study area, already poses direct threat for both human security and vegetation cover. The main forms of environmental degradation in the study area include: resource scarcity; ecosystem degradation; high rate of displacement and migration; and the destruction of natural forests.

In this work, the LU/LC classification scheme is generated representing six land features: i.e. agricultural lands; scattered forests; dense forests; shrubs; scattered trees; and bare land (Fig. 10.7). The history of the study area during the last decades shows that it has been exposed to a series of recurring dry years since 1982. This is very clear from the analysis of 1984 image (Fig. 10.7). The result of the case studies reveals that the area has been severely hit by recurrent droughts, the time and space variability of rainfall, increased temperatures and evaporation and sand encroachment.

On the other hand, the analysis of images 1994 and 2004 indicated that, there is expansion in the agricultural lands. Meanwhile, the investigation of the image in 2014 showed that there is a considerable recover due to the abandonment of agricultural lands during the war period.

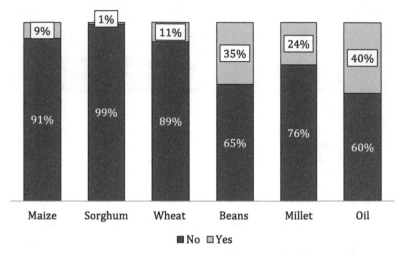

Fig. 10.6 Availability of commodities in 2014 local markets

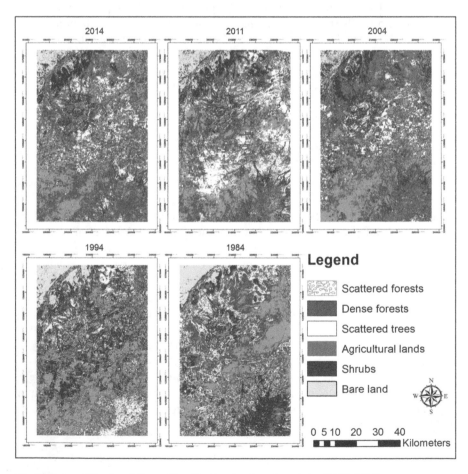

Fig. 10.7 Land Use/Land Cover Classification for the selected site (1984–2014)

Furthermore, to evaluate the results of conversions, PCD technique was applied to quantify and locate the changes. The result reveals that, the natural vegetation has been reforested and modified during the last years, were forest lands were increased from 15.3% in 1984 to 22.9% in 2014. Fig. 10.8 below shows the LU/LC change dynamics in the study area.

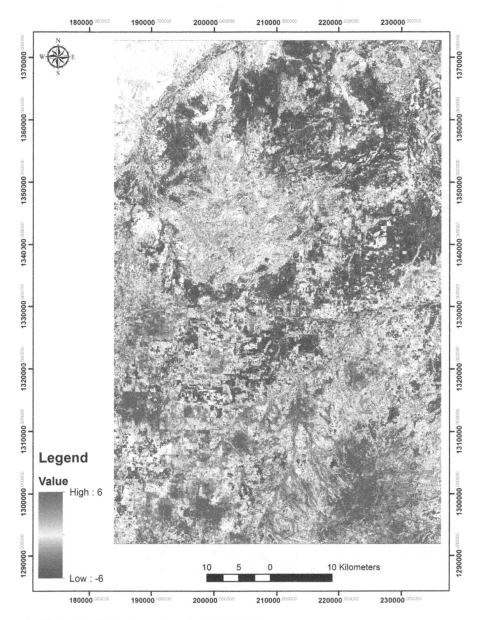

Fig. 10.8 The LU/LC change dynamics (1984–2014)

5   Conclusions

The current study underlines the need for mainstreaming environment into other relevant policy areas, such as poverty alleviation, health, and science. This will provide strategic directions for policy making to reduce vulnerability to EC and enhance human well-being. Indeed, more information exchange is needed to inform actors and decision makers regarding specific experiences, capacity gaps and knowledge to address the EC. Furthermore, new policy and strategies are required to focus on how to deal with consequences of longer term EC rather than with the impacts of sudden natural disasters. Within this perspective, implementing adaptation strategies effectively and at multiple scales may help avert environmental/climate change impacts which are triggers for conflict. However, adaptation strategies must take into account existing social, political, and economic tensions and avoid the exacerbation (Brown and Crawford 2009).

References

Bannon I, Collier P (eds) (2003) Natural resources and violent conflict: options and actions. World Bank, Washington, DC. ISBN 0-8213-5503-1, pp 20–366

Barnett J, Matthew RA, O'Brien KL (2008) Global environmental change and human security. In: Brauch HG, Oswald Spring Ú, Mesjasz C, Grin J, Dunay P, Behera NC, Chourou B, Kameri-Mbote P, Liotta PH (eds) Globalization and environmental challenges reconceptualizing security in the 21st century, vol 24. Springer ISBN 978-3-540-75977-5, pp 355–361

Barnett J, Matthew RA, O'Brien KL (2010) Global environmental change and human security. ISBN 978-0-262-01340-6, pp 3–32

Behnassi M, McGlade K (eds) (2017) Environmental change and human security in Africa and the Middle East. Springer, pp 29–353. https://doi.org/10.1007/978-3-319-45648-5

Brown O, Crawford A (2009) Climate change and security in Africa, a study for the Nordic-African foreign ministers meeting, published by the international institute for sustainable development, manitoba, Canada pp 1–20

Central Bureau of Statistics (CBS) (2009) Fifth sudan national population and housing census. Census tables. Available online at www.cbs.gov.sd. Accessed on 4 Jan 2011

Chuvieco E (1996) Fundamentos de teledetecciónespacial, 3rd edn. Ediciones RIALP, Madrid, pp 21–43

Collier P (2003) Natural resources, development and conflict: channels of causation and policy interventions. World Bank, Washington DC, pp 1–13

Deafalla THH (2012) Non Wood forest Products and Poverty alleviation in semi arid region. LAP LAMBART Academic Publishing GmbH and Co, KG, Germany ISBN: 978-3-659-16411-8, p 22

Deafalla THH, Csaplovics E, El-Abbas MM (2014) Analysis of environmental change dynamics in arid and semi arid climatic zones in International Journal of Performability Engineering, Sustainability: Balancing Technology Environment and Lifestyles. ISSN: 0973-1318, Vol. 10, No. 4:401–410 DIVA-GIS http://www.diva-gis.org/gdata

Ekbom A (2009) Africa Environment and Climate Change Policy Brief, Sida's regional Climate Change and Development-training workshop, May 5–7 2009, Nairobi Kenya. pp 1–27

El Tahir BA, Fadl KM, Fadlalmula AGD (2010) Forest biodiversity in kordofan region, Sudan: effects of climate change, pests, disease and human activities. Biodivers Conserv 11(3–4):34–44

Faki H, Nur ME, Abdelfattah A, Hassan AAA (2011) Poverty Assessment Northern Sudan ICARDA, Aleppo, Syria. ISBN: 92-9127-261-2, pp. vi + 62

FEWS NET (2011) Sudan Food Security Outlook, April to June 2011, Significant household-level food deficits likely due to poor harvest, conflict, and rising prices. http://www.fews.net/east-africa/sudan/food-security-outlook

FEWS NET (2012) Sudan Food Security Outlook, July to December 2012, Season to be impacted by conflict, fuel prices, and atypical livestock movements. http://www.fews.net/sites/default/files/documents/reports/Sudan_OL_2012_07_final.pdf

FEWS NET (2015) Sudan Food Security Outlook, Conflict, poor harvests to result in a deterioration of food security in South Kordofan by March. http://www.fews.net/east-africa/sudan/food-security-outlook-update/december-2015-0

FEWS NET (2016) Harvests to improve food security in most areas, but conflict likely to drive continued high needs. http://reliefweb.int/sites/reliefweb.int/files/resources/SD_OL_2016_10_final.pdf

Intergovernmental Panel on Global Environmental Change (IPEC) (2003) Discussion document for the expert think tank meeting. Oslo, Norway, pp 1–16

International Federation of Red Cross and Red Crescent Societies (IFRC) (2013) The Disaster Relief Emergency Fund (DREF) Sudan/North Kodofan and South Kordofan population Movement. http://www.ifrc.org/docs/Appeals/13/MDRSD017.pdf

Johnson DL, Ambrose SH, Bassett TJ, Bowen ML, Crummey DE, Isaacson JS, Johnson DN, Lamb P, Saul M, Winter-Nelson AE (1997) Meanings of environmental terms. J Environ Qual 26:581–589. https://doi.org/10.2134/jeq1997.00472425002600030002x

Lillesand TM, Kiefer RW, Chipman JW (2008) Remote sensing and image interpretation. Wiley, New York, pp 585–607

Lisk F (2009) The current climate change situation in Africa, CIGI special report: climate change in Africa: adaptation, mitigation and governance challenges, the centre for international governance innovation. pp 8–15

National Geographic Society (NGS) (2005) Human Migration Guide (3–5), Marcopolo Xpeditions, National Geographic Society. pp 1–3

Notaras M (2009) Does climate change cause conflict? United Nations University, Tokyo http://ourworld.unu.edu/en/does-climate-change-cause-conflict

O'Brien K, Barnett J (2013) Global environmental change and human security. Annu Rev Environ Resour J 38:373–391

Richards JA (1993) Remote sensing digital image analysis. Springer-Verlag, Berlin

Singh A (1989) Digital change detection techniques using remotely-sensed data. Int J Remote Sens 10:989–1003

Skoufias E, Rabassa M, Olivieri S, Brahmbhatt M (2011) The poverty impacts of climate change, the World Bank, poverty reduction and economic management (prem) network, Number 51, NE 010. Number 18

Thomas H, Blitt J (1998) Ecoviolence: links among environment, population and security. Rowman and Littlefield Publishers, New York

United Nations Development Programme (UNDP) (2006) Pastoral production systems in south Kordofan (study 2), Sudan pp 8–13

United Nations Environment Program (UNEP) (2007) The fourth global environment outlook–environment for development (GEO-4). ISBN: 978-92-807-2836-1(UNEP paperback). In: Vulnerability of people and the environment: challenges and opportunities, pp 33–342

United Nations Office for the Coordination of Humanitarian Affairs(UNOCHA) (2014) South Kordofan and Blue Nile: Population Movements Fact Sheet. http://reliefweb.int/sites/reliefweb.int/files/resources/South%20Kordofan%20and%20Blue%20Nile%20Population%20Movements%20Fact%20Sheet%20-%2019%20May%202014.pdf

UNOCHA (2016) Humanitarian Bulletin Sudan, Issue 09, pp 22–28

World Food Programme (WFP) (2008) Interagency Workshop Report WFP - FAO. Measures of Food Consumption - Harmonizing Methodologies, Rome, Italy

Chapter 11
Spatial Assessment of Environmental Change in Blue Nile Region of Sudan

Mustafa M. El-Abbas, Elmar Csaplovics, and Taisser H. H. Deafalla

Abstract Nowadays, innovative technologies are becoming progressively inter-linked with the issue of environmental change. They provide a systematized and objective strategy to document, understand and simulate the change process and its associated drivers. In this context, the main aim of this work is to develop spatial methodologies that can assess the environmental change dynamics and its associated drivers. To achieve this objective, optical multispectral satellite imagery, integrated with field survey data, were used for the analyses. Object Based Image Analysis (OBIA) was applied to assess the change dynamics within the period 1990 to 2014. Broadly, the above-mentioned analyses include: Object Based (OB) classifications; post-classification change detection; data fusion; information extraction; and spatial analysis. The dynamic changes were quantified and spatially located as well as the spatial and contextual relations from adjacent areas were analyzed. The study concludes with a brief assessment of an 'oriented' framework, focused on the alarming areas where serious dynamics are located and where urgent plans and interventions are most critical, guided with potential solutions based on the identified driving forces.

Keywords Environmental change · Land-use policy · Remote sensing · Spatial analysis · Dry land

M. M. El-Abbas (✉)
Researcher Professor, Faculty of Environmental Sciences, TU Dresden, Dresden, Germany

Faculty of Forestry, University of Khartoum, Khartoum, Sudan

E. Csaplovics
Researcher Professor, Faculty of Environmental Sciences, TU Dresden, Dresden, Germany
e-mail: elmar.csaplovics@tu-dresden.de

T. H. H. Deafalla
Faculty of Forestry, University of Khartoum, Khartoum, Sudan

© Springer International Publishing AG, part of Springer Nature 2019
M. Behnassi et al. (eds.), *Human and Environmental Security in the Era of Global Risks*, https://doi.org/10.1007/978-3-319-92828-9_11

1   Introduction

The international concern with natural habitat changes is motivated not only by the irretrievable imbalance in the natural environment, but also from the perspective that it is a destructive process in which the economic feasibility is lower than the environmental losses. Environmental change is a complex ecological and socioeconomic process caused by a number of anthropogenic and climatic factors. Nowadays, global land cover is altered principally by humans and their direct activities – such as agricultural expansion, fuel wood collection, forest harvesting, mismanagement, urban and suburban construction and development (Robinson et al. 2011). There are several challenges facing natural resources globally, and in dry land in particular, which impose the need for well-designed information systems and management plans. One of these challenges is Land Use/Land Cover (LU/LC) changes, mainly deforestation and land degradation. Unsustainable wood harvesting and uncontrolled expansion of mechanized rain-fed agriculture inside forestlands, as well as a growing pressure on lands used for shifting cultivation, have been proposed as an approximate cause which led to large-scale deforestation and land degradation (El-Abbas and Csaplovics 2012).

The changes in natural resources have gained attention as a result of the potential effects on desertification, erosion, increased runoff and flooding, increasing CO_2 emissions, climatological effects and biodiversity loss, in addition to other indirect impact such as migration and poverty (Mas 1999; Williams 2006). Moreover, the identification of these driving forces, which is indispensable to understanding ongoing LU/LC processes, constitutes one of the dominant challenges in international LU/LC changes research and in the formulation of regional land/forest management policies.

Several challenges are facing the forest sector in Sudan, which impose the need for well-designed information systems and management plans. One of these challenges is changes in LU/LC, particularly due to deforestation and land degradation. In spite of the fact that LU/LC and its dynamics serve as one of the major input criteria for sustainable development programs, currently, unregistered land covers almost about 85–90% of the land area in Sudan. Meanwhile, the remaining part is a private ownership, which is restricted to the rights offered before the implementation of the Unregistered Land Act in 1970. Before that time, the majority of the area was forest and grassland, which used communally for pasture and traditional farming under customary land laws (Agrawal 2007). In this form of use, the right to cultivate an area of newly opened land became vested to the farmer who cleared it for use. This condition no longer exists, especially when the government disregarded the lands used by customary laws, which were unregistered lands, forest and waste lands managed by government.

Subsequently, large areas were leased to individuals to be used for mechanized rain-fed agriculture, while majorities, who used the land in the past for traditional farming, were not allowed to use the land. Since then, most of the natural land cover was dramatically destroyed as the issued act greatly influenced the exploitation of the land during the past decades (Elsiddig 2004). Accordingly, a clear, accurate, cost-effective, and up-to-date knowledge about LU/LC patterns, in terms of their distribution, magnitude, and changing proportions, is highly demanded by legislators, planners, and local to national officials for the objective to elaborate better land-use policy.

Remotely sensed data are inherently suited to provide an accurate, cost-effective and up-to-date source of information about LU/LC and its dynamics. However, the paucity of information on LU/LC is particularly related to the scarcity of efficient extraction methods (Foody 2002; Estes and Mooneyhan1994). From the above-mentioned perspective, it appears that an innovative approach for the interpretation of increasing availability of advanced remotely sensed data to provide valuable spatial information on change (Lu et al. 2004; Rogan and Chen 2004), and ultimately, this information into usable knowledge, is the highest priority for remote sensing applications (Franklin 2001).

In addition to the spectral properties of image pixels, the spatial extent of the feature under investigation is of great importance to the classification scheme (Flanders et al. 2003; Hay and Castilla, 2006; Platt and Rapoza 2008). A typical application observed in the study area involves extraction of features that are encompassing of multiple pixels from one hand (e.g. agricultural fields and Blue-Nile valley), and mixed with various entities from the other hand (e.g. residential areas and scattered forests). These features require classification of aggregative homogenous pixels that make up a feature. This is often defined as a segmentation process which is applied just before the classification (Martin et al. 2001; Baatz et al. 2004). Therefore, this study utilized the object-based paradigm in order to group pixels of analogous spectral and spatial response, based on predefined criteria to extract features of interest.

It is not an easy task to understand the complex structure of natural systems. This structure, described as a 'mosaic of patches' (Hay and Marceau 1999), requires a conceptual model identified by complex system theory. Complex system theory is a way to explain that the interactions in ecological systems are composed of large numbers of components which interact in a non-linear way and exhibit adaptive spatio-temporal properties (Hay et al. 2003). These components of complex systems are linked to each other in a hierarchical way in that they are built of sub-objects and super-objects (e.g. trees, compartment, stand, forest, vegetation type, etc.) (Hay and Marceau 1999). In the perspective of a remote sensing as a primary data source to fully understand, monitor, model and manage the various interactions between these components, three elements are needed: 1) remote sensing data which contain a sufficient detail to identify surface features in hierarchical manner; 2) methods and theories which provide a capability to discover pattern components, real-word features at their respective scale of representation (Woodcock and Strahler 1987); and 3) the potentiality to connect these features in appropriate hierarchical structure (Hay et al. 2003).

Meanwhile, the development of the applications and interpretation approaches should follow the acceleration of advanced technologies and data available. A variety of studies have addressed the post-classification comparison presented the advantage of describing the nature of the changes (Mas 1999; Yuan et al. 2005; El-Abbas and Csaplovics 2012). In this work, change objects derived from the post-classification comparison were spatially analyzed to address the dynamic change process and the proximate causes.

In this context, the main goal of this work, conducted in the Blue Nile region of Sudan, in which most of the natural habitats were dramatically destroyed, was to develop spatial methodology to assess the dynamic change and its associated factors during the period 1990 to 2014. In different words, spatial analysis of LU/LC change can be aimed at understanding the spatial pattern of changes, tackling the question 'where do LU/LC changes occur?' as well as the rate of change, addressing the question 'at what rate are LU/LC changes likely to progress?' These questions have been referred to as the *location* issue versus the *quantity* issue (Williams 2006), which it has been addressed in this work towards the better understanding of the proximate causes of LU/LC changes.

2   Methods

2.1   Study Site

The rural areas of Sudan, as well as much of its urban areas, rely on forests. Blue Nile region constitutes an important area of forest resources in the Sudan, where the forests play significant roles in the economy through provision of a variety of goods and services. They are main source of sawn timber, fuel wood, charcoal, and construction materials from local to national levels, in addition to many other non-wood forest services such as providing a main source of income and livelihood for the rural poor people (Deafalla 2012).

Sennar state is one of the economically important states of the Republic of Sudan. It has an area of 1.084.600 hectares, an estimated population of 1,270,504 capita according to the last census in 2008, and is growing at the rate of 2.6% per annum. The main economic activity is cultivation within the irrigated schemes besides seasonal rain-fed agriculture. The state is located in the southeast of Sudan, which shares its borders with the Gezira state in the North, White Nile and south Sudan in the west, Gadarif state in the east and Blue Nile state in the south. Selected state, shown in Fig. 11.1, mainly consists of: forests, horticultural land and settlements which lie near the bank of the Blue Nile River. Most of the area is grassland and rain-fed agriculture which was dominated by natural forests till a few decades ago (Estes and Mooneyhan 1994). In spite of the fact that dry-land forests are not well-known for their export-oriented timber production, Sudan is a major producer of Gum Arabic for decades, producing more than 80% of the world market (Abdelgalil 2005). Moreover, forests provide a varied range of environmental services (e.g. protection against desert creep, agricultural land deterioration, protection of the rivers and their tributaries against erosion, soil amelioration, wildlife

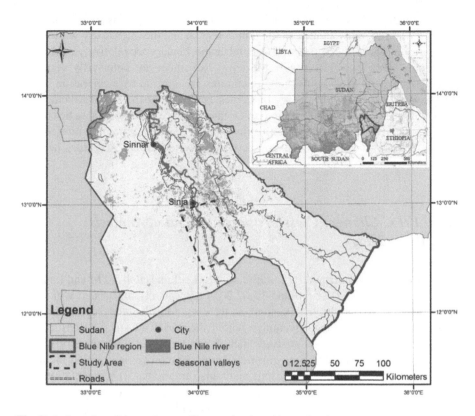

Fig. 11.1 Location of the study area (Source: developed by authors)

sanctuaries, etc.), and are as well important for cultural and religious heritage and traditional practices.

The vegetation cover within the study area mainly consists of:

- Bushland, mainly *Acacia nubica* and *Acacia mellifera* ranging from 2 to 4 m height generally open associated with woody species separated by stretches of short grasses. Wood land ranging from light and open *Acacia Senegal* in the north to *Acacia seyal*, *A. fistula* and *Balanites aegyptiaca*, woodland with areas of tall annual open grassland which has been occurring broadly to the south of the state.

- Open grassland where occurs in wide shallow depressions grey clay soil. *Acacia nilotica* forest. Single storey pure stands of *Acacia nilotica* about 15–20 m high, lie in seasonally flooded basins along the bank of the river and are sometimes found on similar flooded areas, such as drainage channels (*Khours*) and shallow surface catchment areas (*Hafirs*), on the clay plain inland from the Blue Nile. These forests, many of which are managed or regulated by the Forest National Corporation (FNC) for conservation objectives, are as well of considerable economic importance as they provide hard wood round logs and sawn timbers.

The climate of the region is tropical and continental. The year is sharply divided into a very humid, rainy summer and autumn season and the intense dry winter and spring season. The winter months of December and January are relatively cold, with the average of 16°c to 35°c, while March to November are potentially very hot (20°c- 41°c) except in so far as the temperatures are reduced by evaporation in the rainy season. The rainfall in the region comes as a result of the South Atlantic and Congo air masses, with little or no Indian Ocean influence. The study area lies in the zone in which rainfall increases to the south-east. The annual precipitation varies between 300 mm to 500 mm occurring between June to October, and much heavily in August.

2.2   LU/LC Change: Classification and Spatial Analysis Methodology

A mosaic of optical multispectral satellite scenes (i.e. Landsat 4–5 TM and Landsat 8 OLI), acquired in 04.01.1990 and 15.02.2014 respectively, integrated with field survey in addition to multiple data sources, were used for the analyses. OBIA was applied to assess the change dynamics within the period of study (1990–2014). Broadly, the above-mentioned analyses include Object Based (OB) classifications, that can be summarized into four phases: (i) image preprocessing which includes georeferencing, resampling, corrections and transformations; (ii) segmentation and creation of the substantial homogenous units for the classification; (iii) rule sets development based on the expert user's knowledge and intensive field visit to collect training data; and (iv) validation and assessment of the results. Subsequently, post-classification change detection was applied to generate categorical change classes (El-Abbas and Csaplovics 2012), while the focused work in this research includes data fusion, information extraction, and spatial analysis.

Hierarchical multi-scale segmentation thresholds were applied and each class was delimited with semantic meanings by a set of rules associated with membership functions. Consequently, the fused multi-temporal data were introduced to create detailed objects of change classes from the input LU/LC classes. The classification produced a total of nine LU/LC classes from each, i.e. agriculture, bare-land, crop-land, dense-forest, grassland, orchard, scattered-forest, settlements, and water body. The method adopted in this research uses cross operation of multi-temporal classi-fied maps and subsequent reclassification of the overlaid maps to generate the change objects. New layers of segments were created representing the change areas (from-to) as well as the overlapped areas (no change) of the classified maps. In order to achieve a comprehensive description of the LU/LC changes during the period of study, the post-classification change analysis was applied based on OBIA approach. The images were fused and the segmentation result generates a new level of objects that differ from those in the original images. However, the intersections

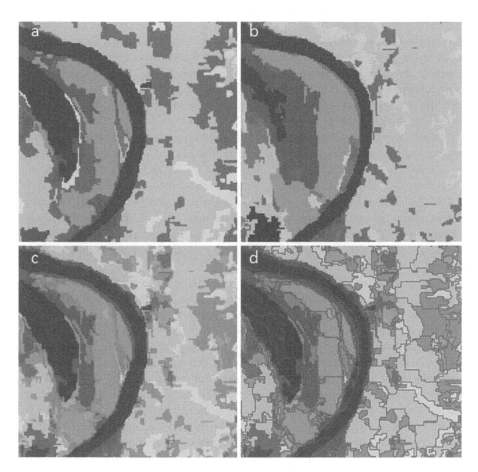

Fig. 11.2 Post-classification change detection procedure: (**a**) and (**b**) illustrate a selected sample of the classified maps used as input layers, (**c**) fused layers, and (**d**) the segmentation result of the fused images

between the change 'from-to' information classes were automatically delineated with controlled scale and shape parameters (Fig. 11.2).

Important aspects which must be considered in change detection analysis techniques are detecting the changes that have occurred, identifying the nature of changes and measuring the spatial extent of the change (Macleod and Congalton 1998). Thematic maps describing LU/LC changes 'from-to' information and its dynamics in the study area were obtained for the periods 1990–2014, by means of the OB post-classification change detection approach. This change map not only shows where LU/LC changes occurred, but also illustrate the spatial location of the changes between each of the nine LU/LC classes. Furthermore, associated matrices quantify the magnitude and shift 'from-to' each class category, which were derived from the fused t_0 and t_1 classified maps.

For the purpose of change detection with OB approach, the adopted novel procedure applied in this study is the fusion of classified thematic maps for t_0 and t_1 as input layers in OBIA project when performing the segmentation. Following the available classified maps generated for the study area for the years 1990 and 2014, segmentation was performed for the fused maps (i.e., 1990 and 2014). The images were fused to generate a new level of objects differing from those in the original images, as shown in a Fig. 11.2.

Several datasets were introduced for this research work. However, in addition to the earth observation data as a main source of data used in this research, Global Positioning System (GPS) to collect field check points and ground control points were also used. The field survey data were collected during the field visit in April 2010, which coincided with the season of acquisition in order to eliminate the variation in natural phenological status and allow for same atmospheric and environmental conditions. The collected data are used for various purposes, including Training and Test Area (TTA) masks that were created as a reliable guidance in the analytical process of social survey of household.

In the present research, stratified random sampling areas were used among the deforested areas to understand the proximate causes of change in which 30 segments were allocated to examine the spatial relations from the adjacent classes. The layout of all sampled areas were the segments derived from the post-classification change analysis based on scale parameters and homogeneity criteria. Subsequently, the spatial relation features were calculated and extracted utilizing *eCognition* software. Remotely sensed data and GIS provide numerous spatially explicit abiotic and anthropogenic variables that might be considered as underlying forces, which can be used to test specific hypotheses about the vegetation dynamics. Prior to model and simulate these explanatory variables, this work examined the correlation among the change classes which provide a framework to extrapolate the proximate causes of deforestation based on spatial relations between the deforested areas and the main LU/LC change classes.

Aggregated to the community-level, social survey of household data provides a comprehensive perspective in addition to the earth observation data, investigating predetermined hot spot degraded and successfully recovered areas. Hence, the study utilized a social survey of household through a well-designed questionnaire based on multi-choice and open ended questions to address the factors affecting LU/LC dynamics based on local community's perception. Population Proportional to Size (PPS) sampling technique was applied to collect 120 questionnaires distributed in 6 villages in the study area. The interviews were carried out through personal contact, which was enumerated by the author himself, and the interviews considered the local community language, which was conducted in Arabic language. Data collected from interviews were coded, interred and analyzed using the SPSS software. And the descriptive statistical analysis was applied to explain the local community perceptions about different aspects of changes dynamics.

3   Results and Discussion

The spatially explicit Pearson's correlation coefficient equation estimates the significance, nature, and magnitude of influence of the spatial relation features on deforestation. Table 11.1 summarizes the association between the randomly selected sample areas of deforestation and the LU/LC change classes, i.e. agricultural fields, urbanization activities, water body, and reforested areas as well as the no-change areas.

As hypothesized, spatial relation feature (the distances to the deforested areas) is associated with some of the land-use activities in the region; specifically, mechanized rain-fed agriculture was observed as a higher feature that is positively correlated with deforestation during the time period examined (0.76). The result is also coinciding with the fieldwork which indicates that the occupancy of the mechanized rain-fed agriculture mainly was expanded over the natural forest land (El-Abbas and Csaplovics 2012). The dynamics observed in the region was greatly attributed to the policy and land ownership system since unregistered land covers almost about 85–90% of the land area. Whereas, the majority of that area was forest and grassland, which used communally for pasture and traditional farming under customary land laws (Agrawal 2007). This condition no longer exists as the government disregarded the land used by customary laws. Consequently, large areas were leased to individuals for mechanized farming. Since then, most of the natural habitats were dramatically destroyed as the issued act greatly influenced the exploitation of the land during the past decades (Fig. 11.3).

Although forest areas were drastically decreased, as clearly observed in Fig. 11.4 below, nevertheless the potentiality of the area to recover is obviously shown in the abandon agricultural field, where the reforested areas were clearly associated with the agricultural field as indicated by a positive correlation shown in the Table 11.1 (0.56). In fact, there is no correlation observed between the residential areas and the

Table 11.1 Results of the associations between the change in forests and the other LU/LC classes (Source: Computed by authors)

Class	Deforestation	Reforestation	No-change	Agriculture	Water	Urbanization
Deforestation	1.00	0.33 0.04	−0.12 0.26	0.76[a] 0.00	−0.16 0.20	−0.35[b] 0.03
Reforestation	–	1.00	−0.02 0.46	0.56[a] 0.00	−0.20 0.15	0.22 0.12
No-change	–	–	1.00	−0.15 0.22	0.69[a] 0.00	−0.16 0.20
Agriculture	–	–	–	1.00	−0.04	0.11 0.29
Water	–	–	–	–	1.00	0.46[a] 0.01
Urbanization	–	–	–	–	–	1.00

[a]Correlation is significant at the 0.01 level (2-tailed)
[b]Correlation is significant at the 0.05 level (2-tailed)

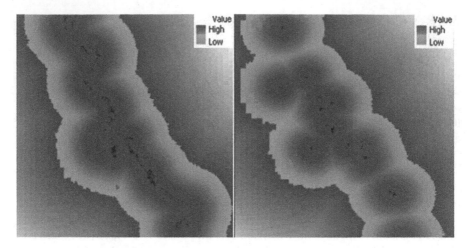

Fig. 11.3 Spatial relation feature (distance to) of the water body on the left and the settlements on the right in relation to neighborhood objects of other LU/LC change classes (Source: Computed by authors)

reforested areas, meanwhile a negative correlation is shown between the urbanization activities and deforestation (−0.35). This negative correlation can be attributed to the contribution of local communities in conservation of natural habitats. In different words, forests are in the far distance to urbanization activities and existing settlements, or those with lower frequency of residential area objects in their neighborhoods are at greatest risk of deforestation.

For the LU/LC change analysis, the dark-orange areas in the map (Fig. 11.4) reveal an intensive clearance of natural forest cover, and as shown in Table 11.2, about one third from the total area were converted from forest to other LU/LC class categories. The occupancy of the mechanized rain-fed agriculture expanded over the natural forest land represents 23636.19 ha, which is 35.53% of the total gained areas obtained from the forest land. The recovered area represents only 7.88%, mainly from abandon agricultural field (5.10%). Therefore, evidence is clear for the potentiality of the area to recover if manmade impact is well-controlled. It is worth mentioning that an area which constitutes 8.35% was converted from Scattered-forest to grassland. This change could be explained by the practice of shifting cultivation in forest land. This issue was supported by the coincided statistics of the areas that were converted from agriculture to grassland as well as the areas that were re-utilized from grassland (Table 11.2).

Interestingly, most of the no-change objects were observed near the Blue Nile valley, representing an area of about 38.93%, rain-fed agriculture occupied an area of about 29.20% of this portion (Table 11.2). Meanwhile, the deforestation increased perpendicularly as the distance increased along with the valley (Figs. 11.3 and 11.4). For instance, the highest value of commercial *Acacia nilotica* pure stand forests, where located in form of strips alternately along with the banks of the Blue Nile valley, were highly protected and well managed under clear property rights and

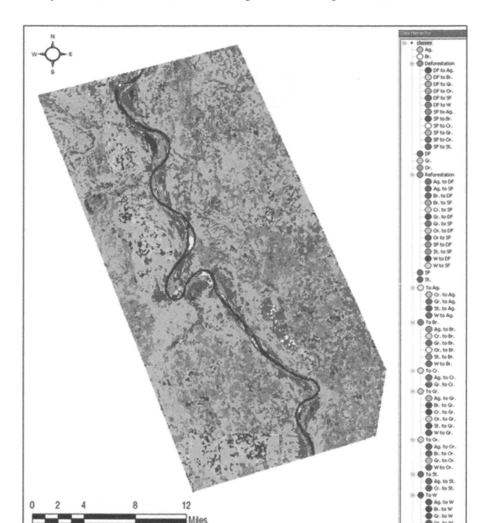

Fig. 11.4 Change map "from-to" classes generated from the overlaid classified images 1990 and 2014 (Source: Developed by authors)

ownership conditions, while the forests with lower commercial value (government owned) were ignored which led to high rates of deforestation in that area. Within this context, high associations were observed between the no-change areas and water body as well as urbanization activities and water body, as shown in the Table 11.1 by the values 0.69 and 0.46 respectively. Therefore, land tenure could play an indispensable role for the stability of these areas (Rogan and Chen 2004).

Aggregated to the community-level, social survey of household data provides a comprehensive perspective additionally to earth observation data. Thus, geospatial determination of hot spots degraded and successfully recovered areas were

Table 11.2 LU/LC change matrix of the study area (ha). (Source: Computed by authors)

Class	1	2	3	4	5	6	7	8	9	Total Area 2009
1-Agriculture	32649.75	–	5825.88	361.86	4199.56	–	23274.33	113.35	96.95	66521.68
	29.20%		5.21%	0.32%	3.76%		20.81%	0.10%	0.09%	59.49%
2-Bare-land	1325.64	111.73	217.97	119.48	237.61	98.94	1282.93	36.22	398.36	3828.88
	1.19%	0.10%	0.19%	0.11%	0.21%	0.09%	1.15%	0.03%	0.36%	3.43%
3-Crop-land	609.39	–	–	–	8.11	–	240.58	–	–	858.08
	0.54%				0.01%		0.22%			0.77%
4-Dense-forest	33.97	2.16	–	1445.03	72.99	202.20	283.29	–	178.23	2217.87
	0.03%	0.002%		1.29%	0.07%	0.18%	0.25%		0.16%	1.98%
5-Grassland	6930.85	61.18	1473.41	197.24	1465.93	119.93	9336.76	12.80	24.87	19622.97
	6.20%	0.05%	1.32%	0.18%	1.31%	0.11%	8.35%	0.01%	0.02%	17.55%
6-Orchard	129.84	26.22	–	497.83	10.99	865.47	389.17	–	162.64	2082.16
	0.12%	0.02%		0.44%	0.01%	0.77%	0.35%		0.15%	1.86%
7-Scattered-forest	5704.69	36.94	1069.47	614.70	853.75	74.34	5104.49	45.95	257.07	13761.4
	5.10%	0.03%	0.96%	0.55%	0.76%	0.07%	4.56%	0.04%	0.23%	12.30%
8-Settlements	554.69	–	50.19	–	–	–	68.93	220.85	–	894.66
	0.50%		0.04%				0.06%	0.20%		0.80%
9-Water	96.41	24.06	–	201.30	11.35	27.93	–	–	1678.04	2039.09
	0.09%	0.02%		0.18%	0.01%	0.02%			1.50%	1.82%
Total Area	48035.23	262.29	8636.92	3437.44	6860.29	1388.81	39980.48	429.17	2796.16	111826.8
1990	42.97%	0.23%	7.72%	3.07%	6.13%	1.24%	35.75%	0.38%	2.51%	100%

Table 11.3 Local community's perception about the causes of forest cover loss. (Source: Computed by authors)

Factors	Number of responses	Total respondents (%) (N = 117)
Agricultural expansion	113	97
Harvesting of woody products	88	75
Climatic factors	34	29
Manmade fire	14	12

Table 11.4 Local community's perception about the mission that it may stimulate the recovery process. (Source: Computed by authors)

Mission	Number of responses	Total respondents (%) (N = 108)
Land tenure	86	80
Afforestation and reforestation	69	64
Job opportunities	77	71
Awareness programs	14	13

investigated. The study utilized a well-designed questionnaire to address the factors affecting land-cover dynamics and the possible solutions based on local community's perception.

Table 11.3 illustrates community perception of the underlying forces of forest cover loss. From the analysis of this result, land clearance for mechanized farming seems to be the major cause of deforestation in the study area, which was indicated by 97% of the total respondents. Cultivation expansion leads to removal of natural woody vegetation and this conversion on land leads to deforestation. Accordingly, vast areas were converted to agricultural land under intensive cultivation. Moreover, as a result of high profitability gained from selling woody products from a new opened land, high percentage of respondents indicated that the harvesting of woody products is considered as the second force of deforestation. Many authors have agreed on those subjects, e.g. Elmoula (1985) and Abdallah (1991) noted that the illegal cutting for energy purpose (fuel wood, charcoal making), as well as excessive clearance for mechanized rain-fed agriculture, has decimated the forest resources in the study area. Meanwhile, as shown in Table 11.3, 29% of the respondents indicate that the climatic factor was responsible for this deterioration.

Before the 1970s, majority of the area was forest and grassland, which was used communally for pasture and traditional farming under customary land laws (Agrawal 2007). In this form of use, the right to cultivate an area of newly opened land became vested in the farmer who cleared it for use. Since 1980 the government disregarded the land used by customary laws and then the unregistered lands became government owned. Subsequently, vast areas were leased to individuals to be used for mechanized rain-fed agriculture, while majority of the local farmer are not allowed to use the land. This might be one of the main causes that lead to illegal degradation. As shown in Table 11.4, 80% of the respondents declared that land ownership system is considered as a prime factor of forest cover loss. The difference between the two systems is that, by customary law, the farmers maintain the land for sustainable use by shifting cultivation, which gives chance for the forest to recover, while the leaser looks for high profitability ratio with intensive cultivation. As a result, the seed bank becomes bankrupted and the ability for the natural forest to recover may be lost forever (Miller 1999).

4 Conclusions

This approach may be applied and expanded to model and simulate the dominant socioeconomic and socio-ecological drivers of such changes, which include population pressures, transmigration, land tenure, levels of poverty occupation and educational level, meteorological information, elevation, ease accessibility and infrastructure (Secrett 1986; Dennis and Colfer 2006). Therefore, earth observation data aggregated to the GIS layers of explanatory variables might provide a comprehensive perspective for and insight into change dynamics and their driving forces as well as to provide possible solutions to restore exposed areas. Identifying the LU/LC rates and distribution, as well as a clear identification of its proximate drivers should comprise the first step toward the analyzing of environmental change. The integrated analyses of post-classification change detection and spatial analysis, reveal that the crucial human intervention was found to be the main cause of changes, quantitatively as well as qualitatively. More precisely, land clearance for mechanized farming is considered as the direct cause of LU/LC changes, as fairly confirmed by this work. In contrast, the local community was found to be negatively correlated to the deforested areas, which means they contribute somehow in the conservation of natural resources.

Meanwhile, the improper policy and legislation of land management system was the major underlying cause of change, since most of the stable areas on the change maps were observed where the land managed under clear property rights and ownership system (i.e. *Acacia nilotica* pure stand forests, orchards, and settlements). During the period covered by this study, there has been an increase in the magnitude of natural habitat loss. This loss is a function of the change in policy and human activities that led to more pressure on the natural resources in the region. The practices of the mechanized agriculture, as a main economic activity in the region, has very significant impacts on natural ecosystem degradation. We expect that the remote sensing and the spatial analysis adopted during this research will prove to be a very valuable tool for policy and decision-making processes in the region based on the valuable information that can provide.

References

Abdallah H (1991) Energy potential from economically available crop residues in the Sudan. Oxford 16(8):1153–1156

Abdelgalil EA (2005) Deforestation in the dry lands of Africa: quantitative modelling approach. Environ Dev Sustain 6:415–427

Agrawal A (2007) Forests, governance, and sustainability: common property theory and its contributions. Int J Commons 1(1):111–136

Baatz M, Benz U, Dehghani S, Heynen M, Oltje AH, Hofmann P, Lingenfelder I (2004) *eCognition* professional userguide, version 4.0. Definiens Imaging GmbH Munchen, Germany: Definiens

Dennis RA, Colfer CP (2006) Impacts of land use and fire on the loss and degradation of lowland forest between 1983–2000 in East Kutai District, East Kalimantan. Singap J Trop Geogr 27(1):39–62

El-Abbas MM, Csaplovics E (2012) Spatiotemporal object-based image analyses in the Blue Nile region using optical multispectral imagery. Proceedings of the SPIE, Vol 8538, doi:https://doi.org/10.1117/12.974546

Elmoula A (1985) On the problem of resource management in the Sudan. Environmental monograph series no. 4. Institute of Environmental Studies, University of Khartoum, Sudan, 131

Elsiddig EA (2004) Community based natural resource management in Sudan, In: Awimbo J, Barrow E, Karaba M (2004) Community Based Natural Resource Management in the IGAD region. IGAD; IUCN

Estes JE, Mooneyhan DW (1994) Of maps and myths. Photogramm Eng Remote Sens 60:517–524

Foody GM (2002) Status of land cover classification accuracy assessment. Remote Sens Environ 80:185–201

Flanders D, Mryka H, Joan P (2003) Preliminary evaluation of *eCognition* object based software for cut block delineation and feature extraction. Can J Remote Sens 20:441–452

Franklin SE (2001) Remote sensing for sustainable Forest management. CRC press LLC, p 448

Hay GJ, Marceau DJ (1999) Remote sensing contributions to the scale issue. Can J Remote Sens 25(4):357–366

Hay GJ, Blaschke T, Marceau DJ, Bouchard A (2003) A comparison of three image object methods for multi-scale analysis of landscape structure. J Photogramm Remote Sens 57:327–345

Hay GJ, Castilla G (2006) Object-based Image analysis: Strengths, Weaknesses, Opportunities and Threats (SWOT), 1st International Conference on Object-based Image Analysis (OBIA 2006), Salzburg University, Austria

Lu D, Mausel P, Brondizio E, Moran E (2004) Change detection techniques. Int J Remote Sens 25(12):2365–2407

Macleod D, Congalton G (1998) A quantitative comparison of change-detection algorithms for monitoring eelgrass from remotely sensed data. Photogramm Eng Remote Sens 64:207–216

Martin D, Fowlkes C, Tal D, Malik J (2001) A database of human segmented natural images and its application to evaluating segmentation algorithms and measuring ecological statistics. In: Proc. 8th Int'l Conf. Computer vision, vol 2, pp 416–423

Mas JF (1999) Monitoring land-cover changes: a comparison of change detection techniques. Int J Remote Sens 20(1):139–152

Miller M (1999) Effect of deforestation on seed bank in a tropical deciduous forest of western Mexico. Trop Ecol 15(2):179–188

Platt RV, Rapoza L (2008) An evaluation of an object-oriented paradigm for land use/land cover classification. Prof Geogr 60:87–100

Robinson BE, Holland MB, Naughton-Treves L (2011) Does secure land tenure save forests? A review of the relationship between land tenure and tropical deforestation, CCAFS Working Paper no. 7. CGIAR ResearchProgram on Climate Change, Agriculture and Food Security (CCAFS). Copenhagen, Denmark

Rogan J, Chen DM (2004) Remote sensing technology for mapping and monitoring land-cover and land-use change. Prog Plan 61(4):301–325

Secrett C (1986) The environmental impact of transmigration. Ecologist 16:77–88
Williams M (2006) Deforesting the earth: from prehistory to global crisis, an abridgment. University of Chicago Press, Chicago, p 543
Woodcock CE, Strahler AH (1987) The factor of scale in remote sensing. Remote Sens Environ 21(3):311–332
Yuan F, Sawaya KE, Loeffelholz BC, Bauer ME (2005) Land cover classification and change analysis of the Twin cities (Minnesota) metropolitan areas by multitemporal Landsat remote sensing. Remote Sens Environ 98:317–328

Chapter 12
Vulnerability to Climate Change and Adaptive Capacity of Social-Ecological Systems in Kenitra and Talmest, North and Central Morocco

Rachida El Morabet, Mohamed Behnassi, Mostafa Ouadrim, Mohamed Aneflouss, Said Mouak, and Zhar Essaid

Abstract Climate change alters the conditions of communities and regions worldwide, thus becoming a global concern. In Morocco, which no exception, many communities and regions are currently suffering from climatic extreme events such as high temperatures, severe storms, floods, and irregular rainfall patterns. Among the devastating impacts of climate change, floods are causing losses of lives and property in both Moroccan rural and urban areas. This problem is also exacerbated by unsustainable development practices and poor landuse planning in the country. This chapter analyses the vulnerability of social-ecological systems to climate change in two regions in Morocco: Kenitra (Sebou watershed); and Talmest (Tensift watershed).

R. El Morabet (✉)
Department of Geography, Research Laboratory on Dynamics of Space and the Society (LADES), Faculty of Arts and Human Sciences, Hassan II University of Mohammedia; Researcher at Center for Research on Environment, Human Security and Governance (CERES), Rabat, Morocco

M. Behnassi
Faculty of Law, Economics and Social Sciences, Center for Research on Environment, Human Security and Governance (CERES), Ibn Zohr University, Agadir, Morocco
e-mail: behnassi@gmail.com; m.behnassi@uiz.ac.ma

M. Ouadrim · M. Aneflouss · S. Mouak · Z. Essaid
Department of Geography, Research Laboratory on Dynamics of Space and the Society (LADES), Faculty of Arts and Human Sciences, Hassan II University of Mohammedia, Rabat, Morocco

© Springer International Publishing AG, part of Springer Nature 2019   221
M. Behnassi et al. (eds.), *Human and Environmental Security in the Era of Global Risks*, https://doi.org/10.1007/978-3-319-92828-9_12

The vulnerability assessment has been performed through focusing on components, emerging from interactions between physical and human elements: their exposure, sensitivity, and adaptive capacity to climate change. We used a mixed-methods approach based on both quantitative (data analysis and modeling) and qualitative (interviews and investigation) data. Increased vulnerability to climatic changes recently observed (in Talmest compared to kenitra) can be explained both by its high exposure to the perturbation and by its socioeconomic situation that intensifies sensitivity and hinders adaptive capacity.

Keywords Climate change · Vulnerability · Adaptation · Social-ecological system approach · Sustainability analysis

1   Introduction

The vulnerability of a socioecological system to an extreme event appears as the propensity of the system to be damaged, in the first analysis, by its exposure to this event. However, multiple and varied links may affect this vulnerability, forcing beyond the exposure conditions to take into account the sensitivity of the considered system and its resilience. Sensitivity to climate change is not reduced to the problem of flooding, regardless of its importance. Awareness must be focused on the conditions of vulnerability that could determine exceptional responses from exposed systems, and complex and varied bonds these systems have with their social and biophysical environment, forcing them to consider other spatial, temporal and functional scales in the analyses. Vulnerability to climate change depends on the context in which these events occur. The heterogeneity of the dynamics requires a spatial approach across a territory (catchment area). The demarcation of the territories in which the study took place faces a double constraint: that of human communities and that of climate change, whose influences are not the same. This study covers both the environmental change dimension related to ecosystems' services and the human-induced perturbations. The links between the different compartments are expressed. From a temporal point of view, the existence of different timescales requires the implementation of specific study devices dedicated to slow or fast dynamics, depending on the characteristics of the system under study. Such an approach corresponds perfectly to the study areas, which are respectively adapted to slow and fast dynamics.

The study areas share a common conceptual framework highlighting the integrating and iterative processes of social-ecological interactions. This offers a variety of different situations regarding both the environment and human communities. The questions of land-use evolution (urban and rural areas) and operation and use of hydro-systems are at the center of the studied areas. The two study areas had been assessed by reference to their physical component, biological component, physico-biological interactions, social component, and finally as social-ecological systems.

2 Kenitra 'Gharb Region' (Sebou Watershed)

Floods are more or less brutal events that induce rapid changes with a very high environmental impact. The Sebou basin (Fig. 12.1), which is an extremely important area from a socio-economic point of view, is a striking example. An analysis of the situation is needed to explicitly address the social and ecological issues, violently induced by these flooding situations in terms of exposure, sensitivity, and resilience. For example, pollution has an economic, social and environmental impact on human communities at the scale of a city or a region. The healthcare system is directly related to the water pollution and flooding is regarded as the major vector in space and time. The event will be floods, with all the social and environmental consequences.

2.1 The Significant Impact of Population Dynamics on the Environment

The Sebou Basin is the second most populated basin of Moroccan Kingdom with 7.4 million inhabitants according to recent statistics (RGPH 2014), thus representing 21.7% of the total population. In addition, the population in this region is growing faster than the national average, with an inter-annual average growth rate of 1.8%, compared to 1.4% for the rest of the country between 2004 and 2014 (Figs. 12.2a, 12.2b).

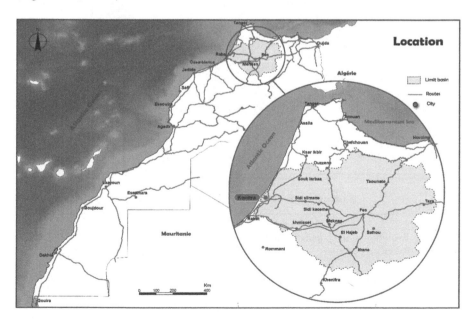

Fig. 12.1 The study area (Sebou watershed)

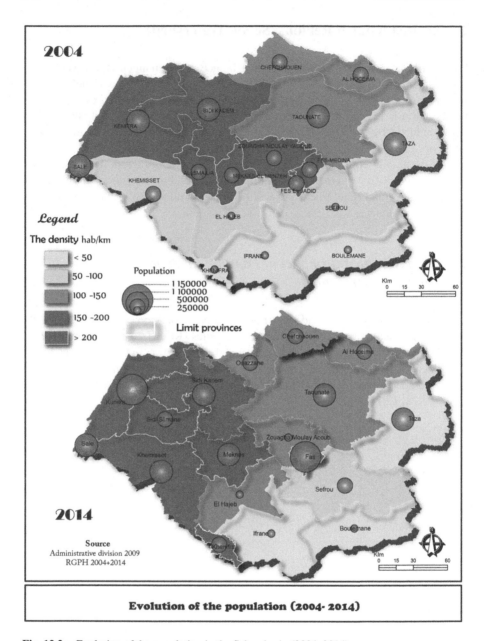

Fig. 12.2a Evolution of the population in the Sebou basin (2004–2014)
Source: HCP data processing

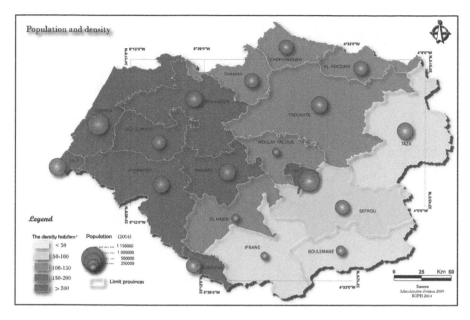

Fig. 12.2b Population and density in the Sebou basin according to the statistics for 2014 (Source: Processing of HCP data)

2.1.1 Urbanization Versus Environmental Regulations

Population dynamics has a significant impact on the environment. The size of the population itself is only a variable, certainly important in this complex relationship, but other demographic dynamics, such as changes in flow and density of the population, can seriously impact the environment. The urbanization exceeds the pace of infrastructure development and environmental regulations, resulting in a high degree of pollution and environmental degradation (Fig. 12.3). There is no simple relationship between population size and environmental change, however, as the population continues to increase, the availability of natural resources becomes a matter of concern. The distribution of the population also influences the environment (Fig. 12.4). Such pressures can be equally important in regions where the income level is lower compared to regions where the level is high. Poverty can lead to unsustainable resource use patterns which focus on the short term satisfaction basic subsistence needs. In regions where per capita income is high, the over-consumption is often associated with a significant production of waste.

In the plain of Gharb, the total volume of domestic wastewater discharges is estimated at over 200,000 m³/day, of which 86% is dumped in rivers, 2% applied to soils, and 12% discarded at sea (JSMO 2006). This pollution is responsible for approximately 60% of the organic pollution and 80% of the nitrogen and phosphorus pollution. The current organic pollution in the Sebou basin is estimated at about 76,000 tons of BOD5 per year, accounting for 25% of the national total, of which

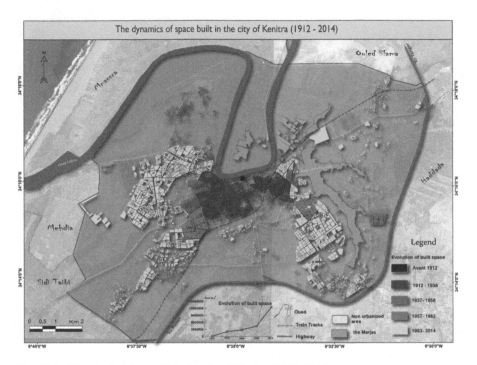

Fig. 12.3 Urbanization dynamics in the city of Kenitra (1912–2014)

nearly 47% comes from domestic pollution. This pollution is mainly carried by the Sebou river (Fes and Kenitra), the impact of the Fes rejections on the latter remains measurable over 100 km, Rdom (Meknès) and Beht (Khémisset), Inaouene (Taza), Ouerega Taounate), Aggay (Sefrou), Tizguit (Ifrane) whose quality is generally poor to very bad (ABHS 2013) (Fig. 12.5).

BOD5, is one of the parameters of water quality. It measures the amount of bio-degradable organic matter contained in water. This biodegradable organic matter is evaluated through the oxygen consumed by the microorganisms involved in the natural purification mechanisms. DBO5 is the biochemical oxygen demand for 5 days.

The growing production of household and industrial wastes causes critical pollution problems. The increasingly complex and heterogeneous nature of these wastes causes difficulties in their treatment and management (collection, sorting, storage, recovery, etc.). The fact that much of these wastes are land filled without precautions constitutes a real and permanent threat to the environment and public health.

The production of household waste in urban areas in the Sebou basin is estimated at around 840,000 tons per day. This waste production is steadily rising due to population growth and urbanization dynamics in big cities of the basin, and changes in consumption patterns. Uncontrolled landfills also represent a significant factor behind the pollution of surface and ground waters. The treatment of household waste is not well developed outside of the wild landfills, which is a common practice

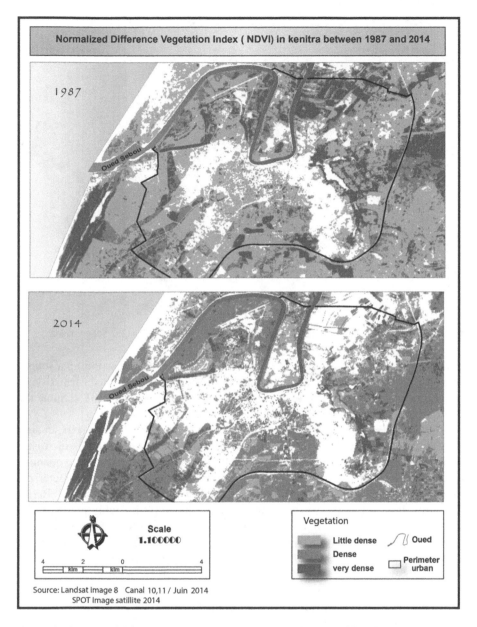

Fig. 12.4 Normalized difference vegetation Index (NDVI) in Kenitra (1987–2014)

for major cities in the country. The maximum pollution from landfills generally comes from the cities of Fez, Meknes, Kenitra, Sidi Kacem and Taza.

By their location, in general, adjacent to the cities and sometimes not far from water environments (river, aquifers), they release leachates which join the surface or

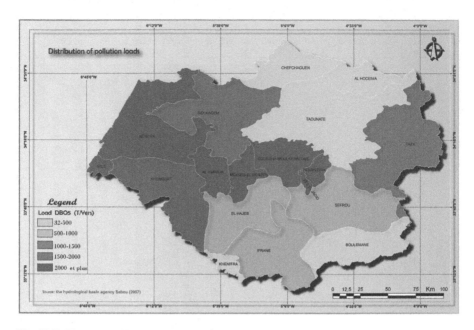

Fig. 12.5 The current biochemical oxygen demand for 5 days (BOD5) in the Sebou basin

underground waters according to the geology of the site, thus constituting a source of pollution not negligible. The Taza landfill, for example, presents a real risk of pollution because it is located in the Larbaa river bed (El Haji et al. 2012). Same for the discharge "Ouled Barjel" in Kenitra.

The Ouled Berjal discharge (Fig. 12.6), which exists since 1973, covers an area of 20 hectares and receives an average of 329 tons per day of all kinds of waste: domestic, industrial, hospital, slaughterhouses, trade, and roads. This quantity reached in 2015: 510 tons per day or some 186,000 tons per year. The characterization of leachate generated by the open uncontrolled discharge of Ouled Berjal (Kenitra) showed that the surrounding of the discharge pit is highly polluted. The microbiological component is characterized by heavy loads of fecal coliforms, staphylococcus, AFMAT and Salmonella. This leachate, with high pollution load, could contaminate groundwater flowing at shallow depths (from 4 to 15 m) (Fig. 12.7) under a permeable bedrock (Abed et al. 2014). In result, these highly toxic effluents permanently damage to the health of the local population and the surrounding environment, and require a specific treatment before being discharged into the receiving environment.

This method of waste wild landfill can no longer continue because of the serious damage it causes to the environment (pollution of ground and surface water resources, soil pollution, impact on human health, emission of greenhouse gases) (El -Fadel et al. 1997).

The exception of the city of Fez, which has a well-equipped landfill, the majority of other cities are at the stage of planning and selecting new sites.

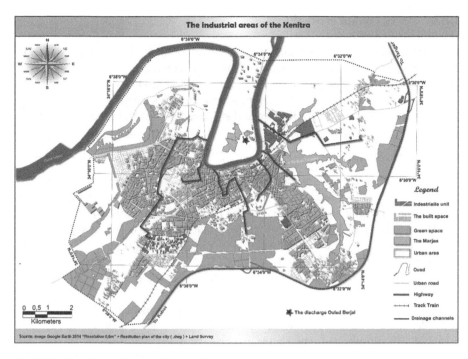

Fig. 12.6 The industrial areas of the kenitra

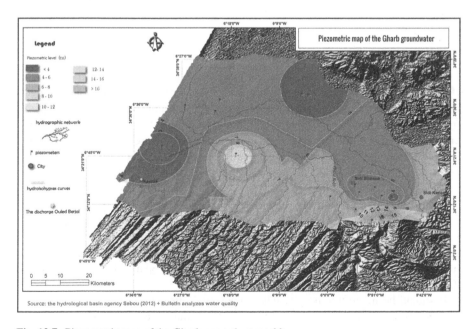

Fig. 12.7 Piezometric map of the Gharb groundwater table

2.1.2 The Economic Activity

One of the specific areas illustrating the complexity of the influence of demographic dynamics on the environment consists of land-use patterns. To meet the needs of a growing population through the intensification of food production in already cultivated lands or the infrastructure development to support such growth, land-use patterns have deeply changed. These forms of changes have several ecological impacts that increase the severity of flooding.

• Agricultural activity:

The Sebou Basin is one of the most important agricultural regions of Morocco. Nationally, this region has one of the highest irrigation potential. The land use is diverse with a dominance of cereals (60%), the rest is occupied by fruit plants (14.4%), legumes (6.6%), industrial crops – mainly sugar beet and sugar cane – (4.2%), oilseed crops (3.6), vegetable crops (3.1%), and forage crops (1.7%) (ABHS 2013). The Sebou basin constitutes nearly 20% of the useful irrigated agricultural area (i.e. 357,000 ha), and 20% of the useful agricultural area of Morocco (ie 180,000 ha).

The agricultural practices in the basin are being intensified with an increased dependence on irrigation, chemical fertilizers and pesticides. Pollution loads consist essentially of nitrates and phosphates and are estimated as follows: 6294 tons per year of total nitrogen; and 1499 tons per year of phosphate (MEMEE 2013). Groundwaters are specifically exposed to this type of pollution due to the infiltration of agro-chemicals. Pollution due to agricultural activities is mainly due to seepage of water that contains fertilizers and pesticides. The groundwater of Gharb is wickedly affected (ABHS 2013).

• Industrial activity:

Industrial activities are also diverse in the Sebou basin and cover many areas such as the activities prosecutors (applicants) in water like: food processing (sugar factories, oil mills, dairies, canneries, etc.), paper mills, tanneries, textiles, oil refining, and alcohol production. In addition, the production of margins is an important part of the overall pollution. It is mainly concentrated in the regions of Sefrou, Taza, Fez and Taounate. This pollution is very sensitive and has extremely strong seasonal peaks (reaching three times the load of untreated organic effluents in the city of Fez).

Industrial activity is concentrated in some cities of the basin (Fig. 12.8). Cost of organic pollution from industrial sources in the basin are estimated at 2.75 million population equivalent, of which about 70% come from sweets and paper mills. Industrial pollution is responsible for 40% of the organic pollution, 20% of the nitrogen and phosphorus pollution, and 100% of heavy metal releases (ABHS 2013).

It should be noted that no city has a functional purification plant, which means that the sewage treatment rate of urban waste is virtually zero until 2009.

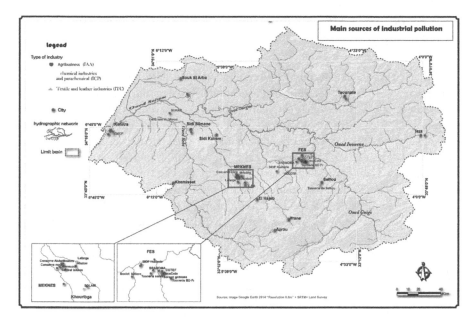

Fig. 12.8 Main sources of industrial pollution

Commissioning of the treatment wastewater plant in the city of Meknes June 27, 2009, who is part of the liquid sanitation project of the city. In the first phase, the effluent is subject only to primary treatment by anaerobic lagoon process, and in the second phase, the effluent will undergo secondary and tertiary biological treatment (activated sludge load average) to allow the reuse of the treated water for irrigation of agricultural areas located downstream of the wastewater plant. To ensure proper operation of the station, the Régie Autonome d'Eau et d'Électricité (RADEM) conducted, as part of a multidisciplinary committee, numerous awareness campaigns among the most polluting industries in the city, especially the food industries (oil mills, flour mills, and canneries), textile, paper and certain activities, including municipal abattoirs, hospitals, and municipal waste (RADEM 2009).

A wastewater plant that was established in the city of Fez in 2014, and it's partially operational since 2015, has aimed to improve the quality of Oued Sebou water through the treatment of household waste and industrial pollutants. The first of its kind in North Africa, it is likely to have positive impacts on public health (since the impacts of pollution on health are disastrous, see below), and on irrigation, besides uplifting the socio-economic conditions of local populations. However, this station will only be fully operational in late 2016.

The autonomous agencies are planning to set up 18 new wastewater plant in many locations such as Taza (where the public health is more vulnerable).

2.2 Floods as a Major Challenge to the Basin: Increasing Pollution

According to Stour and Agoumi (2008), the climatic data recorded in the region during the twentieth century indicate a warming estimated at more than 1 ° C during this century, specially the last 40 years and a clear increase in the frequency of droughts and floods. Thus, we have gone from a drought frequency of 1/10 years to the beginning of the century to 1/5 years in the last three decades. This new situation is accentuated by the scarcity of wet years with very high rainfall and spread over a short period of the year: hundreds of millimeters of water can be seen falling into arid regions in a few days and nothing for the rest of the year.

The Gharb experienced a peak of floods during 1996, 2002, 2009 and 2010 in the majority of territories of the region (OREDD 2014). Driouech (2010) established that the climate of the country, and consequently that of the region, tends towards a decrease in precipitation ranging from −10% to −20%, an increase in temperatures of the order of 1.05 ° C and an increase in evapotranspiration of the order of 0.05%.

A new type of pressure which, according to scientists, seems possible is the exacerbation of the violence of rainy episodes. This would mean that even if the amount of rain falling annually is decreasing, each rain event may be shorter and more intense than in the past (OREDD 2014).

The study area – Sebou Basin – is a region characterized by its dense hydrographic network. 30% of Morocco's surface water resources are drained by the Sebou (Fig. 12.9). The flood phenomenon has repeatedly threatened the riparian populations causing significant physical and human damage (Fig. 12.10). The climate assessment allowed the characterization the study area climate with its spatiotemporal variations. Interestingly, global warming appears to be reflected in the study area by a sudden jump of rainfall amounts and not by drought. Through this assessment, it is observed that the rainfall intensity in recent years explains clearly the exceptional floods from which the center of the study area has recently suffered. Ultimately, due to the climate of this region rainfall amounts and the geomorphological structure present favorable conditions for the acceleration of pollution processes. It is useful to recall here that all pollutants, regardless of their types (organic or metallic) and their origins (household, artisanal or industrial), are discharged almost without prior treatment into the watercourse.

The analysis of 'landsat' satellite image of thermal emissivity (Fig. 12.11) shows a strong thermal emission detected in the Sebou river along with its spatial distribution. This can be interpreted in many ways: 1) the existence of high night radiative temperatures of water at this river results from the combination of topographic, climatic, and geomorphological considerations; 2) the presence, within the thermal emission, of the particles which have a high emissivity. The sample analysis shows that they correspond to the dumping of industrial textile factories, tanneries, pottery wastes, and heavy metals found in the region of Fez, which is upstream of Kenitra.

The study of the role of topography on the surface and sub-surface flow requires the definition of the flow system. Each system has connected and continuous charg-

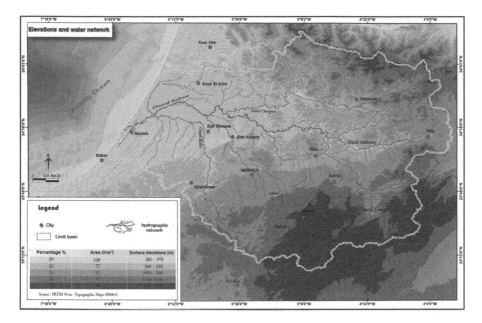

Fig. 12.9 Sebou basin hydrographic network

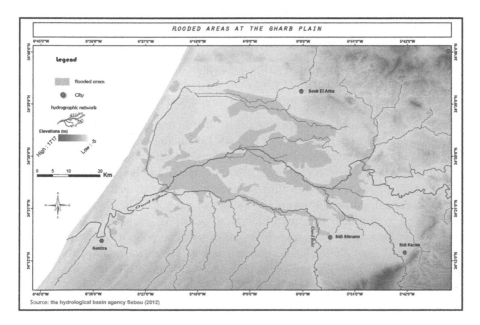

Fig. 12.10 Flooded areas at the Gharb plain

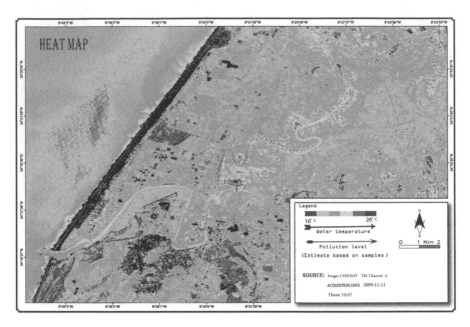

Fig. 12.11 'Landsat' satellite image of thermal emissivity

ing and discharging zones. The water flow system in our case where the surface is undulating slope forms many flow systems near the surface. On the topographic section (Fig. 12.12), we clearly differentiate:

- Local flows (flow from the hill to the adjacent hollow),
- Intermediate flows (separated by local flows)
- Regional flows (which flow from one end of the basin to the other).

The main consequence of this water flow distribution lies in water mixture of very different origins in the discharge regions. Depending on the size of the amplitudes, the relative importance of drainage systems can vary widely.

The analysis of physico chemical, microbiological and biological data of the sampling stations of Sebou basin allows the production of two distribution circles of physicochemical, microbiological, and biological qualities. The distribution circle of the physico-chemical and microbiological quality of the overall Sebou basin (Fig. 12.13) shows that the quality level varies between poor, medium, good, and excellent. The high pollution load could contaminate groundwater circulating at low depths (from 4 to 15 m), under permeable bedrock (Fig. 12.14); this will have a detrimental effect on human health.

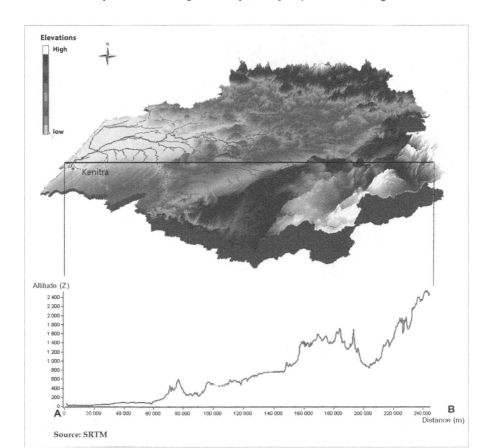

Fig. 12.12 Topographic section of the Sebou basin

In recent years, a sharp deterioration in water quality was observed upstream of the drinking water supply outlets. This degradation reached very critical levels during the campaigns of activities of oil mills, due to the importance of the pollution load.

The increase in industrial facilities, the development of intensive agriculture, and the extension of urban zones cause rapid and uncontrolled degradation of water quality in the basin, thus questioning the sustainability of both current and future water use practices.

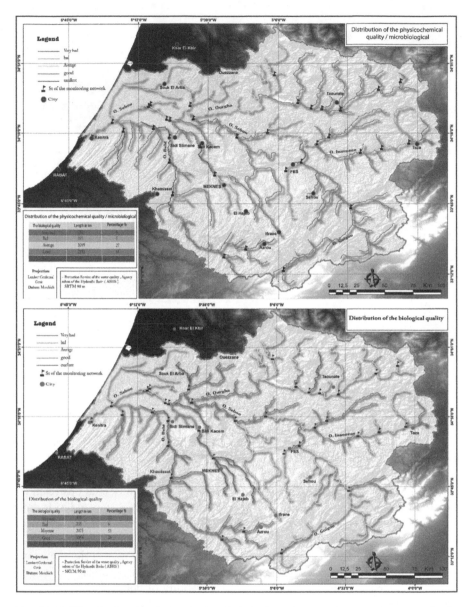

Fig. 12.13 The physico-chemical and microbiological quality of surface water resources in the Sebou basin (ABHS Data Processing, 2013)

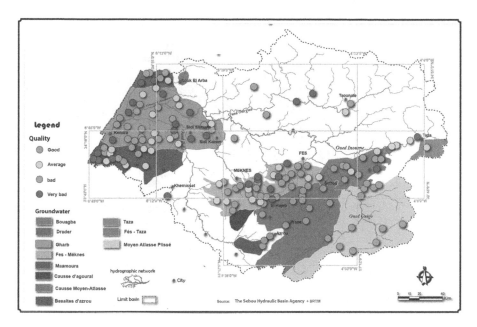

Fig. 12.14 Groundwater quality in the Sebou basin. (ABHS Data Processing, 2013)

2.3   Impacts on Social–Ecological Systems

The Sebou river represent a large part of the country's available water resources. However, it is among the most polluted rivers in the entire region. Pollution in the Sebou basin has reached critical levels and largely exceeds national standards. Flooding is the major pollution vector in space and time, with all its social and environmental consequences. Its negative effects on biodiversity and agricultural development are a potential vector for spreading epidemics.

2.3.1   Health Impact

Waterborne diseases – such as cholera, typhoid, polio, meningitis, hepatitis A and E, dysentery, diarrheal disease, guinea worm disease, and giardia – are often caused by drinking contaminated (due to human, animal or chemical wastes) or dirty water. These diseases have killed more children in the last 10 years than all armed conflicts since the end of World War II (Vedura 2015).

Surveys conducted in 2004 (Shaker 2004) in the centers of Sebou region show that the gastroenteritis prevalence rate near the Sebou river is around 98%, compared to 5% in health centers away from the Sebou river. Similarly, the prevalence of typhoid and cholera are 4 to 6 times higher in the region. These data are confirmed by studies published by the Department of Planning and Financial Resources

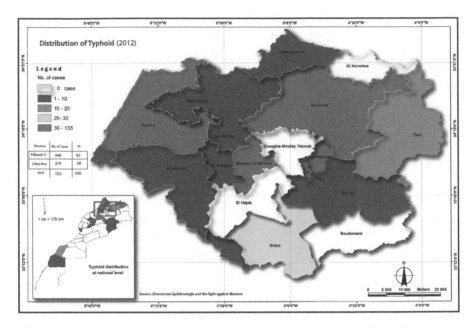

Fig. 12.15 Distribution of typhoid (2012) Typhoid distribution in the Sebou basin, 2012
(Source: Data processing of the Directorate of Planning and Financial Resources on the health situation of Morocco)

on the health situation of Morocco (62% of reported cases of typhoid concern the Sebou basin) (Fig. 12.15). The current situation can be considered as catastrophic.

The results of the investigation conducted in 2015 in the main hospital of the Gharb region shows the extent of the spread of water-borne diseases (Fig. 12.16) which have certainly many repercussions on economic and human development.

This part of the work aims to highlight the need to understand the dynamics of waterborne diseases that are expected to change in terms of geographical distribution and incidence due to climate change. More specifically, climate change will likely affect the distribution of water-related disease, and this scenario is relevant to the Gharb context (Downstream part of the Sebou basin). These diseases represent a major human health burden in the coming years; a burden that is unevenly distributed across many regions. Besides, they are closely linked to many risk factors, including climate change, ecological dynamics and socioeconomic, urbanization, and agricultural practices. The most vulnerable regions are currently in a difficult situation because they have limited adaptive capacities, thus unable to cope with the changing conditions in efficient way.

It is currently recognized that with the constant increase of the planet's temperature, there are, and will be, more frequent and severe floods, droughts, storms, and heat waves. These changes to the Earth's biophysical system will exacerbate and extend the rates and ranges of many diseases and overall contribute to poor health among all populations (Friel et al. 2011). For Wiley (2010), climate change acts

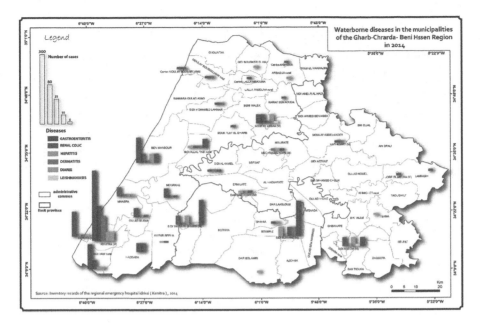

Fig. 12.16 Waterborne diseases in the municipalities of the Gharb-Chrarda-Beni Hssen region (2014)
(Source: Field Survey 2015)

primarily as an intensifier (or risk multiplier), and to some extent a redistributor of existing threats to health. More precisely, climate change affects many environmental determinants of health, mainly water and air (Portier et al. 2010). Similarly, Boxall et al. (2009) asserts that weather and climate factors are known to affect the transmission of water and vector-borne infectious diseases.

Undertaking a vulnerability and adaptation assessment with regard to the protection of public health from climate change impacts should be responsive to the country context and needs in order to appropriately determine what steps to be taken and according to which priorities. The basic components of an assessment are shown in Fig. 12.17.

2.3.2   Environmental Degradation

According to a study about the estimated cost of environmental degradation in the Sebou basin, damages related to the loss and degradation of biodiversity in this basin were estimated at nearly 0.56 billion dirhams (ABHS 2013). Much of this loss is mainly due to the degradation of water quality. There are several classes of water pollutants:

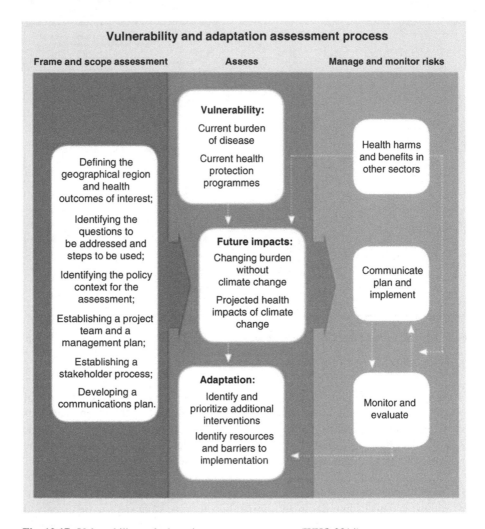

Fig. 12.17 Vulnerability and adaptation assessment process (WHO 2014)

- **Category 1**: pollutants linked to the waste that can be decomposed by aerobic bacteria. The activity of these bacteria results in low oxygen level of the water, causing the death of many water species, such as fish.
- **Category 2**: concerns the water-soluble inorganic pollutants such as acids, salts, and toxic metals. Large amounts of this compound make the water unfit for consumption and leads to the death of aquatic life.
- **Category 3**: nutrients that consist of water-soluble nitrates and phosphates with the potential to cause eutrophication (excessive growth of algae and aquatic plants, which reduces the amount of oxygen in the water). Besides, this type of pollutants, when concentrated in drinking water, can cause death of young chil-

dren (the infant blood met-hemoglobin level also called the blue baby syndrome).

- **Category 4**: composed of organic compounds such as oil, plastics, and pesticides that are harmful to humans, plants, and all animals living in the water.
- **Category 5**: composed of water-soluble radioactive compounds that can cause cancer, birth defects, and genetic changes, and are therefore very dangerous water pollutants.
- **Category 6**: composed of very dangerous class of contaminants linked to suspended sediment, because they reduce the absorption of light by water. In addition, these particles scatter the dangerous compounds – such as pesticides – in water.

Like many other areas in Morocco, Kenitra faces heavy human pressure from a variety of uses. Indeed, industrial and agro-alimentary activities result in substantial pollutions which make the ecosystems fragile and vulnerable. The anthropogenic ecological and socioeconomic impacts of such activities may be immediately observable or predictable due to the short time lags between human-nature interactions and the appearance of such impacts. For instance, the impacts on surface and ground water resources affect the quality these resources, and this appears at the 'downstream' level very rapidly.

The issue of water security has been the object of increased academic and policy interest (Bakker 2012; Cook and Bakker 2012). Accordingly, the good water governance has been increasingly recognized as a critical contributor to the long-term sustainability of water resources (Pahl-Wostl et al. 2012; Huntjens et al. 2011). Similarly, public policies, one of the main factors influencing the dynamics of social-ecological systems, makes decisions which are made in one place and affect another. Achieving water security requires the coordination of different actors' interventions in line with national standards and targets without neglecting local specifities and customs.

According to the Moroccan Economic, Social and Environmental Council (CESE. 2015), in terms of water resource management, the governance system structured around the autonomous unit of the hydraulic basin is a major asset in establishing a localized approach to water resource management, while adopting principles of solidarity and inter-regional support for the state subsidy in order to manage a manifest spatial heterogeneity. However, in the absence of a structured overall framework, this idea of water resource governance, did not find the favourable environment for its implementation. Indeed, despite the existence of a Delegate ministry in charge of water and a High Council for Water and Climate (designed according to Article 13 of law 10–95 on Water and tasked to make useful recommendations for the elaboration and implementation of the national water policy), their power and legitimacy have not yet reached the needed level of maturity and efficiency to ensure the effectiveness of their interventions. In addition, the only operational regulatory body that forms the Inter-Ministerial Water Commission has ceased to be active for years, thus considerably limiting the scope of active actors.

If these organizational deficiencies have generated marginal impacts in the past, the current situation is different for many reasons. Firstly, water supply is increasingly threatened by emerging risks such climate change and the depletion or degradation of conventional water resources due to unsustainable practices aggravated by the existence of an inefficient governance system which should be reformed. Ideally, and in application of the principles of participatory democracy, the choice of projects and implementation of technical solutions must be done according to a bottom-up approach, which enables a multilevel participative involvement of concerned local communities, civil society actors and users in relevant decision-making processes. Furthermore, concerned actors should also project themselves into the future through: the implementation of an inter-generational solidarity in response to the needs of future generations; and the consolidation of knowledge management of water-related risks to allow anticipatory action and optionally a previously planned reactive adaptation.

The diagnosis of the current situation concerning the water resource management shows that the Delegate Ministry in charge of Water and the Higher Council for Water and Climate are not in fact efficient coordination entities. In addition, the Inter-Ministerial Commission for Water is inactive and the Hydraulic Basin Agencies are weak because of their feeble decision-making autonomy and insufficient financial and human resource capacities. From a regulatory point of view, although the law 10–95 on Water confirms the situation above, it remains out of step with current realities and needs and it suffers from weak effectiveness in terms of implementation. Overall, the planning and implementation modes, which cause excessive delays and phase shift situations between sub-projects, should be revised. Dialogue and convergence must be the master words in any schematic layout of agents working in the water sector.

In this regard, the action to be taken is double:

- Firstly, it should mainly cover the water resource assessment taking into account the critical situation in many ways, including the risks associated with over-exploitation of aquifers, theoretical dams storage capacity and the actual capacity due to siltation, the differences between the dam storage time and irrigation and evaporation loss resulting, finally, the low reuse of treated wastewater.
- Secondly, the action should aim at protecting the water resources by implementing the texts relating to the "polluter pays", then developing sanitation and water treatment plants, and specifying the criteria for protection zones to reduce pollution that would be associated with them.

A vulnerability analysis must take into account the concept of resilience alongside those of exposure and sensitivity. Resilience, another component of vulnerability, is the ability of socio-ecological systems to absorb disturbances while retaining their essential structures as well as processes at the origin of these structures (Walker et al. 2002). In fact, the concept of resilience challenges the traditional view of ecological systems – which always seeks to return to the same state of stable equilibrium – by adopting an approach which permits to an ecological system to shift from one situation to another while conserving its balance. This ability is highly needed to face unexpected, sometimes irreversible, and recurrent disturbances through varying strength intensities (Scheffer et al. 2001). Thus, resilient systems have various mechanisms that allow them to cope with changes and crises (Gunderson and Holling 2002; Allenby and Fink 2005; Adger et al. 2005). It can then help explain the possible responses to diverse and varied disturbances and identify what measures may ensure sustainability in an uncertain environment.

3 Talmest 'Chiadma Region' (Tensift Watershed)

In the Tensift basin (Fig. 12.18), the flood phenomena and the exceptional hydrological pulses have gained magnitude in recent decades, in relation on the one hand to the climate change, and on the other hand to local geomorphological and climatic characteristics. The seasonal contrast is very well marked and the rains – often concentrated during fall and winter – are irregular, intense, and violent. During the rest of the year, drought takes considerable magnitude. The previous catastrophic events in the basin had been largely triggered by weather events with remarkable intensity (Table 12.1).

While this area is experiencing a development of farms (Figs. 12.19 and 12.20), soil suffers adverse harm due to flooding and water erosion. Indeed, floods have caused catastrophic damage to farmers. If the protection of natural resources is a major concern for the environment, soil protection remains a high priority

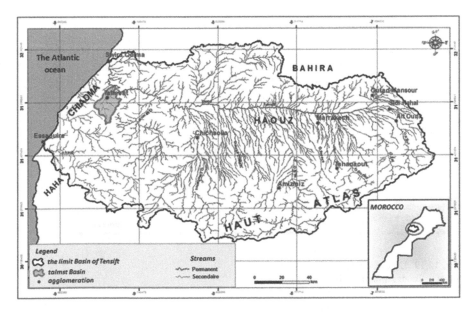

Fig. 12.18 The study area. Source: SRTM 90 m Digital Elevation Database

Table 12.1 Value of peak flows recorded before and during the exceptional flood of autumn 2014

Station	Before flood		During flood Autumn 2014			
	Peak flow (m³s)	Date	Peak flow (m³s)	Hour	Date	Return priod
Tatmest	1275	10/11/1988	3500	11h30	30/11/2014	1/1000

Source: Agence de bassin hydraulique de Tensift (2015)

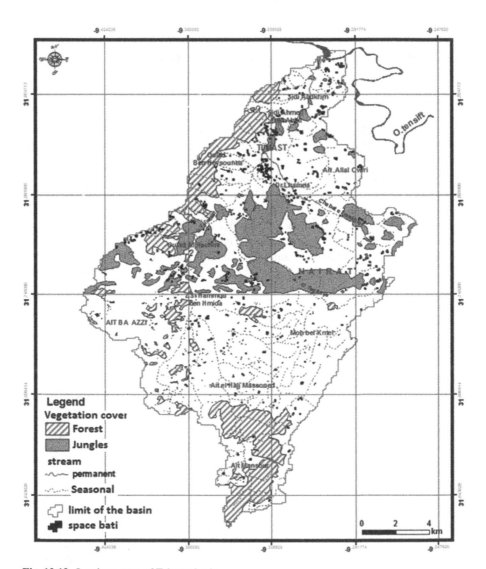

Fig. 12.19 Land use map of Talmest basin
Source: Topographic map of Scale 1/50000

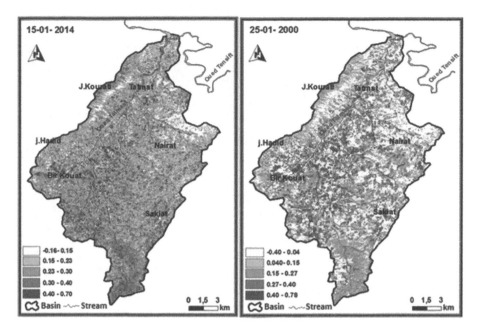

Fig. 12.20 Land cover change (Normalized Difference Vegetation Index (NDVI) in Talmest Basin)
Source: Landsat satellite photo

compared to other resources threatened by floods and water erosion, and additionally exacerbated by climate change and human practices.

3.1 Vulnerability of Physical Environment to Climate Change

The hydrological basin of Tensift is a hydro-system with two major global morphological entities (Fig. 12.21): a very high mountainous set and a vast alluvial plain. A clear morphological asymmetry between the two banks and real mobile meanders appear in different places; their sinuosity index gradually increases following the decrease in the topographic slope. The physiography of watershed provides a conducive environment to the development of intensive floods: exposure favorable to precipitation, steep slopes (Fig. 12.22), substrata of low permeability (Fig. 12.23), low and discontinuous plant cover, and hierarchical water system (Fig. 12.24). The morphological contrast between the two sides gets the alluvial plain located on the left bank much more conducive to overflows of rharbo-soltannian terraces by runoff water during floods and to flooding of corresponding farmland. The result is a side undercutting by migration of active bed of the river. This morphological configuration contributes to amplify the observed peak flows at the outlets. The reliefs are dominated mainly by soft formations that oppose the very heavy rains. The plain

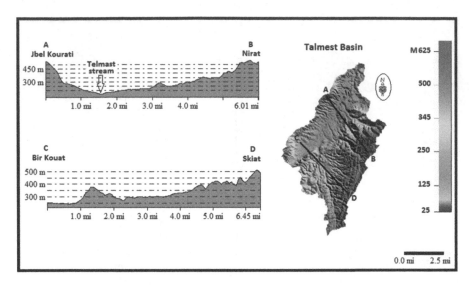

Fig. 12.21 The nature of the basin 'Talmst'
Source: SRTM 90 m Digital Elevation Database

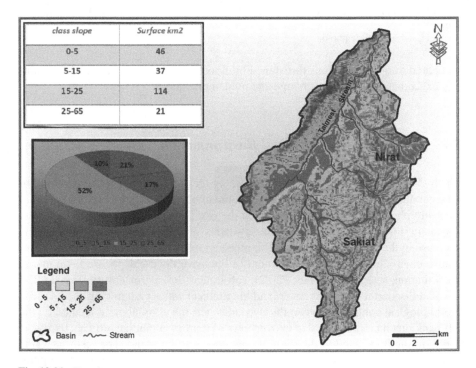

Fig. 12.22 The slope
Source: SRTM 90 m Digital Elevation Database

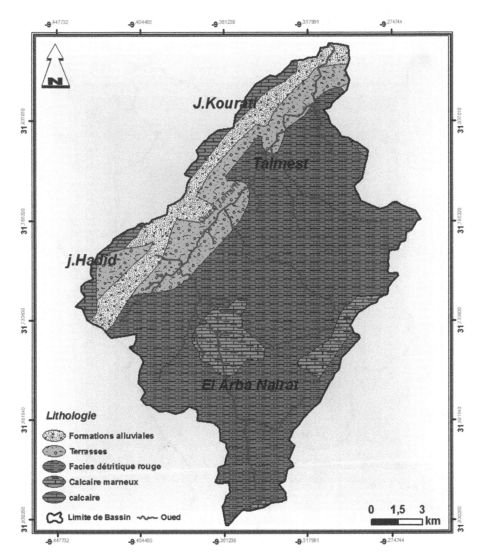

Fig. 12.23 Lithology map of Talmest Basin
Source: Geological Morocco of Scale 1/1000000

area is dominated by carbonate layers and quaternary alluvium. The weakness of the plant cover has the potential to largely promote mechanical erosion, which will determine a large amount of terrigenous particles carried to the sea.

Floods in the Tensift basin are very violent and irregular. The differences between annual average flows and instant maximum flow rates are very high; demonstrating therefore the brutality and magnitude of flood flows. The consequences of these floods are potentially worrying for a large part of the population of the region whose livelihoods depend on the alluvial plain (housing, use of agricultural land, routes).

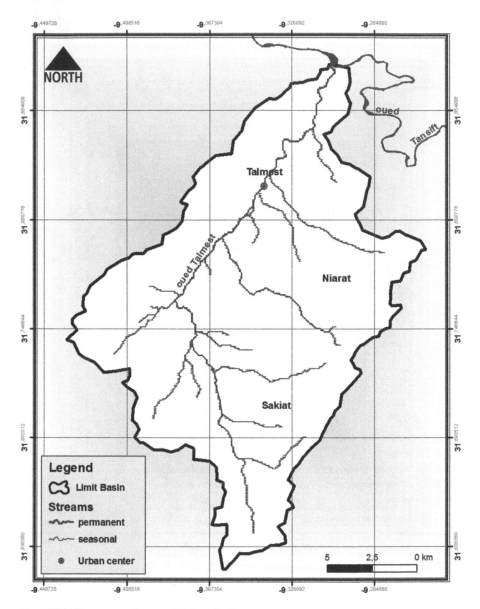

Fig. 12.24 Water network map of Talmest basin
Source: SRTM 90 m Digital Elevation Database

In addition, an increased loss of fertile soil is observed as removed suspended material to the ocean. Based on direct observation of the field and the survey of the local population, the region seems experiencing a continuous degradation. Hydric erosion (Fig. 12.25) continues to dismantle the ground, in the form of gullies in a spectacular way (Fig. 12.26). It should be noted that there is currently no precautionary

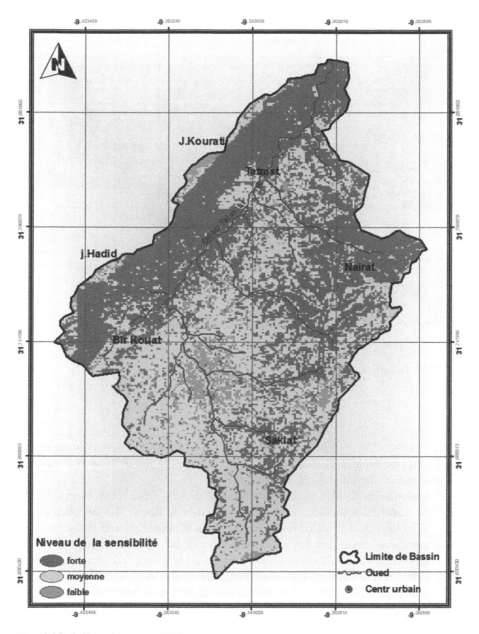

Fig. 12.25 Soil erosion susceptibility map
Source: SRTM 90 m Digital Elevation Database and Field work

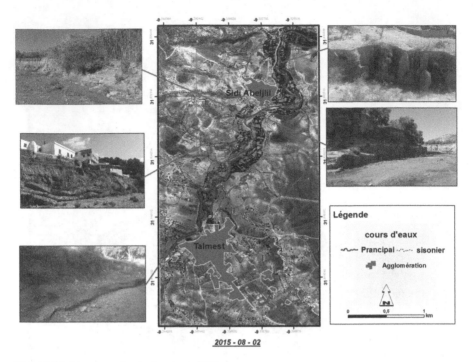

Fig. 12.26 Linear erosion in the river of Talmst
Source: Fieldwork

management of soil fertility. This phenomenon constitutes a major threat to the
ecological and biological value of this site. Indeed, the floods have revealed the
fragility of the alluvial plain of Tensift.

Furthermore, landslides occur when the slope moves from a stable to an unstable
situation. Landslides should not be confused with mud flows. The landslide in the
case of Talmest, is a displacement on a slope, a coherent ground mass, variable
volume and thickness. This surface is generally curved (circular sliding/ rotational
landslides) (Fig. 12.27). It occurs more often in areas where soft ground is found,
such as clay. It is usually produced by an excess of water in the soil. Water softened
the soil and quickly slipped down the slope. This phenomenon can involve large
volumes of materials, and sometimes, in some large landslides, we can even find
whole fields with the houses, which will move without suffering damage. The land
issue is in front of the stage. Strong questions remain about the policy options
available to deal with this phenomenon, the possibilities and the ways to implement
them.

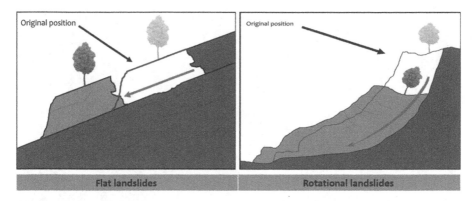

Fig. 12.27 Landslides types

3.2 Transformation and Commoditization of the Floristic Composition

Our planet is currently entering a mass extinction phase, the sixth one since life has diversified on Earth 500 million years ago. Humans beings are engaged in a massive destruction of ecosystems on a global scale and our development model probably condemns a species every 20 min (Billé et al. 2014). Both natural and anthropogenic actions constitute threats to the Moroccan flora. The loss and degradation of habitat is the main driver (90%) of biodiversity loss, especially in countries like Morocco who are rich in biodiversity but conservation measures are not well defined and lack necessary attention.

In forest ecosystems, natural processes and observation approaches are extended in time. The behavior of forest plantations spreads over a long period (from 5–25 years). During this time, we can characterize a complex dynamic constituted of a constant process of change of living matter present In Morocco, the forest degradation has become a serious problem due to the transformation of thousands hectares from primary vegetation to the secondary vegetation.

The study area is part of the Arganeraie Biosphere Reserve of UNESCO,[1] and it is representative of the low mountain Arganeraie. The oldest argan tree would be located in a cemetery of the Talmest. Forest cover is marked by clear forests and an almost continuous grass cover in years of normal rainfall. The region is composed of two distinct vegetation formations: the Arganeraie and the Tetraclinaie (Fig. 12.28).

---

[1] Located in the southwest of Morocco, the Arganeraie Biosphere Reserve is a large intra-montane plain home to the Argan tree (Argania spinosa), a species endemic to the reserve in need of conservation. These trees constitute a major economic resource for local inhabitants as the principal source of Argan oil, which has multiple uses in cooking, medicines and cosmetics. The trees are also used as fuelwood for cooking and heating.

Fig. 12.28 Talmest (January 2016)

The Arganeraie At the mountain "Jbel Kourati", the argan forest dominates (*Argania spinosa*) as open forests (scattered). The argan forest is under strong degradation (regression of superficies and density) due to heavy economic exploitation (argan oil). Currently there are:

- The Arganeraie-Orchard which is located in less hilly areas, on mountain plateaus and on the great plains. Its density is low (20–30 strains/ha). The undergrowth has almost disappeared due to the plowing. It occurs around the douars (small villages) on parts with little accused relief; in this sector, olive tree culture is predominant.
- The Arganeraie-Forest that is confined in hilly areas (mountain). Tree density is higher. The floor is covered by either a forest or an important, and sometimes, continued live coverage.

At altitude, the argan forms mixed forests with cedar.

The Tetraclinaie The tree cover of Barbary thuya (*Tetraclinis articulata*), occupies the high slopes and its ecological space is subject to high anthropogenic pressure. This species presents a high degree of degradation due to the profligate cuttings. The region's archives attest that the main factors contributing to the decline of the forest plant cover are mainly: deforestation; overgrazing; forest fires; and hydric erosion (in about 72% of the areas). This last factor has an obvious negative impact often reinforced by the mechanical action of the wind. The soils are often severely degraded making the natural regeneration very delicate. According to the High Commission for Water and Forests and the Fight against Desertification (High Commission for Water and Forests and the Fight against Desertification 2015), hydric and wind erosion cause land degradation: 500 tons/ km²/year; and 40% of the area is classified as high-risk erosion area (1.12 million ha). Excessive water stress may increase tree mortality, particularly those on steep slopes and highly exposed to sunlight (El Abidine 2003).

Loss of forest cover and ground vegetation can also cause erosion which, in parallel to the edaphic degradation, can make soil unsuitable for germination (Nouaim 2005). The expansion of irrigated agriculture has, as a consequence, intensified the over-pumping of ground-water which results ineluctably in the lowering of the water table. As a result, the vulnerability of trees to drought increases on continuous basis. The degradation of the forest area is the consequence of the combination of both lack of regeneration and loss of trees, which itself depends on biophysical and social drivers.

The land use has as well direct implications on the availability of natural resources. A better understanding of the changing land-use patterns and vegetation cover is a major concern which may contribute to the enhancement of our knowledge with regard to the pace and nature of vegetation change. The results allow for the identification of areas at the watershed scale where interventions are needed to limit soil degradation and, consequently, stop the degradation of vegetation covers.

3.3   Social-Ecological Systems

The flood events experienced by the region imply direct and indirect negative effects on people and property. Based on direct observation of the field and the survey of the local population, floods constitute a threat for local residents, land, homes and infrastructure. The impact of devastating floods and heavy rains could not be explained only by physical factors; only a combination of factors that favored the vulnerability of the plain may provide the appropriate explanation.

To manage the magnitude of damage caused by such events, the local population leads undertook continuous action (Figs. 12.29 and 12.30) to protect and restore the damaged soil despite the additional expenses generated by such efforts. The techniques used by farmers to protect and restore their soils were not subject to any consultation among stakeholders. This creates social conflicts and leads to narrowing of the rivers bed. Therefore, the risk of flooding remains imminent, each intervention challenges the previous one, and unfair competition between farmers continues. This will further affect the environment, the economy, and the society.

In addition, socio-ecological factors exacerbating the impact of climate change on the floristic composition in 'Talmest watershed' have been identified. To convert land cover, a systems thinking approach was used to assess the interrelationships between land-cover change, actors' choice, and their motivation.

The communities within the study area (mainly poor rural people) focus their activities on food production (Figs. 12.31 and 12.32), and their weak financial capacity often prevent them from investing in conservation. In addition, and due to land shortage (including limited access to land), these communities tend to over-exploit the marginal lands, without considering the real impact they may have on the environment (Douglas 2006).

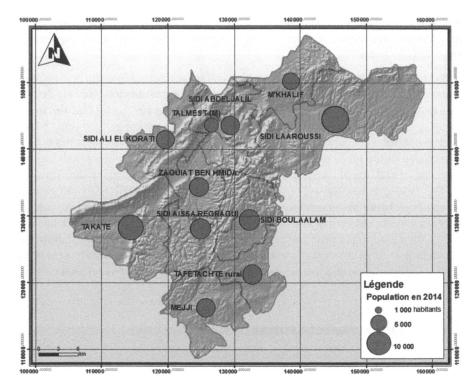

Fig. 12.29 Population and density in the talmest basin according to the statistics for 2014
Source: Processing of HCP data

For example, the adverse impacts of unsustainable agricultural practices are not appropriately perceived by many farmers in the study area. The result is a reduction of the land's capacity to support their livelihood. Surprisingly, endemic argan trees that are 60 years old are continually replaced by these communities to maximize olive oil production, which is their main source of cash income. This link between poverty and environment is often conceptualized as a 'downward spiral' or 'vicious circle' with alarming economic and social effects. The resulting environmental degradation further limits the availability of natural resources and increases poverty (Lestrelin 2010).

It is very important to take into account connections between human activities (social systems) and land use (ecological systems) because the two factors affect the land-cover change in ecosystems. This may improve the policy-makers' understanding of socio-ecological interactions and the various drivers of land-cover change. The Talmest region is no exception.

Current and future land-use changes often occur as the result of human decisions that are influenced by various driving forces (Hersperger et al. 2010). These driving forces are commonly determined by the emergence of complex and non-linear feedback relationships between social and ecological systems (Van Noordwijk et al. 2011).

Fig. 12.30 Continuing struggle

These global changes in tree strata lead to the loss of biodiversity and trivialization of flora (abundance of banal species, with a very wide distribution and progressive reduction of endemic species) and will continue over the coming years if nothing is done to stop this real deterioration. For example, there is a very serious problem of regeneration of the argan tree, which is endemic to Morocco, in natural forests in the total absence of natural regeneration and ecological balance. The cause is linked to the important collect of the fruit pit of the tree by users for the extraction of oil (used in cosmetics and main source of income for local populations). Moreover, the few seedlings, grown from seeds that have escaped the harvest, are routinely grazed by animals (M'hirit et al. 1998).

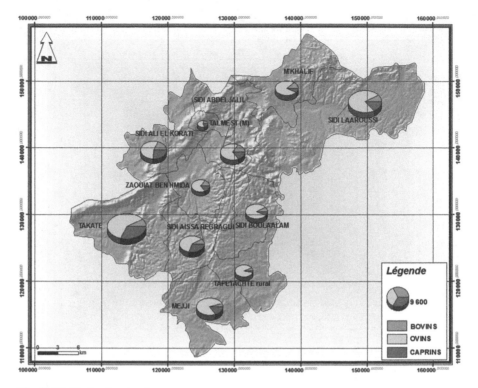

Fig. 12.31 Payload pastoral map
Source: Marrakech's regional direction's data, 2015

Some of the key factors playing a major role in the destruction of the biological community structure, species loss, and thus loss of biodiversity are, inter alia: changes in land-use patterns; abandonment of local agricultural practices; pollution due to herbicides and fertilizers; and intensification of agriculture along water courses and habitat loss.

Climate, abiotic environment and disturbance history can provoke changes in vegetation dynamics and the installation of the spread of invasive species and newly introduced species (example of olive tree). In favorable ecological conditions, the invasive species become more competitive and could affect the endemic flora (Argan tree and Thuya). The risks caused by these species to endemic flora are apparent and vary from competition for natural remedies to disappearance of species. They cause a significant loss of biodiversity, due to the underestimation of the issue and the lack of information and awareness.

These perspectives will help to fulfill the commitments and objectives adopted by Morocco under international agreements, in particular the Global Strategy for Plant Conservation (GSPC 2020). In a country like Morocco where biodiversity is the main source of local livelihoods, biodiversity loss will continue unless signifi-

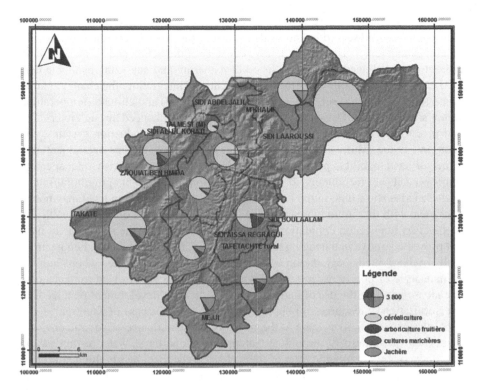

Fig. 12.32 Types of culture
Source: Marrakech's regional direction's data, 2015

cant conservation measures are quickly implemented in full coordination with the local populations that need to realize the extent to which this biodiversity is important for their well-being, identity, and future livelihoods.

4 Results Discussion

The durability of social-ecological systems depends on the interactions of humans with space and resources and the relations between social viability and ecosystems. Therefore, the assessment of potential risks of disruption between social and ecological viability based on present land use, demographic parameters, climate change, natural and human dynamics, the categories of actors and situations which should promote their innovative approaches, and failure in governance may help support the elaboration of future efficient policies.

Social-ecological systems are interactive sets of entities, each of them characterized by a pace of evolution. Complex interrelations are generated by their specific diversities and by interdependence of their respective functional scales (global-local). Management of these systems proceeds on a sector-based approach linked to compartmentalized scientific knowledge. The study of socioeconomic and environmental parameters to assess the adaptive capacity of the Sebou and Talmest basin to current risks and uncertainties consisted of gathering and analyzing available data on the changes of conditions in the studied zones by integrating: the changing landscape (land use/land cover, houses and settlement areas) and climate (temperature, water, soil, vegetation) based on satellite data analysis, enhanced through interviews and investigations, evolution of population densities and production systems, and economic tissue. Multivariate analysis suggests an unsatisfactory overall status in terms of sustainability, particularly at border areas at river. Also, management improvements are recommended. When, projections of different agricultural/industrial scenarios show that, in the long-term, the current exploitation level may not be consistent with the future increase of local demand and a loss of human health and biodiversity may occur. Population and economic dynamic in the studied areas have not proved operative in managing natural resources, most of which are being chronically overexploited. Degradation of natural resources is exacerbated when planning is inadequate or poorly executed, and when financial incentives lead to inadequate decisions regarding the use of natural resources or when development plan leads to over-utilization of resources for short-term production (the case of Kenitra:

Anthropogenic activities are at the origin of the disturbance of aquatic environments both quantitatively, through the intensive exploitation of water resources, and qualitatively, by the modification of their physico-chemical and biological properties. Indeed, economic and industrial development is accompanied by an increase in waste streams containing pathogens and various organic and inorganic substances. The main problems associated with releasing these pollutants into aquatic environments are eutrophication and contamination. Eutrophication-related problems lead to water depletion in oxygen, resulting in an increased risk of mortality of fish and other aquatic organisms with eventual destruction of the environment and a hazard to human health and the cattle. In addition to anthropogenic activities, the dynamics of pollutants in aquatic systems are also influenced by the climate. In arid or semiarid regions, low flow periods limits the dilution process of anthropogenic inputs, which leads to intense deposition of contaminated particles in the storage areas. These dry periods are generally followed by violent and rapid rainy periods provoking heavy surface washing, remobilization of cumulative pollutants during the low water period and also dilution of various rejected substances) (Fig. 12.33).

Such short-sighted approaches should be replaced by an integrated resource planning and management in which the local user is placed at the center. This will ensure: the long-term quality of natural resources to meet the human needs; and the prevention or resolution of social conflicts related to land use and conservation of high biodiversity ecosystems. The need for an integrated approach is widely affirmed, particularly in the context of a region like Talmest whereabouts the small-scale plots, often multispecies, are very important in terms of production, employment and food security. However, these plots and the exploited resources are often characterized by a social-ecological interactions complexities. Given the extent of the damage caused by the events such as: mud flows, increased loss of fertile soil,

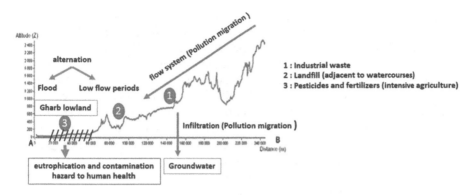

Fig. 12.33 Remobilization of cumulative pollutants in Sebou basin

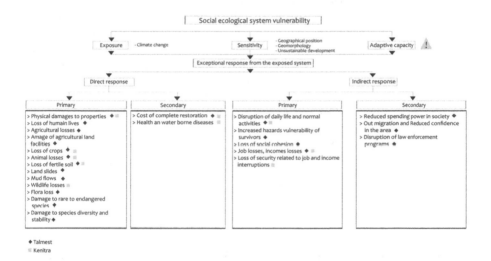

Fig. 12.34 Vulnerability to climate change of social ecological systems in Kenitra and Talmest

landslides with sometimes whole fields, the harmful situation affects both the environment, the economy, and the society. The diagnosis made for the two case studies (Kenitra and Talmest), allows for the investigation of exploitation conditions of natural environment and human activity, complex mechanisms such as livelihood interactions, competition, population dynamics, and economic investment processes.

Trends in environmental change and other social and demographic processes make individuals and social systems always vulnerable to unexpected events and susceptible to unforeseen consequences of action (Fig. 12.34). While policy-makers often express their surprise while unexpected events occur, many of these even are predictable or at least 'imaginable'. Yet, vulnerability persists due to inherent unpredictability in some physical systems and ideological barriers to the perception of certain risks. If the key objective of sustainable development consists of eliminating risks to the most vulnerable, the application of the precautionary principle should be central to relevant decision-making processes (Adger 2006).

The produced maps facilitate the identification and analysis of vulnerability. The vulnerability analysis was the outcome of an integrated analysis of identified external and endogenous risks and the availability or lack of resources to cope with the situation (capacity). The adaptive capacities depend on past and present environmental policies and local rules that has constrained or facilitated the social ecological coviability. Climate change, population growth, economic development, etc. are factors which generate increasing pressure on natural resources in terms of availability and access. Resilience of social-ecological systems refers to the magnitude of disturbance that can be absorbed before a system changes to a different state, as well as the capacity to adapt to emerging circumstances. According to the Community and Regional Resilience Institute (CARRI 2013): the community resilience is the capability to anticipate risk, limit impact, and bounce back rapidly through survival, adaptability, evolution, and growth in the face of turbulent change. With reference to Adger (2006), vulnerability is the state of susceptibility to harm from exposure to stresses associated with environmental and social change and from the absence of capacity to adapt which has contributed to present formulations of vulnerability to environmental change as a characteristic of social-ecological systems linked to resilience. The challenges for vulnerability research are linked to the ability to develop credible measures, to incorporate diverse methods that include perceptions of risk and vulnerability, and to incorporate governance research on the mechanisms that mediate vulnerability and promote adaptive action and resilience.

5 Conclusion

In Morocco, the flood risk has become a serious problem for sustainable development. The analysis of local and regional context shows an evolution of the human influence and recent developments, more than that of climate. Certainly, extreme weather events are clearly involved in the genesis of floods, but their occurrence is nothing new in such an environment where the irregularity is structural in all its aspects. The factors that may explain most of the damage caused by floods and torrential rains are related to the characteristics of the affected sites and relevant anthropogenic actions.

From a comparison of the data from both case studies, the most striking fact is that they provide important insights about the diverse complex characteristics that cannot be observed in a single study. Both have demonstrated varied effects (consisting in the interactions between human and natural systems). Because of the independent nature of these two case studies, information from one study is not necessarily available in or transferable to other studies.

A challenge is posed by the vulnerability assessment both for the analysis of governance structure and for the implementation of governance solutions to environmental change. Vulnerable people and places are often excluded from decision-making and from access to power and resources (Dow 1992; Pelling 2003; Adger 2003; Stockholm Environment Institute 2001). Therefore, policy interventions to reduce the vulnerability social-ecological systems should be able to identify these vulnerabilities, to recognize the mechanisms behind this vulnerability, and to address marginalization as a cause of social vulnerability. Further, policy interventions need to recognize the plural forms of knowledge and governance systems that are used throughout the world to manage risks and promote resilience (Ostrom 2001; Brown 2003). In this perspective, vulnerability may challenge the capacity of a governance system to promote resilience and minimize exclusion, thereby it is useful to reduce both the severity of perceived vulnerability and its structural causes. Overall, the coviability of social-ecological system requires a spatial approach across a territory (catchment area), a governance system adapted with different scales of decision-making, and necessarily the insurance of local population livelihoods.

References

Abed H, Esamil A, Barrahi M, Chahboun N, Khadmaoui A, Ouhssine M (2014) Evaluation analytique de la qualité microbiologique des eaux de lixiviats de la décharge publique de Kénitra. International Journal of Innovation and Applied Studies. ISSN 2028–9324 Vol 7 No 3, pp 1225–1231 © Innovative Space of Scientific Research Journals. http://www.ijias.issr-journals.org/

Adger WN (2003) Social capital, collective action and adaptation to climate change. Econ Geogr 79:387–404

Adger WN (2006) Vulnerability. Glob Environ Chang 16:268–281 www.elsevier.com/locate/gloenvcha

Adger WN, Hughes TP, Folke C, Carpenter SR, Rockstrom J (2005) Social-ecological resilience to coastal disasters. Science 309:1036–1039

Agence du Bassin Hydraulique de Sebou (ABHS) (2013) Étude d'actualisation du plan directeur d'aménagement intégré des ressources en eau de bassin hydraulique de Sebou. Note de synthèse, Agence du Bassin Hydraulique du Sebou

Allenby B, Fink J (2005) Toward inherently secure and resilient societies. Science 309:1034–1036

Bakker K (2012) Water security: research challenges and opportunities. Science 337:914–915. https://doi.org/10.1126/science.1226337

Billé R, Maris V, Cury P, Loreau M (2014) Biodiversité : vers une sixième extinction de masse

Boxall Alistair BA et al (2009) Impacts of climate change on indirect human exposure toPathogens and chemicals from agriculture, environmental health perspectives, vol 117: 4 April, 508–514

Brown K (2003) Three challenges for a real people-centred conservation. Glob Ecol Biogeogr 12:89–92

Community and Regional Resilience Institute Report (CARRI) (2013) Definitions of community resilience: an analysis. Meridian institute

Conseil Economique, Social et Environnemental (CESE) (2015) Gouvernance des ressources en eau au Maroc. Gouvernance par la gestion intégrée des ressources en eau au Maroc. Levier fondamental de développement durable. http://www.cese.ma/Pages/Auto-saisines/AS_15_2014-Gouvernance-par-la-gestion-integree-des-ressources-en-eau-au-Maroc-Levier-fondamental-de-developpement-durable.aspx

Cook C, Bakker K (2012) Water security: debating an emerging paradigm. Glob Environ Chang 22:94–102. https://doi.org/10.1016/j.gloenvcha.2011.10.011

Driouech F (2010) Distribution des précipitations hivernales sur le Maroc dans le cadre d'un changement climatique: descente d'échelle et incertitudes (Doctoral dissertation, INPT)

Douglas IAN (2006) The local drivers of land degradation in South-East Asia. Geogr Res 44:123–134

Dow K (1992) Exploring differences in our common future(s): the meaning of vulnerability to global environmental change. Geoforum 23:417–436

El Abidine A (2003) Le dépérissement des forêts au Maroc: Analyse des causes et stratégies de lutte. Sécheresse 14(4):209–218

El -Fadel M, Findikakis AN, Kekei JO (1997) Modeling leachate genrationnad transport in solid wast landfills. Environ Technol 18:669–686

El Haji M, Boutaleb S, Laamarti R, Laarej L (2012) Qualité des eaux de surface et souterraine de la région de Taza (Maroc): bilan et situation des eaux. Afrique Science: Revue Internationale des Sciences et Technologie 8(1):67–78

Friel S, Trevor H, Tord K, Gordon M, Monge P, Joyashree R (2011) Urban HealthInequities and the added pressure of climate change: an action-oriented research agenda. J Urban Health., Oct 88(5):886–895

Gunderson LH, Holling CS (2002) Panarchy: understanding transformations in human and natural systems. Island Press, Washington, DC

Hersperger AM, Gennaio MP, Verburg PH, Bürgi M (2010) Linking land change with driving forces and actors: four conceptual models. Ecol Soc 15:10

High Commission for Water and Forests and the Fight against Desertification (2015) The National Forest Council meeting Tuesday 26 May 2015 http://www.eauxetforets.gov.ma/fr/contenu.asp x?detail=yesandRubrique=9andid=1328

Huntjens P, Pahl-Wostl C, Rihoux B, Schlüter M, Flachner Z, Neto S, Koskova R, Dickens C, Nabide Kiti I (2011) Adaptive water management and policy learning in a changing climate: a formal comparative analysis of eight water management regimes in Europe, Africa and Asia. Environ Policy Govern 21:145–163. https://doi.org/10.1002/eet.571

Journal de la Santé de la Méditerranée Orientale (JSMO) (2006) "Pollution par les margines et production d'eau potable. Cas de l'Oued Sebou au Maroc", dossier 8, 1

Lestrelin G (2010) Land degradation in the Lao PDR: discourses and policy. Land Use Policy 27:424–439

M'hirit O, Benzyane M, Benchekroun F, El Yousfi SM, Bendâanoun M (1998) L'arganier, une espèce fruitière-forestière à usage multiples. MARDAGA, Belgique, 150 p

Ministre de l'Energie, des Mines, de l'Eau et de l'Environnement - Département chargé de l'Eau (MEMEE) (2013) Les bassins hydrauliques du Maroc

Nouaim R (2005) L'arganier au Maroc: entre mythes et réalités. Paris: L'Harmattan. 227 S

Observatoire Régional de l'Environnement et de Développement Durable (OREDD) (2014) Fiche changements climatiques dans la région du Gharb chrarda beni hssen. www.4c.ma/medias/ le_changement_climatique_dans_la_region_de_gharb_chrarda_beni_hssen.pdf

Ostrom E (2001) Vulnerability and polycentric governance systems. IHDP Update 3/01, pp 1–4; http://www.vedura.fr/social/sante/deces-pollution-eau

Pahl-Wostl C, Lebel L, Knieper C, Nikitina E (2012) From applying panaceas to mastering complexity: toward adaptive water governance in river basins. Environ Sci Pol 23:24–34. https:// doi.org/10.1016/j.envsci.2012.07.014

Pelling M (2003) The vulnerability of cities: natural disasters and social resilience. Earthscan, London

Portier CJ, Thigpen Tart K, Carter SR, Dilworth CH, Grambsch AE, Gohlke J, Hess J, Howard SN, Luber G, Lutz JT, Maslak T, Prudent N, Radtke M, Rosenthal JP, Rowles T, Sandifer PA, Scheraga J, Schramm PJ, Strickman D, Trtanj JM, Whung P-Y (2010) A human health perspective on climate change: a report outlining the research needs on the human health effects of climate change. Environmental Health Perspectives/National Institute of Environmental Health Sciences, Research Triangle Park, NC https://www.niehs.nih.gov/health/materials/a_human_ health_perspective_on_climate_change_full_report_508.pdf

Recensement Général de la Population et de l'Habitat (RGPH) (2014) Report (Statistics) of the High Commission to the plan (HCP, 2014)

Régie autonome d'eau et d'électricité (RADEEM) (2009) Mise en service de la station d'épuration des eaux usées de la ville de Meknès. http://mediterranee.typepad.fr/marketing/2009/06/mise-en-service-de-la-station-d%C3%A9puration-des-eaux-us%C3%A9es-de-la-ville-de-mekn%C3%A8s.html

Scheffer CS, Foley JA, Folke C, Walker B (2001) Catastrophic shifts in ecosystems. Nature 413:591–596. https://doi.org/10.1038/35098000 http://www.nature.com/nature/journal/v413/n6856/abs/413591a0.html

Shaker SM (2004) "Pollution physico-chimique et parasitologique du bas Sebou par les eaux usées urbaines de la ville Kénitra", Diplôme d'études supérieures approfondies en : Eaux Usées et Santé. Département de Biologie, Faculté des Sciences, Université Ibno Tofail de Kénitra

Stockholm Environment Institute (2001) Strategic environmental framework for the greater Mekong subregion: integrating development and environment in the transport and water resource sectors. Stockholm Environment Institute and Asian Development Bank, Stockholm

Stour L, Agoumi A (2008) Sécheresse climatique au Maroc durant les dernières décennies. Hydroécologie Appliquée 16:215–232

Van Noordwijk M, Lusiana B, Villamor G, Purnomo H (2011) Feedback loops added to four conceptual models linking land change with driving forces and actors. Ecol Soc 16:702

Vedura (2015) Décès dus à la pollution de l'eau. Santé et développement durable. Source : http://www.vedura.fr/social/sante/deces-pollution-eau. Droits de reproduction : http://www.vedura.fr/legal/droit-reproduction-contenu (à lire impérativement avant toute reproduction de contenu)

Walker B, Carpenter S, Anderies J, Abel N, Cumming GS, Janssen M, Lebel L, Norberg J, Peterson GD, Pritchard R (2002) Resilience management in social-ecological systems: a working hypothesis for a participatory approach. Conserv Ecol 6(1):14 [online] URL: http://www.consecol.org/vol6/iss1/art14/

Wiley LF (2010) Mitigation/adaptation and health: health policymaking in the global response to climate change and implications for other upstream determinants. J Law Med Ethics 38(3):629

World Health Organization (WHO) (2014) WHO guidance to protect health from climate change through health adaptation planning. WHO Library Cataloguing-in-Publication Data http://apps.who.int/iris/bitstream/10665/137383/1/9789241508001_eng.pdf

Chapter 13
Participation of Female Farmers Groups in *Kai* Algal Processing and Production in Northern Thailand

Seksak Chouichom and Lawrence M. Liao

Abstract The role of female farmers has increased in recent years to supplement family income in many rural households. This research involved female farmers engaged in processing *kai* algae in Chiang Rai Province, Thailand studied by the population parameters technique. An interview schedule with open-ended and closed questions and descriptive statistics were used. The responses were scored on a five point Likert's scale ranging from 'very strongly agree (5)' to 'very strongly disagree (1)' in order to assess the level of participation among respondents. Results of this study revealed that female farmers were mostly between 51 and 60 years old (33.2%), and most of them were members of the locally organized *kai* algal processing group (86.6%) of the village. Their educational attainment was low with 60.0% of respondents completing primary education and comprising a family of 3 to 4 members. Monthly income for most of them (43.3%) ranged between 5001 and 10,000 Thai Baht (100 THB = 2.83 US$). The participation of female farmers in group activities got established strongly within five areas, namely: 1) planning process (budget planning, mean = 4.27); 2) group production and implementation process (all production responsibility, mean = 4.47), 3) group activities evaluation process (group activities' suggestion and improvement for the management, mean = 4.34); 4) selling activities (booth set up and exhibition, mean = 4.40); and 5) benefit distribution (benefit management for group member, mean = 4.53). The respondents indicated that the raw material was insufficient for the whole year production due to the highly seasonal nature of the algae. Moreover, the group needed more logistic and financial support for marketing and promotion both inside and outside the province.

Keywords Algal products · Cooperative · Countryside development · Women farmers

S. Chouichom (✉)
Kasetsart University, 50 Ngamwongwan Rd., Ladyao, Chatuchack, Bangkok, Thailand
e-mail: seksak.ku@gmail.com; joys-tistr@hotmail.com

L. M. Liao (✉)
Graduate School of Biosphere Science, Hiroshima University, Higashi-Hiroshima, Japan
e-mail: lliao@hiroshima-u.ac.jp

© Springer International Publishing AG, part of Springer Nature 2019
M. Behnassi et al. (eds.), *Human and Environmental Security in the Era of Global Risks*, https://doi.org/10.1007/978-3-319-92828-9_13

1   Introduction

Recently, more and more female farmers have played vital roles and gained more empowerment within the Thai agricultural, economic and social sectors. In terms of agricultural expansion and development work, female farmers have shown more enthusiastic participation in extension work than their male counterparts (Chouichom 2014). Moreover, female farmers have gained more advancement compared to Thai male farmers in their agricultural knowledge and abilities, a fact that can be obviously observed in many other sectors of society such as in many kinds of Thai media like the printed media, television, radio (Sakaowrat et al. 2012).

Thai agrarian communities have typically been male-dominated, but owing to changing demographics and urban migration in recent years, there is increasing need for female family members to assume roles traditionally performed by male members. In fact, the growing participation of Thai females in farming activities involved a long process that is anchored on their knowledge of people and society and their problem-solving abilities (Tripakornkusol 2002). While females cannot always make up for decreased income arising from the lack of manpower especially of male members, there is a need to search for alternative sources of farm income.

Kai algae (*Cladophora glomerata*) are freshwater green algal species distributed in many parts of the world. In the rivers of northern Thailand and neighboring Laos, they grow abundantly albeit seasonally and are harvested to make dried sheets consumed as snacks (Peerapornpisal et al. 2006; Ajisaka and Wakana 2004). Along the Mekong River passing through Chiang Rai province in Thailand, *kai* algae are particularly common where some folk medicinal uses have been documented, such as anti-pyretic and stomach ulcer cures (Peerapornpisal et al. 2006). In Vieng Subdistrict, Chiang Khong District, Chiang Rai province, female farmers have processed fresh *kai* algae for making some commodities in support of the One Tambon (Village) One Product (OTOP) government livelihood program and which can potentially increase family income and support their farm expenditures (Sriwan and Prasert 2001). They processed *kai* algae into various kinds of products – such as biscuits, cookies, steamed fish with curry paste, crisp fried sheets– as a supplementary source of income. Nevertheless, some activities within the female farmers group involving *kai* production are still facing some production obstacles causing activities to slow down and weakening the cooperation among the members.

Based on these, the specific objectives of the study were: (1) to survey some socio-economic background of female farmers group engaged in *kai* production; (2) to examine their participation in the production of processed *kai* commodities; and (3) to identify some obstacles in their participation and suggest the best ways to improve the situation.

2 Methodology (Study Area, Data Collection and Analysis)

This study was conducted among female farmers who processed *kai* products. In April 2014 30 members of a group in Vieng Sub-district, Chiang Khong District, Chiang Rai province, Thailand using the population parameters technique following the pretest conducted one month earlier. The study employed a semi-structured and structured questionnaire. In order to complement both quantitative and qualitative data, more information was collected through focus group discussion by face-to-face interviews. This analysis used population-based survey to determine the participation of female farmers in group activity involving *kai* commodity production. Interviews were conducted on farm sites or in the farmers' households. The interview schedules were composed of open-ended and closed questions but some questions elicited quantitative data as well. The modified interview schedule includes seventeen statements of participation regarding *kai* commodity production. The received responses were scored on a five-point Likert's scale (1932) ranging from "strongly agree (5)" to "strongly disagree (1)". The data were analyzed with the Statistical Package for the Social Science (SPSS) for Windows. Descriptive statistics was applied to analyze percentage, arithmetic mean, minimum and maximum.

3 Results and Discussion

3.1 Demographic Characteristics of Women Farmers

As shown in Table 13.1, all respondents of *kai* commodity production in this study were between 51 and 60 years old. Sakaowrat et al. (2012) mentioned female farmers in the age of 45.12 years in average mostly graduated from elementary school (46.50%), while Wirote et al. (2001) showed younger respondents in their study in northeastern Thailand with ages between 41 and 50 years (26.2%) and mostly graduated from elementary school (67.1%). These implied that Thai female farmers are ageing, with lots of group working experiences even with the low level of education. Most of them could read and sign their names. Female farmers often cited economic reasons as constraints towards attaining higher educational level at that time, inconvenient location of schools and difficult access, among others. Their family members in their households were 3–4 persons (60.0%). Most of them were farmers (46.7%), got a monthly income between 5001and 10,000 Thai Baht (43.3%), and were actively engaged in *kai* production (86.6%).

In contrast, Chouichom (2015) found that members of female farmers group in some parts of northeastern Thailand got higher than 20,000 Thai Baht monthly income. Sakaowrat et al. (2012) said that the total agricultural income of female farmers group in Surin province, northeastern Thailand was around 140,171 Thai Baht a month. This discrepancy can be explained by the different nature of agricultural products and market situation there. Additionally, they also mostly received

Table 13.1 General profile of female farmers involved in *kai* production

Data	Frequency (Persons)	Percentage
1. Gender		
Male	0	0.00
Female	30	100.00
Total	30	100.00
2. Age (years)		
40–50	5	16.70
51–60	10	33.20
61–70	8	26.70
71–80	5	16.70
>80	2	6.70
Total	30	100.00
3. Status		
Group leader	1	3.30
Group committee	3	10.00
Group member	26	86.60
Total	30	100.00
4.Educational level		
Elementary school	18	60.00
Junior high school	4	13.30
High school	4	13.30
Diploma	1	3.30
Bachelor 's degree	3	10.00
Total	30	100.00
5. Household members (Persons)		
1–2	8	26.70
3–4	18	60.00
5–6	4	13.30
Total	30	100.00
6. Main occupation		
Farmers	14	46.70
One Tambon one product (OTOP) member	2	6.70
Laborer	5	16.70
Seller/vendor	8	26.70
Company employee	1	3.30
Total	30	100.00
7. Monthly income (baht/ThB)		
<5000	12	40.00
5001–10,000	13	43.30
10,001–15,000	3	10.00
15,001–20,000	1	3.30
>20,001	1	3.30

(continued)

Table 13.1 (continued)

Data	Frequency (Persons)	Percentage
Total	30	100.00
8. Agricultural media adoption		
TV	19	63.33
Radio	5	16.67
Printed media	4	13.33
Internet	2	6.67
Total	30	100.00

agricultural information via the television (63.3%). Recently, agricultural media has become important for female farmers particularly the television that has motivated them to acquire new agricultural knowledge. Chouichom (2016) mentioned that 48.0% of rice farmers of both sexes in the central region of Thailand received agricultural information through the television. Doss and Morris (2001) found a slight disparity in the adoption of agricultural technology between sexes but it can partly be attributed to difference in resource access. Wheeler (2007) and Chouichom (2014) observed that agricultural information and knowledge can also be obtained from cooperatives, unions and other organizations. In contrast, it was apparent from this survey that Thai farmers have not much accessed to digital information by using Internet (6.7%) when in developed countries, such use of digital media is on the rise for the management of farms (Mishra et al. 2009).

3.2 Participation of Female Farmers' Group in Kai Production

In this part, the study explored the female farmers' participation and involvement in group activities in *kai* production in order to measure their actual participation in five stages of the production activities as shown in Table 13.2. Briefly, a very strong agreement in participation levels (mean score = 4.21–5.00) was shown in almost every factor measured describing their participation in *kai* production. For instance, the participants were involved and participated in group benefits distribution at the most highest (strongest) level of participation (mean = 4.50), and involved in production and implementation as well as in the selling process at similarly strong agreement levels as well (means = 4.40 and 4.40, respectively). Most female farmers reasoned that benefit distribution for all members was the most important issue which supported group cooperation work among all members. Chouichom (2015) also revealed in his research study that most female farmers in northern Thailand showed their active participation in the selling process by showing their products during exhibition activities, similar to the results obtained in the present study at very strong agreement level (mean = 4.57). Supachai et al. (2013) have shown that female farmers were strongly involved in the project performance activities because they have a clear idea about the need to increase income and decrease expenditure.

Table 13.2 Levels of Participation Among Female Farmers in Various Steps of *Kai* Algae Production

Participation Details	Participation Levels					Mean
	5(%)	4(%)	3(%)	2(%)	1(%)	
1. Participation in planning process						
1.1 Material preparation	2 (6.70)	23 (76.70)	5 (16.70)	0 (0.00)	0 (0.00)	4.00 (Strongly Agree)
1.2 Production period	7 (23.30)	23 (76.70)	0 (0.00)	0 (0.00)	0 (0.00)	4.23 (Very Strongly Agree)
1.3 Group laborer	6 (20.00)	24 (80.00)	0 (0.00)	0 (0.00)	0 (0.00)	4.20 (Strongly Agree)
1.4 Group budget	8 (26.70)	22 (73.30)	0 (0.00)	0 (0.00)	0 (0.00)	4.27 (Very Strongly Agree)
Total						4.18 (Strongly Agree)
2. Participation in production and implementation						
2.1 Collaboration in all group activities	14 (46.70)	16 (53.30)	0 (0.00)	0 (0.00)	0 (0.00)	4.47 (Very Strongly Agree)
2.2 Act like one's own business	12 (40.00)	18 (60.00)	0 (0.00)	0 (0.00)	0 (0.00)	4.40 (Very Strongly Agree)
2.3 Group decision making and sharing	11 (36.70)	19 (63.30)	0 (0.00)	0 (0.00)	0 (0.00)	4.37 (Very Strongly Agree)
2.4 Assuming group activities responsibility	11 (36.70)	19 (63.30)	0 (0.00)	0 (0.00)	0 (0.00)	4.37 (Very Strongly Agree)
Total						4.40 (Very Strongly Agree)
3. Participation in group activities evaluation						
3.1 Having description about weak activities of group to other members	11 (36.70)	19 (63.30)	0 (0.00)	0 (0.00)	0 (0.00)	4.37 (Very Strongly Agree)
3.2 Suggesting improvement of group activities to group leader	13 (43.30)	17 (56.70)	0 (0.00)	0 (0.00)	0 (0.00)	4.43 (Very Strongly Agree)
3.3 Always implemented groups' suggestion and implementation	12 (40.00)	18 (60.00)	0 (0.00)	0 (0.00)	0 (0.00)	4.40 (Very Strongly Agree)
Total						4.40 (Very Strongly Agree)
4. Participation in selling process						
4.1 Join selling process every time	12 (40.00)	18 (60.00)	0 (0.00)	0 (0.00)	0 (0.00)	4.40 (Very Strongly Agree)

(continued)

Table 13.2 (continued)

Participation Details	Participation Levels					
	5(%)	4(%)	3(%)	2(%)	1(%)	Mean
4.2 Exhibition and booth showing every time	12 (40.00)	18 (60.00)	0 (0.00)	0 (0.00)	0 (0.00)	4.40 (Very Strongly Agree)
4.3 Knowing price of each algal products	11 (36.70)	19 (63.30)	0 (0.00)	0 (0.00)	0 (0.00)	4.37 (Very Strongly Agree)
Total						4.39 (Very Strongly Agree)
5. Participated in benefit distribution						
5.1 sharing opinion in benefit distribution for all members	14 (46.70)	16 (53.30)	0 (0.00)	0 (0.00)	0 (0.00)	4.47 (Very Strongly Agree)
5.2 decision making in benefit distribution for all members	15 (50.00)	15 (50.00)	0 (0.00)	0 (0.00)	0 (0.00)	4.50 (Very Strongly Agree)
5.3 searching for further benefits for members	16 (53.30)	14 (46.70)	0 (0.00)	0 (0.00)	0 (0.00)	4.53 (Very Strongly Agree)
Total						4.50 (Very Strongly Agree)

Besides, in term of the benefit and distribution stages, female farmers also had very strong agreement particularly in decision making for benefit distribution towards all members (mean = 4.50) which can accrue to more group advantages in the future. Dusiya (2011) found that the participation in decision making had the highest score among her respondents who expressed strong involvement in the management of the village fund while the participation in the evaluation or monitoring of the group obtained the lowest score. Chouichom (2015) found that the high participation in production and implementation stage (mean = 4.67) was due to the inclination of group members to act like business owners. But Prasert (2006) remarked that in general, female farmers participated in every stage of problem solving, problem analysis, planning, performing and evaluation with equal intensity. This is in contrast with a more selective group of farmers studied by Wirote et al. (2001) who were involved more in evaluation activities (65.8%) but lesser so in investment decision making (58.4%).

In the case of planning processes, respondents expressed the lowest agreement level (mean = 4.18). This implies that group members still lack the initiative and stimulus to participate in the planning activities even in short, mid and long-term planning process. This could be explained by general lack of technical know-how and low leadership confidence. This is particularly evident in traditionally male-dominated societies like Japan where the gender gap is still wide (Kawate 2010). In many developing countries, many forms of gender segregation in the workplace still exist with women often accorded the least stable positions (Saptari 1991). Nonetheless, Dolisca et al. (2006) have suggested that policies designed to improve technical assistance are essential to strengthen farmers' participation in the program. In addition, female farmers still needed some more suggestions from agricultural extension workers. Interestingly, Lahai et al. (1999) mentioned that female farmers who dealt with female extension agents had relatively higher level of awareness and participation of extension activities organized, adoption of and technological knowledge of recommended technologies/practices and higher satisfaction with the quality of agents' service and credibility.

3.3   Women Farmers' Constraints and Suggestions from this Study

According to the survey, female farmers said that raw material particularly the *kai* algal raw material was not enough for year-round production due to the seasonality of the algae. A tentative solution was to store the raw materials in dried form obtained during the harvest season. This creates a storage problem and besides, dried material may not be suitable for dried algal sheet production. Moreover, they needed more support for marketing like dedicated budget for exhibitions or product shows both inside and outside the province. Chouichom (2015) had suggested that more attractive packaging should be provided for a better image.

4   Conclusions

The female farmers involved in the production of the *kai* algal final product indicated their participation and support in the benefit distribution stage through a very strong agreement (mean = 4.50), particularly in the search for further economic benefits to members (mean = 4.53). The other indicators also pointed to high Likert scale values suggesting a high level of participation and interest in all stages of production and operations. The current set-up should be maintained but improvements must also be introduced for the industry to maintain its viability. For example, the problem about raw material shortage and better marketing strategy should be addressed with the help of agricultural extension workers and government technical agencies.

Kai algal production represents a unique and recent economic augmentation activity for female famers, and only in the extreme northern regions of Thailand along the tributaries of the Mekong River where *Cladophora glomerata* grows abundantly albeit on a highly seasonal basis. Other supplemental income sources for female farmers have traditionally included basket making and silk weaving

(Chouichom 2014) but these are low-scale and have encountered stiff market competitions. In addition, other options for new livelihoods for female farmers are limited considering their low educational attainment and lack of schooling opportunities.

In areas of northern Thailand where *kai* algal production is a growing cottage industry, the majority of these rural industries involved female farmers' groups. Our study revealed that female farmers viewed their participation as a means for supplementing their family income, representing an important support for the rural economy. The emergence of new economic activities like *kai* algal production is also important in a region notoriously known as The Golden Triangle where opium poppy (*Papaver somniferum*) planting is a lucrative but dwindling activity. The maintenance of women-centered industries is also very important in this region especially when the emigration of rural men into larger cities is increasing, thereby entrusting greater responsibility on women for farm management and income generation. To achieve these goals, technical support and trainings in entrepreneurship among rural female farmers must be intensified to sustain their participation and interest, to insure the viability of their industries and to sustain rural development.

References

Ajisaka T, Wakana I (2004) Researches of Cladophoraglomerata (Linnaeus) Kuetzing and Spirogyra spp. in Laos. 2004 Human Ecology Project Report, Kyoto University. pp. 338–344. (in Japanese, with English abstract)

Chouichom S (2014) Some socio-economic factors affecting farmers' participation of agricultural extension education effort: a case study in northeastern Thailand. In: Behnassi M et al (eds) Science, policy and politics of modern agricultural system: global context to local dynamics of sustainable agriculture. Springer, New York, pp 47–60

Chouichom S (2015) Participations of female farmers' group in Kai algae products production in Tha Wang Pha District, Nan Province. 53rd Kasetsart University Annual Conference, February 3–6, 2015, Bangkok, Thailand

Chouichom S (2016) Rice farmers' expectation towards the single fertilizer utilization on rice production in Suphan Buri Province, Thailand. 54th Kasetsart University Annual Conference, February 2–5, 2016, Bangkok, Thailand

Dolisca F, Carter DR, McDaniel JM, Shannon DA, Jolly CM (2006) Factors influencing farmers' participation in forestry management programs: a case study from Haiti. J For Ecol Manage 236(2–3):324–331

Doss CR, Morris ML (2001) How does gender affect the adoption of agricultural innovations? The case of improved maize technology in Ghana. Agric Econ 25:27–39

Dusiya W (2011) The management participation of people to village fund Donducyae in Tambon Phang-Tru Sub district, Thamuang District, Kanchanaburi Province, Thailand. Bansomdejchaopraya Rajabhat University master's thesis. (in Thai, with English abstract)

Kawate T (2010) Change and problems regarding women farmers in Japan. In: Tsutsumi M (ed) A turning point of women, families and agriculture in rural Japan. Gakubunsha, Tokyo, pp 22–34

Lahai BAN, Goldey P, Jones GE (1999) The gender of extension agent and farmers' access to and participation in agricultural extension in Nigeria. J Agric Educ Ext 6(4):223–233

Likert R (1932) A technique for the measurement of attitudes. Archives of Psychology 140:5–55

Mishra AK, Williams RP, Detre JD (2009) Internet access and internet purchasing patterns of farm households. Agricultural and Resource Economics Review 38:240–257

Peerapornpisal Y, Amornledpison D, Rujjanawate C, Ruangrit K, Kanjanapothi D (2006) Two endemic species of macroalgae in Nan River, northern Thailand, as therapeutic agents.*Science Asia* 32. Suppl 1:71–77

Prasert K (2006) A study of participation in good hygiene practice of Kanom Chun (Thai dessert): A case study of Satripattana agricultural wives club, Thailand/Rachapatrachanakarin. University master's thesis

Sakaowrat M, Sunun S, Jinda C (2012) The operations of farm women groups in Surin Province. The 2nd STOU Graduate Research Conference. September 2–5, 2012. Sukhothai Thammathirat Open University, Nonthaburi Province, Thailand

Saptari R (1991) The differentiation of a rural industrial labour force: gender segregation in east Java's *kretek* cigarette industry, 1920-1990. In: Alexander P et al (eds) In the shadow of agriculture: non-farm activities in the Javanese economy, past and present. Royal Tropical Institute, Amsterdam, pp 127–150

Sriwan C, Prasert C (2001) The study of kai (green algae) in ecosystems. The Thailand Research Fund (TRF). (Completed Report)

Supachai J, Bumpen K, Jinda K (2013) Participation and role of member in farm woman group cooperation in Mueang District of Mae Hong Son Province, Thailand. The 3rd STOU Graduate Research Conference. September 3–4, 2013. SukhothaiThammathirat Open University, Nonthaburi Province, Thailand

Tripakornkusol S (2002) Application of participation concept to solve health problems in community. Khon Khean University master's thesis

Wheeler SA (2007) What influences agricultural professionals' views towards organic agriculture? Ecol Econ 65:145–154

Wirote I, Sithichai K, Pote B, Somejate J, Am-On A (2001) Participation in community development of farmer housewives, Potipisal Sub-district, Kusumal District, Sakhon Nakhon Province, Thailand. Kasetsart University, Bangkok. (Completed Report)

Part III
Environment and Development: Case Studies

Chapter 14
Risks and Opportunities of Sustainable Biomass and Biogas Production for the African Market

Olaf Pollmann, Szilárd Podruzsik, and Leon van Rensburg

Abstract Almost all important natural resources – e.g. oil, diamonds, gold, platinum, coal, copper, ore, phosphate etc. – including rare earth metals for industries functioning in the international arena, are available on the African continent. Roughly 15% of the total world market of resources is in Africa. International interest in Africa is via the relatively political and economic stable countries of the 'Africa 7' – South Africa, Botswana, Morocco, Ghana, Nigeria, Egypt and Kenya. According to Doing Business Report and the World Bank, statistically, about 60% of the resources from agriculture in Africa are still reserves. Based on the international trends to support developing countries by establishing a market for biomass production, with the aim of advanced energy production, resource efficiency, emission reduction, it has to be balanced with the internationally supervised and important sector of food security. As markets in Africa show, the agricultural sector, including cattle production, to be a very strong market, biomass residues such as cow manure and other organic substances are available as a cheap or free by-product. To turn these natural resources into a valuable commodity, it simply has to be converted by fermentation into biogas usable for cooking or electricity and fertiliser usable for private gardening. Current knowledge has proved how effective and simple biomass and biogas production can be. The output of one ton of silage can produce up to 200 m^3 of biogas with a productivity of 5.0–7.5 kWh electricity per m^3 of biogas. Cow manure as a source of waste product is available at an annual amount of 7.5–21.0 m^3 per animal with a productivity of about 30 m^3 of biogas including 56% of methane (CH_4). A micro-biogas production facility is viable with a capacity of about 1 kWh – this is equivalent to the digestion of cow manure from seven milk cows. Using biomass additionally supports the reduction of greenhouse gases

O. Pollmann (✉) · L. van Rensburg
North-West University, School of Environmental Science and Development,
Potchefstroom, South Africa
e-mail: 20942737@nwu.ac.za; o.pollmann@scenso.de; leon.vanrensburg@nwu.ac.za

S. Podruzsik
Corvinus University of Budapest, Department of Agricultural Economics and Rural
Development, Budapest, Hungary
e-mail: szilard.podruzsik@uni-corvinus.hu

© Springer International Publishing AG, part of Springer Nature 2019
M. Behnassi et al. (eds.), *Human and Environmental Security in the Era of
Global Risks*, https://doi.org/10.1007/978-3-319-92828-9_14

(GHG) particularly CO_2, NH_4, N_2O. Compared to CO_2, NH_4 has got 25-times higher and N_2O 298-times higher Global Warming Potential (GWP). The reduction of these greenhouse gases as a side benefit of biomass production provides an indication of how positive this strategy could be for both the agricultural sector as well and general public.

Keywords Resource efficiency · Biomass · Food security · Biogas production · Emission control and reduction

1 Introduction

Biomass is any material that is or was alive. The trees, crops, garbage, and animal waste all represent potential biomass. The energy in biomass is produced by the sun's rays (through photosynthesis) and stored in plant leaves and roots in various forms of sugar. Technically it is possible to convert this energy from biomass to gas or liquid fuels. As the biomass and its stored energy are renewable, the natural production of biomass takes only short time for reproduction. It is theoretically possible to always grow more plants until the limits of nature's resilience are reached. Some of years ago, biomass served as the major source of energy people used, i.e. by burning wood for heat to cook food and to keep the rooms warm in winter. In many developing countries, especially those in the third world, wood is still used to address most of their energy requirements. Additionally, corn cobs and straw is burned for energy as well. In places without trees, waste from cows and pigs are primarily used.

Biomass can also be used to produce biogas, an energy-rich gas. Biogas is similar to natural gas used in stoves and furnaces. In India and Pakistan, many farmers are already using garbage and animal waste to produce their private biogas for heating, light and cooking. They produce the gas by storing the waste in anaerobic tanks and decomposition of the organic material. Most of the waste that is left after the biomass decomposes can be used as fertilizer for crop production, and hence this practice significantly cuts the demand of chemical fertilizers.

The produced biogas is a mixture of gases that is composed of (based on Kaltschmitt and Hartmann 2001): methane (CH_4): 50–75 vol.%; carbon dioxide (CO_2): 25–45 vol.%; steam (H_2O): 2–7 vol.% (20–40 °C); dioxygen (O_2): < 2 vol.%; nitrogen (N_2): < 2 vol.%; ammonia (NH_3): < 1 vol.%; hydrogen sulfide (H_2S): 20–20.000 ppm; and tracer gas: < 2 vol.%.

The liquid fuel biodiesel can be made by the chemical reaction of alcohol with vegetable oils, animal fats, or greases also from recycled restaurant grease. Nowadays, most of the available biodiesel is produced from soybean oil. Biodiesel exceeds normal general diesel in cetane number, the performance rating of diesel fuel which is important for the superior ignition. It has a higher flashpoint which makes it more versatile and causes safety concern. Horsepower, acceleration, and torque are comparable to diesel. Nevertheless, biodiesel has the highest BTU content (British Thermal Unit) of any alternative fuel but slightly less than diesel.

The desirable characteristics of biodiesel pertained to the facts that it is renewable, nontoxic, and biodegradable. Compared to diesel, biodiesel reduces sulfur oxide (SO_x) emissions by 100 percent, particulates by 48 percent, carbon monoxide (CO) by 47 percent, unburned hydrocarbons by 67 percent, and hydrocarbons by 68 percent. As a side effect, nitrogen oxides (NO_x) emissions, however, increase slightly. Biodiesel blends generally reduce emissions in proportion to the percentage of biodiesel contained in the blend. When biodiesel is burnt it releases carbon dioxide (CO_2), which is a major contributor to climate change. However, biodiesel is made from crops that absorb CO_2 and releases oxygen. This cycle would maintain the balance of CO_2 in the atmosphere, but because of the CO_2 emissions from farm equipment and production of fertilizers and pesticides, biodiesel adds more CO_2 to the atmosphere than it removes.

One concern with the use of biomass, biodiesel, and biogas is that the required farmland for growing the biomass might compete with land needed to grow food (especially staples) for undeveloped countries. This could result in increasing food prices and dearth in local market. It is interesting to note that in South Africa, the "fuel versus food" debate has been stopped by the government by implementing legislation that no agricultural crop can be used for biodiesel production – this led to the initiation of a lot of research into the potential of using algae (waste from the cooling towers of SASOL and ESKOM) i.e. energy suppliers that burn fossil fuels – interesting irony.

2   Materials and Methods

2.1   Resources on the African Continent

International investigations state that most of the natural resources in total are available on the African continent (Fig. 14.1). Besides the international trade of natural resources, it is more important to focus on food safety for the people.

The worst scenario is the competition between food and goods for international export. Because of the discussion of periodically occurring food and hunger crises and draughts, new alternatives have to be adapted to these special circumstances. One possible option of supporting developing and emerging countries to upgrade life conditions could be the use of biomass and biogas production without interfering with the food availability. The generally utilized materials for the production of biomass and biogas are organic residues and animal manure. These materials are normally not suited for any further use and has to be deposited somewhere for decomposition or it will be burnt. The former may also cause an accumulation of nutrients with a combined leaching of water-soluble substances into the ground water.

Africa as a continent, with its huge agricultural potential, also has the possibility to effectively utilize biomass for the production of direct usable gas for cooking or

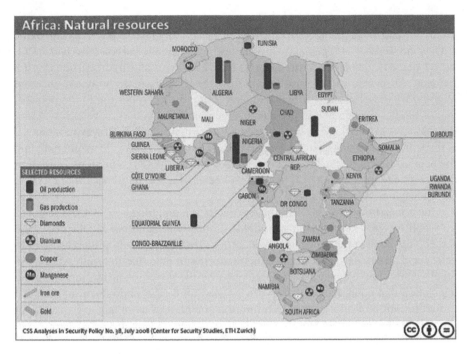

Fig. 14.1 Natural resources on the African continent
Source: Center for Security Studies (2008)

additionally to produce electricity. These additional resources from nature should not be unused and wasted.

2.2 Production of Biogas

Biogas is one of the resulting products of microbial degradation of organic substances in a moist anaerobic condition. In nature, this process of biological decomposition usually occurs on the bottom of lakes, in swamps or in paunches of ruminant animals. The fermentation is normally separated in different interdependent processes: hydrolysis, acidogenesis, acetogenesis and methanogenesis with different microorganisms (Fig. 14.2). During the process of hydrolysis, all complex organic bonds are separated to simple bonds (step 1) which are degraded to organic acids during the acidogenesis (step 2). At this stage, alcohol, hydrogen, and carbon dioxide are generated as basic materials for the methane production. In the following process of acetogenesis (step 2), the organic acids and alcohols are transformed to acetic acid, water, and carbon dioxide. All other products are transformed to methane, carbon dioxide, and water during the process of methanogenesis (step 3).

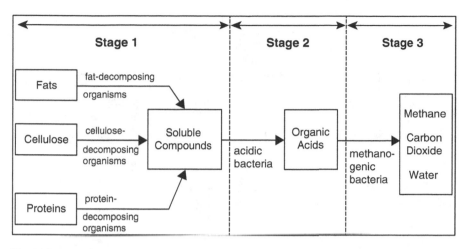

Fig. 14.2 Anaerobic digestion, which takes place in three stages produces biogas. Different kinds of micro-organisms are responsible for the processes that characterize each stage
Source: US Department of Energy (2010)

Generally, all phases occur concurrently. Because of different environmental conditions and different active microorganisms, a compromise of optimal parameters, like temperature, pH-value or nutrient status, must be found. The process of fermentation is sensitive to changing conditions caused by production or inhibitors.

The residue of the fermentation process is called the digestate. This digestate, generally, is recycled and applied to agricultural areas as a soil ameliorant. With this process step, the nutrient cycle can be closed (ISAT, 2011). If the digestate conforms to the required legislative requirements, it can also be used as farm fertilizer or high value organic fertilizer (Fig. 14.3). Most of the pathogenic germs and weed seeds are eliminated by the process and the plant available nutrient content is generally high.

For the biogas production, a wide variety of different organic substrates can possibly be used. Cattle and pig manure, as well as some purposely cultivated energy plants, are primarily used for biogas production. Internationally, a market has been established over the years which are constantly producing new biogas substrates and renewable primary products. The substrate input for biogas production is about 45% animal manure, 46% renewable primary products (energy plants), 7% biological waste as well as 2% residuals from industry and farming.

As a result of criticisms directed towards the use of renewable primary products – e.g. maize, grain, grass – the biogas plant operators have to explain possible negative impacts on soil fertility and biodiversity. An important fact is that not the maize crop itself is used only but the maize silage (blade) is used for the fermentation process. The use of manure also contributes towards the reduction of emissions and, therefore, represents an important step towards climate protection (mitigation).

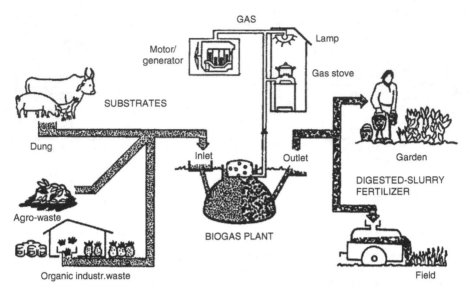

Fig. 14.3 Example of simplified biogas systems
Source: GIZ/GTZ (1989)

The gas yield of different substrates is dependent on different gassing potentials and technical and biological parameters of the production process. The energy production is a direct result of gas yield and specific energy content (Fig. 14.4).

2.3 Possible Risks of Biomass and Biogas Production

Despite all the positive aspects described and benefits of potential biomass use and biogas production, there are also negative aspects. As per usual, this market also depends on supply and demand. If a market for goods is easily accessible and potentially derived benefits have got higher value than the own consumption, the farmers will rather sell products for the benefit. Many studies have substantiated these facts (Susanti and Burges 2012). The most common observed side effect is the long-term degradation of farmland due to over exploitation in striving for greater benefits (Susanti and Burges 2012).

If the use of biomass and biogas production is limited to mainly self-utilization, then these options are more than efficient with no significant unintended negative consequences. However, if it is seen as a commercial product, for the economic market, then the influences could be more negative.

Internationally, the dilemma of competition between biomass for energy production and biomass for nourishment has been exhaustively discussed. Therefore, it is very important that the input for the production of biogas be manure and maize silage (Fig. 14.4) – input substrate without any valorisation potential. If the production is for personal use only, the no criticism can be levied, anyway.

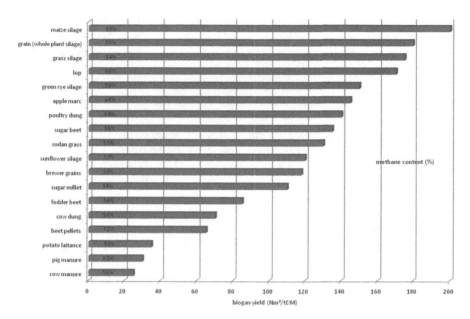

Fig. 14.4 Biogas earnings from different substrates
Source: Leitfaden Biogas (2010)

2.4   Economic Benefits of Biomass and Biogas Production in Africa

There are five main sources of energy consumption in Africa (Kauffmann 2005). The largest portion being biomass that represents 58%. The fractional contribution of petroleum is less than half of that of biomass, being only 25%. With 4 and 5%, coal and gas are the same quantity as the other energy sources. The electricity share is double (8%) that of coal or the gas. In Africa, biomass is mainly used in a traditional manner, although there are several other potential applications for which the biomass can be used such as for bioenergy production (Rutz and Janssen 2012).

Instead of answering the question whether food or energy production has priority, we wish to present a theoretical approach on the production of energy from a part of agricultural waste. Having a closer look at Africa, more precisely at the 'Africa 7' (Botswana, Egypt, Ghana, Kenya, Morocco, Nigeria, South Africa), some of the energy consumption, like electricity and natural gas, should be substituted by renewable energy production. The CIA World Factbook[1] indicates a notable number of animals in the 'Africa 7' countries. Animal manure is available in large quantities with a total of 60 million cattle in the selected countries. Previous studies presented the potential electric production or gas production in quantities from the animal manure (Chen et al. 2010; Cuéllar and Webber 2008).

---

[1] www.indexmundi.com

The animal manure transformation into biogas and electricity may support self-sufficiency, improve trade balance, and contribute to the GDP. The bioenergy production opens new possibilities of knowledge as well as technology market. The electricity production on the basis of the American and European average is 4.2 kWh per day per cow. The quantity produced depends on the animal size, feed, the applied technology, and the regulatory framework of the gas consumption (Mehta 2002).

Clearly, in 2011 all countries among 'Africa 7' consumed electricity in different quantities.[2] Kenya has the lowest consumption with 4.86 billion kWh yearly. The largest consumer is South Africa with its more than 215 billion kWh. Relating to the trade balance, Nigeria is a self-sufficient country in electricity where there is no electricity trade, either export or import. The trade balance position is the worst in Morocco with minus 3.5 billion kWh per year. Morocco is followed by Botswana (−2.1 billion kWh) and Ghana (−186 million kWh). South Africa shows a positive trade balance (3.5 billion kWh) as well as Egypt (563 million kWh) and Kenya (36 million kWh).

The natural gas consumption presents a different picture in these countries in 2011. In Botswana, Ghana, and Kenya there is no natural gas consumption or trade either. Egypt is the largest natural gas consumer (42.5 billion m^3). Nigeria is in the second place with its consumption of 12.2 billion cubic meters. South Africa has 6.4 billion cubic meters and Morocco has only 560 million cubic meter consumption from natural gas yearly. Countries where trade exists – such as Nigeria and Egypt – have positive trade balance (20.5 billion and 8.5 billion m^3), while South Africa has a deficit of 3.2 billion cubic meters and Morocco pointed out a deficit of 500 million cubic meters as well.

The number of cattle and the quantity of manure production should have a positive effect on the above-mentioned economic indicators, if the countries implement a good practice of animal waste management and transform manure into biogas or electricity production. On the basis of the cattle numbers in the 'Africa 7' countries, Kenya has the largest electric power potential of 27.3 million kWh. Botswana should produce 3.9 million kWh. This means that the countries could produce a part of their own consumption or trade with the surplus of the electricity. Kenya could produce 563% of its own consumption from electricity. This figure is 148% in Botswana and 128% in Nigeria. The potential is the smallest in Egypt with 7%.

The total consumption of biogas for substitution of the natural gas could mean 44% of its own consumption for Morocco. South Africa gas potential shows the second place with 18%, Nigeria has 12% and Egypt has only 1%. In the countries where there is no natural gas consumption – such as in Botswana, Ghana, and Kenya – the surplus could be traded (Table 14.1).

Biogas production, independent if it is produced from animal waste or other sources, depends on the cleaning and upgrading process of the raw material. Bio

---

[2] www.indexmundi.com

Table 14.1 Electric power potential and biogas production potential in Africa 7, 2011 (Source: Authors' calculation on the basis of FAO data and CIA Factbook)

	Botswana	Egypt	Ghana	Kenya	Morocco	Nigeria	South Africa
Electricity consumption, million kWh	2650	104,100	5700	4860	20,780	19,921	215,100
Electricity export, million kWh	0	814	249	58.3	0	0	14,160
Electricity import, million kWh	2181	251	435	22.5	3429	0	10,570
Electricity trade balance, million kWh	−2181	563	−186	35.8	−3429	0	3590
Natural gas consumption, million m³	0	42,500	0	0	560	12,280	6449
Natural gas export, million m³	0	8550	0	0	0	20,550	0
Natural gas import, million m³	0	0	0	0	500	0	3200
Natural gas trade balance, million m³	0	8550	0	0	−500	20,550	−3200
Cattle, million head	2.5	4.5	1.5	17.9	2.9	16.6	13.7
Electric power potential, million kWh	3909	6937	2229	27,384	4439	25,414	21,050
Biogas production, million m³ (1020 BTU equivalent)	219	389	125	1534	249	1424	1179
Electric power potential/ electricity consumption	148%	7%	39%	563%	21%	128%	10%
Biogas production potential/natural gas consumption	...	1%	...	...	44%	12%	18%

methane then can be used for natural gas replacement, and transportation fuel represents the final and top quality of the production process. Fixed cost and variable costs determine the production expenses. The production of biogas requires a capital cost investment in the form of infrastructure. Management and operational costs are moderate relative to the capital. In the production activity, the rule of the economy of scale is applicable. Increased production relates to the lower production costs (Chen et al. 2010) (Fig. 14.5).

Cost of externalities can change economic and financial indicators. Animal breading and waste can cause negative externalities. Air pollution of harmful gases such as methane, ammonia, carbon dioxin and hydrogen sulphide has a harmful effect on human health and animal health. These externalities cause extra expenses on health. Proper waste management may contribute to avoiding water and soil quality distortions that has effect on the agricultural product quality as well (El Shol and Wesseler 2010).

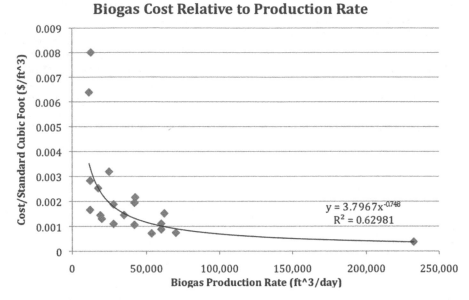

Fig. 14.5 Biogas cost relative to production rate
Source: Chen et al. (2010:10)

3   Importance of Biomass and Biogas for the African Market

As major disparities exist between developed and developing nations as well as within the developing world, both between nations and within nations, the lack of access to energy or the lack of adequate supply of energy is a key aspect of poverty but has to balanced with fair environmental accountability. For example, despite South Africa's modest contribution to global emissions (1.1% in 2005), its per capita emission rate of 9 tonnes CO_2 per person is above the global average of 5.8 tonnes CO_2 per person and about six times higher than the average for sub-Saharan Africa at 1.4 tonnes CO_2 per person. The "food, energy/development and environment trilemma" (Tilman et al. 2009), as it relates to biomass production and organic waste management, is therefore particularly relevant within the African context.

The problem of global warming of anthropogenic origin is intimately tied to the question of energy. Nations of the world continue to be heavily dependent on fossil fuel sources for much of their energy needs. However, major disparities exist with regards to per capita electricity production, provision, and consumption. Thus, as mentioned earlier, the lack of access to energy or the lack of adequate supply of energy is a key aspect of poverty. Energy deficiency is to be seen as an important element of the basket of deprivation that constitutes poverty.

Some 2 billion people around the world, including 89% of the sub-Saharan African (SSA) population, use biomass for cooking and heating. Although energy access is an important aspect of comprehensive development, access alone does not reflect upon sustainability of that development, especially in the use of global atmospheric space. The world has a finite carbon budget which needs to be equitably shared. Clearly, it is incumbent on developed nations that currently over-occupy the carbon space as well as developing nations to transition to a low-carbon economy.

According to Ram (2010), SSA has the world's lowest electricity access rate at 26%, with a rural electricity access rate of only 8%. 85% of the people in SSA rely on biomass for energy. In a quest for modern energy, 70% of household income is spent on energy (diesel, kerosene, charcoal); 0.4 million hectare forests are cleared each year in Africa. Furthermore, SSA largely depends on biomass and its associated products, of which the production of charcoal and use for fuel wood is inefficient. As a result, deforestation is rampant.

It is also worth noting that exposure to indoor pollution is globally responsible for between 1 to 2 million premature deaths and a substantial portion of these deaths occur in SSA. Although properly designed and used, a biogas digester mitigates a wide spectrum of environmental undesirables: it improves sanitation; it reduces greenhouse gas emissions; it decreases demand for wood and charcoal for cooking, and therefore helps preserve forested areas and natural vegetation; it provides a high-quality organic fertilizer; and from the developing world's perspective biogas's greatest benefit may be that it can help alleviate the very serious problem of poor indoor air quality (Brown 2006).

The above numbers indicate a huge energy infrastructure gap and require an urgent fix (Ram 2010). On a positive note, proponents argue that developing countries can, in theory, adopt more renewable energy technologies and gain experience in them more easily, in essence accelerated adopting of the technologies through 'technological leapfrogging', thereby offsetting or decreasing projected carbon emissions (GEF 2006).

4   Results and Discussion

Investigations have shown that the output of one ton of maize silage is up to 200 m^3 of biogas with a productivity of 5.0–7.5 kWh electricity per m^3 of biogas. The source cow manure as a waste product is available at an annual amount of 7.5–21.0 m^3 per animal with a productivity of about 30 m^3 of bio-gas including 56% of methane. With these calculations, it is lucrative to run a micro-biogas production for small private users.

Using this biomass for biogas production additionally supports the reduction of greenhouse gases (GHG) emissions, particularly CO_2, NH_4, and N_2O. Compared to CO_2, NH_4 has got 25-times higher and N_2O 298-times higher Global warming potential (GWP). The reduction of these gases as a side effect of biomass production shows how positive this strategy is for the agricultural sector as well as for the general public.

With the overall calculations, the discussed technology of using biomass for middle size or even small biogas production are a lucrative business for all involved parties from emerging countries as Southern Africa. The resources of cow manure, maize, and grass silage are more or less free available and can be used for direct production. Farmers will have the possibility to use a combination of maize silage together with cow manure for fermentation with an output of about 30 m³ biogas per cow per annum. Profitable will the production be with a capacity of about 300 cows for a middle sized settlement. For small sized private users, biogas from the manure of about 10 cows produce about 300 m³ biogas per annum with an equivalent of more than 1 kWh - which is enough for direct cooking during a year for the entire family.

Especially for countries with impacted areas of mining activities, the fermentation of grass silage will be a perfect solution of biomass usage. Most of grassland areas around mines are heavily polluted by heavy metals and other toxic substances; therefore that biomass source cannot be used for animal feeding or any other use. One hectare of this grass silage harvested from that polluted areas can constantly produce about 170 m³/t biogas. This could be as effective as the production of biogas from cow manure.

This technology is everywhere effective where enough resources for the biogas production are available. The above calculation proved that a biogas production plant is an ideal technique for abattoirs. Cattle have to be stored for about a day before they are slaughtered. With a daily capacity of some thousand animals per abattoir the availability of continuous manure production is ideal for biogas production and power equivalent of the entire abattoir.

5   Conclusion

The discussion of this project of biomass production for the African market has shown that different aspects are in focus and important. The well-functioning technique is economically and ecologically relevant in industrialized countries like Europe, especially because of getting governmental subsidy for leading energy from biogas and biomass in the public electricity grid. Other processes of generating energy like photovoltaic are more efficient even in Europe.

Even if no subsidy is paid, this technique is well suitable for developing and emerging countries. All organic residues and manure can easily be used for fermentation to produce biogas for cooking, warm water and heating or additionally produce electricity for light.

The balance of the biomass / biogas process is entirely positive: the input material is free of charge; and the output is the valuable product 'biogas' and 'fertilizers'. Only the investment of the micro-plant has to be done. The amortisation is – depending on the plant size – between 2–4 years.

References

Advisory Service on Appropriate Technology (ISAT) (2011) Biogas digest. vol 1: Biogas basics, vol 2: Biogas - application and product development

Brown V (2006) Biogas: a bright idea for Africa. Environ Health Perspect 114(5):A300–A303

Center for Security Studies (2008) CSS analzses in security policy No. 38, ETH Zuerich

Chen P, Overholt A, Rutledge B, Tomic J (2010) Economic assessment of biogas and biomethane production from manure. White Paper, Callstart

Cuéllar AD, Webber ME (2008) Cow power: the energy and emissions benefits of converting manure to biogas. Environ Res Lett 3(3)

El Shol M, Wesseler J (2010) The economics and policy of biogas production, Wageningen University, thesis code ENR-80424

GIZ (former GTZ) (1989) Biogas plants in animal husbandry, Vieweg

Global Environment Facility (GEF) (2006) Discussion note for high level roundtable of climate change: mitigation and adaptation. Third GEF Assembly, Cape Town

Kaltschmitt M, Hartmann H (2001) Energie aus Biomasse – Grundlagen, Techniken und Verfahren. Springer Verlag, Berlin/Heidelberg/New York

Kauffmann C (2005) Energy and poverty in Africa, Policy insights no. 8, OECD

Leitfaden Biogas (2010) 5. vollständig überarbeitete Auflage

Mehta A (2002) The economics and feasibility of electricity generation using manure digesters on small and mid-size, January. University of Wisconsin – Madison, USA

Ram B (2010) Renewable energy development in Africa - challenges, opportunities, way forward 21st world energy congress, theme 2: availability - what is the right energy mix for long term stability Montreal, Canada from the 11th to 16th Sep

Rutz D, Janssen R (2012) Keynote introduction: biomass technologies and Markets in Africa. In: Rutz D, Janssen R (eds) Bioenergy for sustainable development in Africa. Springer Science+Business Media

Susanti A, Burges P (2012) Oil palm expansion: competing claim of lands for food, biofuels, and conservation. In: Behnassi M, Pollmann O, Kissinger G (eds) Sustainable food security in the era of local and global environmental change. Springer, Netherlands

Tilman D, Socolow R, Foley JA, Hill J, Larson E, Lynd L, Pacala S, Reilly J, Searchinger T, Somerville C, Williams R (2009) Beneficial biofuels – the food, energy and environment trilemma. Science 325(July):270–271

US Department of Energy (2010) Anaerobic digestion, which takes place in three stages produces biogas. Different kinds of micro-organisms are responsible for the processes that characterize each stage (https://www1.eere.energy.gov/tribalenergy/guide/biomass_biopower.html)

www.fao.org

www.indexmundi.com

Chapter 15
Global Biomass Supply and Sustainable Development

Lucia Beran and Harald Dyckhoff

Abstract Biomass – in form of nutritional energy and energy-rich material – is not accounted for in conventional energy statistics. It constitutes a neglected energy carrier although it has ever since provided the basis for human life and activity. In this work, we assess current global draw on the earth's biomass resources by examining the indicators 'Ecological Footprint' and 'Human Appropriation of Net Primary Production', quantifying humankind's biomass demand and the earth's biomass supply. It is revealed that humankind appropriates about 20–30% of the ecosystem's supplying capacity. Other definitions partly suggest lower and higher values. We then use the energetic metabolism accounting concept to acquire data on biomass supply for the past centuries to complement conventional energy statistics. It is disclosed that the actual energy supply to humankind is about twice as high as conventional energy statistics essentially suggest. Depending on the approach taken, current biomass supply amounts to 10–12 TW or to 14–15 TW. Against the results yielded, ideas like substituting fossil resources with biomass in the future for the provision of energy services to mitigate the current energy and climate crisis might be controversial to the achievement of sustainability.

Keywords Historic primary energy supply · Biomass · HANPP · Ecological Footprint · Energetic metabolism accounting

1 Introduction

When the energy supply of a certain country or region is analyzed, it is customary to refer to energy statistics or balances. However, this refers to commercial energy only, i.e. to commercially-traded energy that is used in technical devices for the provision of energy services (Haberl 2001:11; compare also OECD 2011; BP 2011).

L. Beran (✉) · H. Dyckhoff (retired, 2016)
Lehrstuhl für Unternehmenstheorie, Nachhaltige Produktion und Industrielles Controlling, RWTH Aachen, Aachen, Germany
e-mail: beran.lu@pvw.tu-darmstadt.de

© Springer International Publishing AG, part of Springer Nature 2019   291
M. Behnassi et al. (eds.), *Human and Environmental Security in the Era of Global Risks*, https://doi.org/10.1007/978-3-319-92828-9_15

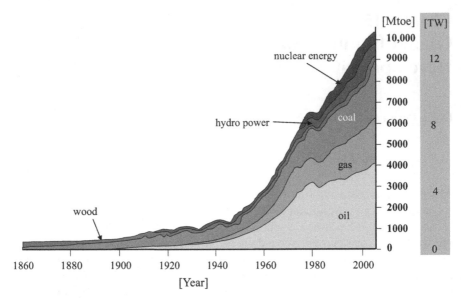

Fig. 15.1 Conventional primary energy supply in the industrial era (Mtoe = million tonnes of oil equivalent; TW = 10^{12} W.)
Source: Adapted from Paeger (2011)

They most notably include fossil energy carriers like coal, crude oil, and gas, as well as alternatives such as nuclear and regenerative energy carriers (compare Fig. 15.1).

Considering the development pictured in Fig. 15.1, it seems like people having lived before 1860 hardly had any noticeable energy at their disposal. Such a conclusion would be incorrect as biomass – organic material produced in the photosynthesis process by the primary producers' transformation of solar into chemical energy – has always provided the vital energy source for almost all life on earth. While the total amount of chemical energy produced by photoautotrophic organisms (like green plants) of an ecosystem within a certain time period is called gross primary production (GPP), part of this energy is used for the primary producers' own metabolic processes. The remaining fixed energy is the net primary production (NPP) and is either used for the buildup of biomass stocks or for the feeding of the ecosystem's consumers (Campbell and Reece 2003:1434; Haberl et al. 2004:280).

The human economic system, being a subsystem of the ecosystem, draws from many resources and services provided by the latter: resource supply, waste absorption and the provision of space to be occupied by human infrastructure. An ecosystem's NPP, the amount of which depends on factors like solar influx, nutrient and water availability, the composition of plant species and soil quality, provides nutritional energy and energy-rich material for the socioeconomic stocks of a human society, which are humans, domesticated animals as well as artefacts (Haberl et al. 2004:280; Haberl 2006:88).

For a comprehensive analysis on global energy supply, energy values from conventional energy statistics must be completed with values on biomass supply. This is particularly important in order to assess humankind's current draw on the earth's NPP, so that ideas like substituting fossil resources with biomass in the future for the provision of energy services to mitigate the current energy and climate crisis can be critically evaluated. Although the need to capture data on biomass quantities supplied to the economic system to complement conventional energy statistics was recognized as early as in 1952 by the United Nations (for instance UN 1952:101), only in the very last decades have scientists/institutions worked at the systematic development of accounting tools to be employed for the acquisition of data on humankind's demand for the ecosystem's NPP.

In this chapter, we pursue two purposes. The first consists of examining two different approaches that quantify humankind's biomass demand and the earth's biomass supply. Such juxtaposition helps to assess the global draw on the earth's NPP. We, therefore, start by taking a macro-perspective approach and examine the metric Ecological Footprint (EF), being followed by the measure Human Appropriation of Net Primary Production (HANPP) and alternative studies on NPP appropriation. The second purpose consists of complementing the conventional energy statistics with values on biomass in order to assess the actual energy supply to humankind. We, therefore, continue by taking a micro-perspective approach and analyze the energetic metabolism accounting concept, being based on energy flow accounting and allowing for the tracing of (per capita) energy flows through a defined societal compartment. The basic questions pervading the examination of EF, HANPP and the energetic metabolism accounting concept are: What is revealed by the defined concept with regard to demand for and supply of biomass? To what extent can the information provided be used to assess global draw on the earth's NPP and to complement conventional energy statistics with biomass estimates?

This analysis is structured as follows. In Sect. 2, approaches to the estimation of biomass supply are presented. In Sect. 2.1, the Ecological Footprint is examined, being followed by Human Appropriation of Net Primary Production in Sect. 2.2. Section 2.3 is devoted to the energetic metabolism accounting concept, the presentation of two diverse approaches to the approximation of global biomass supply and the complementation of conventional energy statistics with biomass supply. The key findings established in Sects. 2.1–2.3 are juxtaposed in Sect. 2.4. Section 3 brings our considerations to a close.

2   Approaches to the Estimation of Biomass Supply

2.1   Ecological Footprint (EF)

The metric Ecological Footprint (EF)[1] was conceived in the early 1990s by the University of British Columbia's academics Wackernagel and Rees (Ewing et al. 2010:9). Both researchers were convinced that human actions cannot be seen as activities that are independent from nature; in contrast, they believed that any form of growth must take place within nature's ecological limits. Following the idea that humankind lives ecologically unsustainably and will experience a decline in well-being in the long run when it consumes more ecological products and generates more wastes than can be provided and absorbed by the ecosystem, the EF was created as "an indicator of human demand for ecological goods and services linked directly to ecological primary production" (Ewing et al. 2010:90). In analogy to bank statements, it juxtaposes "human demand on ecological assets", i.e. the EF and "the ability of these assets to meet this demand", i.e. the biocapacity (Ewing et al. 2010:8). To quantify the EF, the amount of biologically productive land and water area that is required for the generation of all resources currently used and for the absorption of all wastes produced – given the prevalent resource management practices and technology – is measured.[2] The biocapacity, representing the supply of biological materials that are demanded for by human activities is assessed by the quantification of bioproductive areas available to secure provision thereof (GFN 2012). Thereby, it is indifferent where the bioproductive areas are located. The EF thus provides an insight into "how much bioproductive area is needed exclusively to sustain the activities of a given society" (Haberl et al. 2004:284). Both demand and supply are measured in hypothetical area units, namely the global hectare (gha); this means that the global EF can actually exceed the earth's biocapacity, a situation called ecological overshoot (Ewing et al. 2010:104; van den Bergh and Verbruggen 1999:64).

2.1.1   Assumptions Underlying Ecological Footprint Accounts

There are six fundamental assumptions providing the basis for EF and biocapacity calculations, i.e. for the ecological footprint accounts (Kitzes et al. 2007:3; Ewing et al. 2010:8–9):

- It is possible to track all resources consumed and wastes produced in a given territory per time unit in physical form as tons, joules or m^3.
- Most of these material flows can be attributed to the bioproductive areas that are needed for the resources' provision and the wastes' removal. Resource and waste

---

[1] http://www.footprintnetwork.org/en/index.php/GFN/page/at_a_glance/

[2] It should be noted here that carbon dioxide emissions are so far the only waste product included in national footprint accounts (Ewing et al. 2010:14).

flows that cannot be quantitatively captured in terms of biologically productive areas are ignored in the assessment.

- Worldwide diverse bioproductive areas can be converted into a mutual unit, the global hectare (gha). Global hectares represent hectares "with world-average productivity for all biologically productive land and water" (GFN 2012) and are used to express both the EF and the biocapacity.
- The entire demand can be aggregated by the merging of all mutually exclusive areas generating resources and absorbing wastes. Such an addition is guaranteed through the conversion of physical hectares into gha and is also valid for the determination of supply, i.e. biocapacity.
- When expressed in gha, humankind's demand for natural capital (EF) can be directly compared to its supply (biocapacity).
- If human demand for the resources of a specific ecosystem exceeds this ecosystem's regenerative capacity, an overshoot is present and ecological assets are being reduced.

2.1.2   Methodology of Ecological Footprint Accounts

The EF is calculated through the capture of all individual resource requirements over several bioproductive area categories and the conversion into the standardized area unit gha by means of yield and equivalence factors (Kitzes et al. 2007:1–2). The comprehensive data points needed to calculate a territory's footprint or biocapacity are provided by global databases, primarily by the UN's Food and Agriculture Organization FAOSTAT, UN's Comtrade, and OECD's International Energy Agency. Also, satellite imaging is used, for instance with regard to built-up land (Ewing et al. 2010:8–9/12–13).

In ecological footprint accounts, six diverse bioproductive area categories or land types are distinguished (Ewing et al. 2010:13–14/100):

- Cropland, the most bioproductive land type delivering food, fodder, rubber and fiber.
- Grazing land, providing above-ground biomass for livestock to deliver meat and dairy products, wool produces and hide.
- Fishing grounds, delivering catch from continental shelves and inland water areas. Catch estimates are converted into an equivalent quantity of primary production, according to the species' trophic levels.
- Forest land, supplying lumber, timber products, fuelwood and pulp.
- Built-up land, which does not provide any resources but – in contrast – prevents any resource provision given human urban and infrastructural development. It is assumed that built-up land replaces the biologically most productive area category, i.e. cropland.
- Carbon dioxide land, describing the biologically productive area that is necessary to absorb the anthropogenic carbon dioxide emitted into the atmosphere through the combustion of fossil fuels, land-use change, industrial processes, and transport. It is the only EF component that is dedicated to mere waste (i.e. carbon

dioxide) absorption. In ecological footprint accounts, all carbon dioxide emissions not sequestered by oceans are translated into the amount of bioproductive forestland needed to absorb the remaining emissions. Carbon dioxide land is thus actually forestland needed for carbon dioxide uptake.

For the purpose of identifying the original area needed to provide resources for processed products, the latter "are converted into primary product equivalents" (Kitzes et al. 2007:4) and translated into gha by means of extraction rates in order to account for the individual nation's transformation efficiencies. The energy required for processing is also considered in ecological footprint accounts and is eventually represented in the carbon footprint (Ewing et al. 2010:11–12).

In order to compare the various area categories by a common denominator, physical hectares (ha) needed for resource provision and waste absorption are converted into gha; the conversion from a physical ha into a gha is undertaken by means of two factors, which are valid both for the EF and the biocapacity (GFN 2012; Kitzes et al. 2007:6–7).

- Yield factors describe the extent of a national land type's biological productivity in comparison to this area category's global average productivity. They are used to convert one physical ha of any land type into the equivalent quantity of world-average hectares through the multiplication of the physical hectare by the appropriate national yield factor of the area category in question. This implies that a yield factor indicates the amount of world-average ha existent within a physical hectare of the defined national area category.
- Equivalence factors specify the productivity potential of a certain area category in relation to the global average productivity of all area categories. They allow for the conversion of one world-average ha of any land type into the corresponding quantity of gha through the multiplication of the world-average ha by the appropriate equivalence factor. This implies that an equivalence factor indicates the amount of gha existent within a world-average ha of a specified land category.

2.1.3   The EF as a Tool for the Assessment of Humankind's Draw on the Earth's NPP

With regard to the purpose of using global EF and biocapacity data, as provided by the EF concept for an assessment of humankind's draw on the earth's NPP, some adjustments must be made to avoid drawing false conclusions. According to the Global Footprint Network (GFN), humankind needs 1.5 planets to uphold its standard of living: the hypothetical area needed to supply products and services is larger than the bioproductive area needed to meet this demand, or its regenerative capacity; natural capital is depleted as a consequence. The current exploitation of the planet's renewable biological resources thus amounts to 150% (GFN 2012). However, this percentage includes the carbon footprint, which is related to the deficiency of single service accounting:

- The concept of ecological footprint accounting is designed in such a way that any given land type can solely render a single ecological service. This assumption is questionable because various land types do provide multiple services and allocation difficulties are almost certainly predestinated (van den Bergh and Verbruggen 1999:65). This is particularly valid with regard to forestland. Forestland does not only provide wooden products but it absorbs much of the carbon dioxide emissions caused by human activities. The EF distinguishes forestland and carbon dioxide land as mutually exclusive areas. In this respect, the forestland's footprint, being caused by human demand for wooden products, is juxtaposed with the biocapacity provided by forests, the single service of which is the supply of the products demanded for. However, the biocapacity provided by forests also absorbs much of the carbon dioxide. As the EF concept is designed in such a way that each land area can solely render a single service, no "corresponding carbon sequestration biocapacity" (Venetoulis and Talberth 2008:452) is additionally allocated. This implies that the demand for land area to absorb carbon dioxide (characteristically exactly forestland) is included in the acquisition of a territory's footprint and expressed as a carbon dioxide land, the supply of bioproductive area to do so, however, is neglected in the accounting framework (due to the conceptual decision that multiple land uses are not considered). This conceptual decision clearly affects the result of a territory's footprint and biocapacity, making the draw on earth's natural resources seem more intense.

In order to offset this deficiency, we decide to omit the carbon footprint in our assessment of humankind's draw on natural biomass resources. This is target-aimed to the effect that we want to determine the quantity of biological products that is appropriated for the maintenance of global society's socioeconomic metabolism.[3] When the carbon footprint, amounting to 54% of global EF (GFN 2012), is subtracted from the current rate of use of planet earth, namely 150%, the exploitation of the planet's resources is reduced to ~70%. Moreover, the estimate of the carbon footprint, amounting to 54% of global EF, should be seen with certain skepticism, as aggregation inaccuracies might be existent:

- In ecological footprint accounts, all carbon dioxide emissions not sequestered by the oceans are translated into the amount of bioproductive forestland needed to absorb the remaining emissions. However, depending on the kind of forest (mature vs. immature, for instance), the net carbon dioxide uptake can differ and the carbon footprint is likely to be falsely aggregated (Herendeen 2000:357). Most notably, ecosystems other than forests also take up emissions and the "carbon budget from just forest" should be reassigned "to the entire globe" (Venetoulis and Talberth 2008:452).

---

[3] The term 'metabolism' refers to the functioning of living organisms, the internal (chemical and physical) processes of energy-rich material intake to enable sustenance and reproduction as well as output in form of entropy and waste (Ayres 1994:xi/3). Analogous to the biological notion, the socioeconomic metabolism approach examines these processes within certain human societies and between such societies and their natural environment.

Also, we want to contrast human demand for biological resources with the NPP instead of the biocapacity:

• In ecological footprint accounts, the global EF is contrasted with the biocapacity. In contrast to the NPP, the biocapacity only embodies areas that deliver products and render services that are of direct productive use to humankind (Ewing et al. 2010:12). This anthropocentric point of view neglects ~36 bn hectares of land (from 51 bn hectares in total) like mountains, tundra, deserts, ice sheets and most of the ocean, because – according to the EF concept – these areas are too unproductive to provide any economically useful biological materials or services (Venetoulis and Talberth 2008:446/449/452). Although the NPP of land types like deserts is below that of biologically more productive areas like tropical forests or crop land, the complete omission of these land types' productivity does not only disregard their role in global biocapacity provision and carbon dioxide sequestration (Venetoulis and Talberth 2008:449) but also leads to a downward bias of the biocapacity available.

Following these considerations, the draw on biological resources caused by human demand can be deduced from the information given on the EF. When the current rate of use of planet earth amounts to 150% and the biocapacity is indicated to include 15 bn gha, the present EF actually corresponds to 22.5 bn gha. When thereof 54% are subtracted in order to neglect the carbon footprint, the EF is reduced to ~10 bn gha. Setting in relation this quantity with the global bioproductive area, namely the 51 bn ha that generate NPP, the proportion of biological resources drawn from the earth's ecosystems' NPP amounts to ~20%. However, as the additional 36 bn ha of bioproductive area are not weighted according to their NPP's usefulness for humans and the lands' characteristics have a lower NPP than the land area accounted for in the biocapacity, the corresponding value of 51 bn ha expressed in gha would be lower; the appropriation of 20% can thus be regarded as a minimum.

2.2 Human Appropriation of Net Primary Production (HANPP)

The indicator HANPP was elaborated by Haberl (1997) and is based on the belief that human activities disturb the ecosystem's functioning by altering vital ecological energy flows and reducing the NPP available. Measuring in physical units (dry matter biomass, Joule, carbon) the quantity of NPP appropriated by humans in a defined land area, the indicator HANPP discloses the intensity of land use and examines the changes in energy flows that result from human use of the ecosystem's services. It can thus be used for an evaluation of the extent of human domination in a given territory (Haberl et al. 2004:280/286; Krausmann and Haberl 2002:181).

The definition of HANPP was formulated after a concept given by Wright (1990), stating that the biomass appropriated by humans through land use is the energy that would be potentially available without the presence of human beings. More precisely, HANPP is defined as the difference between the potential NPP (NPP_0), i.e. the NPP that would prevail in the ecosystem "in the absence of human intervention" (Haberl et al. 2007a:12942) and the NPP actually remaining in the ecosystem after harvest has been completed (NPP_t). NPP_t is calculated by subtracting the NPP harvested and destroyed in the harvest process by human beings (NPP_h) from the current, actual vegetation (NPP_{act}). HANPP is thus $NPP_0 - NPP_t$ with $NPP_t = NPP_{act} - NPP_h$. If the difference between NPP_0 and NPP_{act} is denoted as ΔNPP_{LC}, namely being "NPP changes induced by soil degradation, soil sealing, and ecosystem change" ($\Delta NPP_{LC} = NPP_0 - NPP_{act}$), then "HANPP becomes equal to $NPP_h + \Delta NPP_{LC}$" (Haberl et al. 2007b:3). ΔNPP_{LC} thus embraces productivity changes (losses or gains) compared with potential vegetation (Haberl et al. 2007a:12945).

Methodologically, NPP_h (i.e. the biomass harvested and destroyed in the harvest process) is quantified on the basis of statistical data on agricultural yields, livestock and wood harvest, mainly delivered by the UN's FAO or other agricultural statistics as well as on the basis of "spatially explicit data on land use in grid-based geographical information systems" (Haberl et al. 2007a:12946). NPP_{act} (i.e. the NPP that currently prevails in the ecosystem) is assessed on the basis of existing statistical datasets on land use and land cover stemming from gridded geographical information systems databases (Haberl 2007a:12942). Also, dynamic global vegetation models (DGVM) are being made use of, more precisely the Lund-Potsdam-Jena DGVM, simulating biogeochemical processes and productivity of global vegetation. If there is a lack of reliable and consistent data, harvest indices are used or assumptions made (Haberl et al. 2007b:5). Data for the determination of potential terrestrial NPP, i.e. NPP_0 is either derived from the Lund-Potsdam-Jena DGVM or from extrapolation of typical NPP values per defined unit and year given by literature sources (Haberl et al. 2007a:12946; Haberl et al. 2007b:4).

In their study, Haberl et al. (2007a) quantify and map HANPP in the earth's above- and belowground terrestrial ecosystems. They reveal that HANPP amounted to 15.6 Pg C/yr. around year 2000; this equals 18.3 TW.[4] It is the sum of NPP_h (9.6 TW), ΔNPP_{LC} (7.4 TW) and NPP appropriated in human-induced fires (1.3 TW), the latter being separately shown in the authors' calculation.

---

[4] Pg C/yr. = 10^{15} g carbon per year. For better assessment of this chapter's contents as well as on grounds of uniformity, all original units used in various studies are converted into the unit of power, namely Watt (W) (1 W = 1 J/s; TW = 10^{12} W). Conversions are undertaken using the following factor: 1 kg dry matter biomass equals 0.5 kg carbon or 18.5 MJ (Haberl et al. 2007b:6). All figures stipulated hereafter derive from exact conversions and are only as precise as the original data.

2.2.1 Further 'HANPP' Estimates

As terms, scope of consideration and methods for the assessment of HANPP are not (yet) standardized – the standardization being the aim of Haberl et al. (2007a:12945) – estimates on the appropriation of NPP generated by other authors cannot be directly juxtaposed to each other because 'HANPP' is defined differently in every study. Therefore, we attempt to systematically classify and synthesize alternative definitions of NPP appropriation with the corresponding estimates, therewith providing an overview of the amount of biomass that is taken from the ecosystem to meet human needs.

a. Classification

From a conceptual point of view, it is Wright's (1990) study that is most comparable to HANPP. This is based on the fact that Haberl (1997) adopts Wright's idea of NPP appropriation being compared "with the amount of photosynthetic energy that would flow through natural ecosystems in the absence of human impact" (Wright 1990:189):

- In his assessment of NPP appropriation, Wright (1990) focuses on long-term effects on the ecosystem's productivity that result from human activities. Such long-term effects are produced by habitat destruction like the conversion of natural vegetation into ecosystems serving human purposes as cropland and urban areas. In contrast to Haberl et al. (2007a), he neglects short-term effects like losses of plant biomass that are caused by burning or timber harvest. Wright argues that such biomass losses only have a minimal impact on the NPP that remains in ecosystems to be consumed by wild species (the implications for species endangerment being Wright's main object of research). Productivity losses caused by habitat degradation, like desertification or the conversion of forest into pasture as well as fodder consumed by domesticated animals, are included in Wright's estimate. His study suggests a NPP appropriation of 20.8 TW from aboveground terrestrial ecosystems around year 1988.

Another study that includes long-term productivity losses in the assessment of NPP appropriation is the one by Vitousek et al. (1986). In addition to ΔNPP_{LC}, they account for the entire NPP that is lost from a defined ecosystem due to its human occupation:

- Vitousek et al. (1986) produce an estimate on the amount of NPP that is appropriated by humankind in above- and belowground terrestrial ecosystems. Their so-called 'high estimate' includes the entire NPP of "lands devoted to human activities" (Vitousek et al. 1986:368) and productivity losses compared to potential NPP as a consequence of human interference in the ecosystem. Productivity losses as defined by Vitousek et al. can be relatively clearly juxtaposed with the HANPP component ΔNPP_{LC}. However, the acquisition of the entire NPP of "lands devoted to human activities" does indeed include the entire NPP that is produced in a defined human-dominated ecosystem – regardless of whether the NPP is actually withdrawn through harvest or logging or whether it actually remains there. In this context, the definition suggests that the estimate also includes onsite backflows of harvested biomass to nature (like killed roots or grazing animals' feces) and the often very productive vegetation in urban and infrastructural areas (like gardens, parks or roadside greening) (Haberl et al. 2007a:12943; Haberl et al. 2007b:5). Haberl et al. (2007a) only capture the NPP that is effectively withdrawn for further processing, thus distinguishing between NPP_h and NPP_t. According to Vitousek et al.'s (1986) 'high estimate', humankind appropriated 34.1 TW of terrestrial NPP in the 1980s.

Other estimates omit the consideration of ΔNPP_{LC} and define NPP appropriation solely as the entire NPP of human-dominated lands. A distinction between NPP_h and NPP_t is thus again not made.

- Besides the 'high estimate', Vitousek et al. (1986) also produce an 'intermediate estimate', which is the result of the subtraction of ΔNPP_{LC} from the 'high estimate'. Vitousek et al.'s 'intermediate estimate' for above- and belowground terrestrial ecosystems suggests an appropriation of 23.8 TW in the 1980s.
- The study by Rojstaczer et al. (2001) is oriented towards the 'intermediate estimate' generated by Vitousek et al. (1986). This means that their approximation accounts for the entire productivity of all human-dominated land, including the biomass directly and indirectly consumed. Rojstaczer et al. reassess the estimate on NPP appropriation by using more recent data – compared to Vitousek et al. (1986) – and by conducting stochastic simulations to incorporate estimates of uncertainty. According to their reassessment, the NPP appropriated in above and belowground terrestrial ecosystems amounted to 22.9 TW, presumably in the 1990s. The 95% confidence interval accounts for ±15.8 TW.
- Imhoff et al. (2004) produce a regular estimate on NPP_h (see below), but point out that this regular estimate would increase to 24.4 TW if all the NPP of land dominated by humans was included. In this case, the amount and content appropriated are similar to the 'intermediate estimate' by Vitousek et al. (1986) and the reassessment estimate by Rojstaczer et al. (2001).

Imhoff et al. (2004) commonly define NPP appropriation as all biomass harvested for further processing. From a content point of view, their estimate compares to NPP_h:

- Imhoff et al. (2004) define NPP appropriation as the sum of the terrestrial NPP needed to produce biomass products demanded for by humankind and the NPP that is lost during harvest, logging and processing and that is consequently no further used economically. Their study suggests that NPP appropriation in aboveground (and partly belowground; belowground NPP of grazing land is not included) terrestrial ecosystems amounted to 13.5 TW around year 1995.

There are further estimates based on the definition that NPP appropriation solely includes biomass that is directly consumed by humans or animals; losses occurring during the process of harvesting or logging are not included:

- In this respect, Whittaker and Likens (1973) produce an estimate accounting for the quantity of food (terrestrial and aquatic) and timber directly consumed by humans. They suggest a direct NPP appropriation of 2 TW. The exclusion of aquatic food from the calculation does not have any noticeable effect; direct NPP appropriation remains at a value of 2 TW. The authors do neither specify whether their estimate refers to above- or belowground ecosystems nor what year they refer to. It might be as early as 1950.
- Vitousek et al. (1986) produce a 'low estimate' that only incorporates the NPP that is directly consumed by humans and animals in form of food (terrestrial and aquatic) as well as timber. Their result reveals a direct NPP appropriation of 4.2 TW in the 1980s; when aquatic food is omitted, direct NPP appropriation is reduced to 3 TW.

b. **Synthesis**

Depending on scientists referred to and the corresponding definitions taken, appropriation estimates differ. In the following, we synthesize the individual estimates on terrestrial NPP appropriation classified above. We start with the estimate accounting for least NPP components and end with the most comprehensive estimate. All estimates but the one by Wright (1990) – and possibly Whittaker and Likens (1973) – refer to above- and belowground ecosystems.

- *Estimates on appropriated NPP through direct consumption*: According to Whittaker and Likens (1973), an amount of 2 TW was appropriated in the 1970s through direct consumption. The 'low estimate' produced by Vitousek et al. (1986) suggests an amount of 3 TW in the 1980s, consumed by humans and domesticated animals.
- *Estimates on appropriated NPP_h*: Haberl et al. (2007a) indicate an appropriation of NPP_h of 9.6 TW around year 2000. Imhoff et al. (2004) suggest an amount of 13.5 TW around year 1995.
- *Estimates on HANPP*: Wright (1990) suggests a NPP appropriation of 20.8 TW around year 1988. Haberl et al. (2007a) quantify a NPP appropriation of 18.3 TW around year 2000.
- *Estimates on appropriated NPP of human-dominated lands*: Estimates range from 22.9 TW presumably in the 1990s (Rojstaczer et al. 2001) to 23.8 TW in the 1980s (Vitousek et al. 1986) to 24.4 TW around year 1995 (Imhoff et al. 2004).
- *Estimate on appropriated NPP of human-dominated lands + ΔNPP_{LC}*: Vitousek et al. (1986) suggest an appropriation amounting to 34.1 TW in the 1980s.

Disregarding the estimates on NPP appropriation through direct consumption on grounds of their relatively limited acquisition of NPP components from a content point of view, it is revealed that humankind appropriates – depending on the definition in question – between 10 TW (the estimate on appropriated NPP_h by Haberl et al. 2007a) and 34 TW (the estimate on appropriated NPP of human-dominated lands + ΔNPP_{LC}; i.e. the "high estimate" by Vitousek et al. 1986) of above- and belowground terrestrial NPP. Vitousek et al.'s (1986) 'high estimate' outranges all others due to the very broad conception of NPP appropriation, accounting for the

Table 15.1 Diverse NPP appropriation estimates

Classification of NPP appropriation definitions	Absolute appropriation [TW]	Terrestrial NPP_{act} [TW]	Appropriation of NPP_{act} [%]	Sources
Appropriated NPP through direct consumption	2	63	3.2	Whittaker and Likens (1973)
	3	77	3.9	Vitousek et al. (1986) "low estimate"
Appropriated NPP_h	9.6	70 (77)*	13.7(12.5)**	Haberl et al. (2007a)
	13.5	70	19.3	Imhoff et al. (2004)
HANPP (NPP_h + ΔNPP_{LC})	18.3	70 (77)*	26.1 (23.8)**	Haberl et al. (2007a)
	20.8	76 (89)*	27.4 (23.4)**	Wright (1990)
Appropriated NPP of human-dominated lands	22.9 (±15.8)	70	32.7 (± 22.6)	Rojstaczer et al. (2001)
	23.8	77	30.9	Vitousek et al. (1986) "intermediate estimate"
	24.4	70	34.9	Imhoff et al. (2004)
Appropriated NPP of human-dominated lands + ΔNPP_{LC}	34.1	77	44.3	Vitousek et al. (1986) "high estimate"

*the figures in brackets refer to NPP_0
**the figures in brackets refer to the appropriated % of NPP_0
Source: Adapted from Whittaker and Likens (1973), Vitousek et al. (1986), Wright (1990), Rojstaczer et al. (2001), Imhoff et al. (2004), and Haberl et al. (2007a)

entire NPP of human-dominated land areas and all "land-use-induced productivity changes" (Haberl et al. 2007a:12945) caused by human existence. It is also shown that estimates on ΔNPP_{LC} range from 7.4 TW (Haberl et al. 2007a) to 9.7 TW (Vitousek et al. 1986).

A systematic overview of the results by absolute numbers is provided in Table 15.1.

2.2.2   The Earth's Supply of NPP

So far, we have disclosed estimates on human demand for terrestrial biomass. What is the earth's supply of terrestrial NPP? This information is necessary to establish approximations on human draw on biological resources, which can be used as additional assessments to the Ecological Footprint considerations.

Haberl et al. (2007a:12942) distinguish between NPP_{act} (i.e. the NPP that currently prevails in terrestrial ecosystems) and NPP_0 (i.e. the terrestrial NPP that would prevail in absence of human civilization).

- *NPP$_0$ of terrestrial ecosystems*: Estimates on NPP_0 are provided by Haberl et al. (2007a:12943) and Wright (1990:189), only. Their corresponding estimates amount to 77 TW and 89 TW, respectively.
- *NPP$_{act}$ of above- and belowground terrestrial ecosystems*: Estimates on NPP_{act} range from a minimum of 63 TW (Whittaker and Likens 1973:358; Smil 1991:52) to a maximum of 77 TW (Vitousek et al. 1986:369), with numerous scientists assuming a NPP of 70 TW (Rojstaczer et al. 2001:2552; Imhoff et al. 2004:872; Krausmann et al. 2008:481; Haberl et al. 2007a:12943).[5]

These approximations seem to be validated by estimates on gross primary production (GPP) that are believed to amount to an absolute maximum value of 150 TW with regard to terrestrial ecosystems (Dyke et al. 2011:155–156). Following the assumption that actual NPP amounts to ~50% of GPP (Kleidon 2012:1030), terrestrial ecosystems produce a maximum NPP of 75 TW.

2.2.3   The Appropriation Percentage of NPP

Again, disregarding the estimates on NPP appropriation through direct consumption, a juxtaposition of the estimates on terrestrial NPP appropriation with the corresponding estimates on terrestrial NPP supply reveals that humankind appropriates a minimum of 13.7% (NPP_h by Haberl et al. 2007a) and a maximum of 44.3% (the "high estimate" given by Vitousek et al. 1986) – depending on the definition. However, it might be more plausible to assume a range of 20–35% due to the following reasons. First, Haberl et al. (2007a) regard NPP_h only as one component of HANPP that must be completed with ΔNPP_{LC}. When the appropriated amount of 18.3 TW is juxtaposed with NPP_{act}, humankind appropriates 26.1% of the NPP produced by earth's terrestrial ecosystems. When the absolute appropriation is juxtaposed with NPP_0, the percentage declines to 23.8%. Wright (1990) has similar results, with an appropriation of 27.4% of NPP_{act} and 23.4% of NPP_0. Second, the estimate on appropriated NPP in human-dominated lands + ΔNPP_{LC}, i.e. Vitousek et al.'s (1986) "high estimate" might be too elevated because the production capacity lost (ΔNPP_{LC}) is not juxtaposed with a corresponding estimate on potential NPP production (NPP_0; due to a lack of data provision by Vitousek et al. (1986)). Also, there is no distinction between the NPP actually harvested (NPP_h) and the NPP actually remaining in the ecosystems (NPP_t). This latter argumentation is in fact also valid for the estimates on appropriated NPP in human-dominated lands, all showing an appropriation percentage of up to 35%. These appropriation estimates are higher than HANPP, suggesting that NPP_t must actually account for a quite substantial

---

[5] NPP_{act} of the entire ecosystem (terrestrial and aquatic) is quantified at a minimum of 94 TW (Whittaker and Likens 1973:358) to a maximum of 130 TW (Vitousek et al. 1986:369).

quantity. If these are neglected and solely the estimates on HANPP are regarded, humankind's draw on the terrestrial ecosystem's NPP amounts to 20–30%. Table 15.1 summarises the diverse definitions classified, the corresponding absolute appropriation estimates, approximations on terrestrial NPP and the appropriation percentage yielded.

2.3 The Energetic Metabolism Accounting Concept

The examination of the EF, HANPP and of the alternative definitions of NPP appropriation has provided a snapshot of current demand for and supply of the earth's biological resources. We have therewith completed the first purpose of our paper, namely the assessment of global draw on the earth's NPP. In order to assess the actual energy supply to humankind, we now turn to the second purpose of our paper, namely the complementation of conventional energy statistics with values on biomass. We thus need estimates on the amount of biomass that has throughout the time been directly and indirectly consumed by humans; the estimates on NPP_h as provided by HANPP deliver a global approximation for today, only. In order to acquire estimates for years in the past, we proceed with the examination of the energetic metabolism accounting concept, which allows for the tracing of the (per capita) energy flows of different societies.

The accounting concept allowing for the tracing of energy flows through a societal compartment (energy flow accounting) was developed by Haberl following the presently applied methods of material flow accounting (Krausmann and Haberl 2002:179). Being based on the same system boundaries and concepts, a methodologically stringent quantification of a society's energetic metabolism is ensured. Thereby, all energy-rich material that enters the socioeconomic unit under consideration is accounted for. In contrast to conventional energy statistics, this includes nutritional energy for humans and domesticated animals and all other biomass inputs regardless of their purpose. The term input, as used in energy flow accounting methodology, can be compared with the term supply: It is the energy supplied to a defined system. The prerequisites of energy flow accounting are summarised in Fig. 15.2 and read as follows.

Total primary energy input refers to the total amount of energy that is mobilised by a societal compartment. Total primary energy input includes hidden flows: Energy flows that are mobilised for the procurement of the direct energy input but not crossing the boundary between society and environment. Hidden flows can thus be compared with the losses occurring during biomass harvest; total primary energy input thus with NPP_h (ΔNPP_{LC} is neglected in the energetic metabolism accounting concept). The direct energy input is the amount of energy actually entering into the societal compartment under consideration. When exports are subtracted from the direct energy input, domestic energy consumption is calculable.

After several conversion processes, primary energy becomes final energy, which includes power and fuels but also human food and the draught animals' nutrition.

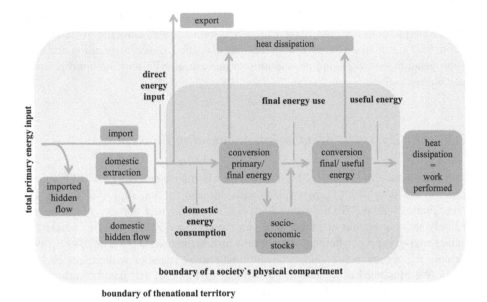

Fig. 15.2 The energetic metabolism accounting concept for a defined societal compartment
Source: Haberl (2002:73), Krausmann and Haberl (2002:180)

Haberl (2001:26) regards these components as final energy because he explicitly accounts for the conversion process from plant biomass (constituting the primary energy) to human food and fodder. In another conversion process also taking place within the societal compartment under consideration, useful energy is generated in the form of animate drive power. Useful energy, i.e. the energy that actually performs useful work eventually delivers energy services. The amount and quality of the energy services derived from energy inputs is often the decisive factor for the utility experienced by the society under consideration. The part of energy that is not immediately used for any energy provisions is often stored and maintained in the form of artefacts and thus contributes to a society's socioeconomic stocks. During every conversion process, energy gets lost in the form of heat, in accordance with the second law of thermodynamics (Haberl 2001:27–28).

2.3.1 Per Capita Estimates Given by Haberl (2002) and his Summation to Global Biomass Supply

On the basis of the energetic metabolism accounting concept, Haberl (2002) determines the average per capita quantity of biomass that is supplied to three archetypical societies. Thereby, losses occurring during harvest, logging or processing are also included, i.e. total primary biomass input per capita is regarded. All societies are characterised by a distinctive mode of subsistence:

- Hunter-gatherer societies: Haberl (2002:73–74) suggests that the primary energy supply to a typical hunter-gatherer society amounts to 10 GJ/cap/yr. Thereof, 3.5 GJ/cap/yr. are allocated to food supply. This corresponds to ~10 MJ/cap/day or 116 W/cap. Boyden (1992: 80) calls this amount of somatic energy, i.e. the energy flowing through the human body, the "human energy equivalent". Another ~10 MJ/cap/day are added by Haberl (2002) in order to account for the losses occurring during food collection and preparation. Further on, he estimates the extrasomatic energy, i.e. the "energy from sources other than human muscles" (Common 1995:68) consumed in form of firewood to be roughly equal to the energy consumed in form of nutritional energy (although the individual fire usage naturally depends upon environmental conditions and the tribes' behaviour). Nevertheless, this arguing lifts the total primary energy supply to hunter-gatherers to ~30 MJ/cap/day, or – as Haberl (2002) puts it – to 10 GJ/cap/yr. This corresponds to 317 W; the energy carrier being solely biomass.
- Agricultural societies: For an estimate on agricultural societies' energy usage, Haberl (2002) refers to a field study carried out by Grünbühel et al. (1999) in the north-eastern Thai village Sang Saeng in 1998. As in any ideal-typical agricultural society, the village's domestic energy extraction is exclusively biomass, which also constitutes the dominating primary energy input into the village. It is used as burning material (charcoal), nutritional energy for humans and domesticated animals and it is stored as energy-rich material in the form of artefacts. Haberl (2002:75) indicates a total biomass supply of 70 GJ/cap/yr. (2220 W/cap).

This estimate is validated by two alternative approximations yielded: On the basis of a case study carried out for year 1875 (Netting 1981), Fischer-Kowalski and Haberl (1997:69–70) estimate the biomass supply to the Swiss alpine village of Törbel to amount to 65 GJ/cap/yr (2061 W/cap). A calculation made for the energetic metabolism of Austria during the nineteenth century suggests an energy supply (being characterised to more than 99% by biomass and to less than 1% by coal) of 72 GJ/cap/yr. (2283 W/cap) (Krausmann and Haberl 2000). As "all these examples refer to societies at an early state of transition from agricultural to industrial society" (Haberl 2002:76), the author suggests believing in a range of biomass supply amounting to 40–70 GJ/cap/yr. (1268–2220 W/cap).

- Industrial societies: For an estimate on industrial societies' energy usage, Haberl (2002:76–78) presents a calculation for Austria in 1995. Austria's domestic biomass extraction with an amount of 59 GJ/cap/yr. is somewhat higher than that of the Thai agricultural village Sang Saeng (48 GJ/cap/yr.). Including domestic hidden flows, imported biomass and imported hidden flows, the total biomass supply rises to 84 GJ/cap/yr. (2664 W/cap). Whereas agricultural societies need a large portion of their biomass supply for fire pits and the nutrition of domesticated animals, industrial societies also consume much biomass in form of meat, for "non-energetic" purposes like the provision of furniture, pulp and paper as well as the realisation of construction work.

Haberl approximates graphically global biomass supply from 1800–2000 in several of his publications (compare Haberl 2000:39–41; Haberl 2006:93) by rather arbitrarily allocating a constant biomass supply of 70 GJ/yr. to each individual and multiplying this estimate by world population. When we numerally calculate global biomass supply according to this methodology and by taking reference to world population estimates given by McEvedy and Jones (1978), Maddison (2010) and the USBC (2011), global biomass supply amounted to 2 TW in 1800, to 3.5 TW in 1900, to 5.7 TW in 1950, to 13.5 TW in 2000 and to 15.3 TW in 2010.

2.3.2 Global Biomass Supply According to a Societal Composition Approach

The estimates produced by Haberl in the manner described are not only very crude (as admitted by the author (Haberl 2006:92) but neglect the complexity of today's societal composition: The earth accommodates all three typical societies, each disposing of a characteristic quantity of energy. We produce alternative global biomass supply estimates that are generated according to the following considerations (Dyckhoff et al. 2010).

First, we generate ranges of per capita supply in order to avoid spurious accuracy. From the information provided by Haberl (2002), it is assumed that typical hunter-gatherer societies disposed of biomass amounting to 200–400 W/cap.[6] It was mainly supplied in form of somatic energy (food) and extrasomatic energy (mostly firewood but also clothing (furs) and tools (throw sticks)). Agricultural societies learned how to produce biomass, providing nutritional energy for themselves and for their livestock. Also, large amounts of wooden biomass were consumed: Private logging was effected in order to gain new soil that could be used for the growing of crops and much wood was needed for the maintenance of fire for cooking, heating, lighting and as base material for any devices. The per capita biomass supply of agricultural societies thus increased to 1200–2300 W.[7] The biomass supply of industrial societies augmented to 2500–3500 W/cap.[8] This augmentation can be explained by rising demand for a larger variety of food evoked by higher incomes. In this respect, the rising demand for animal products plays a major role. Also, so far increased demand has been met by expanded supply, which was facilitated through the industrialisation of the primary sector in the past decades, enabled by artificial fertilisers, agricultural chemicals (pesticides, herbicides), mechanisation and plantbreeding, which was optimised for energy-intensive agriculture following the 1950s (Beran

---

[6] Other estimates on the hunter-gatherer's biomass supply fall into this range and thus seem to validate this spectrum (compare Boyden 1992:80; Malanima 2010:6–7; Cook 1971:136).

[7] Again, other estimates on the agricultural societies' biomass supply seem to validate this range (compare for instance Kumar and Ramakrishnan 1990:331–334).

[8] The higher limit of 3500 W also accounts for the biomass supply to more consumption-oriented societies, like Northern America, having a per capita biomass supply of ca. 3227 W (Krausmann et al. 2008:476–477).

and Dyckhoff 2012). As the range refers to the most advanced societies within each mode of subsistence, i.e. to societies having most sophisticated technologies at hand and thus employing the highest possible amount of primary energy, a 2:1 weighting scheme in favour of the lower value is used to adequately consider those societies that do not (yet) belong to the most advanced ones. Following this method, the per capita mean for hunter-gatherer societies amounts to 267 W, for agricultural societies to 1567 W and for industrial societies to 2833 W.

Second, we assess the share of people that has been living in one of the three modes of subsistence throughout human history. We assume the following process to have taken place:

- For a very long period of time, hunter gatherers were the only society on earth: They dominated earth until 10,000 BC. Braidwood and Reed (1957:23) classify the levels hunter-gatherer societies went through in terms of subsistence patterns as the "food-gathering of full Pleistocene times" and "the more specialised food collecting". Their population increased from 125,000 in 1,000,000 BC to 2 mio in 100,000 BC (Deevey Jr. 1960:196; McEvedy and Jones 1978:14).
- In 10,000 BC, first agricultural societies developed, experiencing with the cultivation of plants and animal husbandry (a level Braidwood and Reed (1957:23) refer to as "incipient agriculture"). We assume that in 10,000 BC, earth was home to 3.8 mio hunter-gatherers and 200,000 agriculturalists; or: 95% of the earth's population were hunter-gatherers and 5% agriculturalist. By 5000 BC, "primary village farming" (Braidwood and Reed 1957:23) communities came into existence, followed by "primary urban" and "various vegecultural-primary village-farming blends" (Braidwood and Reed 1957:23), being an intensification and diversification of the first village farming communities. "Pastoral nomadism" (Braidwood and Reed 1957:23) emerged in arid and semi-arid regions.
- By the Nativity in year zero, the earth was home to 170 mio people (McEvedy and Jones 1978:345) and we presume that the hunter-gatherers' share had meanwhile declined to roughly 0.3%; agricultural societies dominated the earth with a share of 99.7%. In this sense, 500,000 hunter-gatherers remained. However, they did not vanish from our planet but maintained their population size until today. The figure is a rough estimate, deriving from the assumption that there are about 400,000 Pygmies (Vidal 2007), the hunter-gatherers living in Central Africa and Southeast Asia as well as about 100,000 San (Wikipedia 2010), the Bushmen living in Southern Africa on earth nowadays. Pygmies and San represent the two classic hunter-gatherer groups these days (Lee and Hitchcock 2001:257). Although they are still nomadic, living more or less on the products of their environment, their living conditions have changed insofar that the art of trading to acquire different food or material items is probably well-known to them (Murdock 1975:13). Also, there are further groups like the Yanomanis and some native inhabitants, whose food income is, however, majorly secured through the cultivation of crops (Survival International 2011).

- By 1800, agricultural societies have grown to a population of 900 mio (McEvedy and Jones 1978:349). With the process of industrialisation, more and more countries changed from being characterised by agriculture to being characterised by industry and the share of industrial societies consequently increased – first in Europe and then in other regions of the world. Earth was no longer home to two societies but to three: To hunter-gatherers, to agricultural societies and to industrial societies.
- Following 1950, a time when world population amounted to 2.5 bn people (USBC 2011), industrialisation reached many developing countries, especially in Latin America and in Asia. Many agriculturalists migrated to the cities in order to find better living conditions. With this gigantic urbanisation, a new type of hunter-gatherers evolved: The slum hunter-gatherers, subsisting more or less on the spontaneous gathering of energy-rich material in the form of food and artefacts, findable on streets and sites for waste disposal. They majorly settled in large cities in regions such as South-central Asia, Eastern Asia, Sub-Sahara Africa, Latin America and the Caribbean. The slum hunter-gatherers' population increased in the second half of the twentieth century from virtually zero before industrialisation to more than a bn today (UN-HABITAT 2010). Thus, in addition to agricultural and industrial societies, the earth is home to two types of hunter-gatherers, the traditional and the slum hunter-gatherers.
- Today, about 7 bn people reside on earth. About 4 bn of them live in developing countries in agricultural societies. Their population increase is significant, as they have not concluded their demographic transition yet. More than 1.5 bn people live in developed countries in industrial societies (Kapitza 2006:149). Also, more than 1 bn people are of hunter-gatherer type, both traditional (like the Pygmies, San, Bushmen) and slum hunter-gatherers.

Third, estimates on global biomass supply are generated through the multiplication of the quantity of all humans living in one of the defined societies by the appropriate weighted biomass value used by each society under consideration. In contrast to Haberl's (2000:39–41; 2006:93) approach, this method explicitly considers the share of people having lived in each mode of subsistence in the course of time (Dyckhoff et al. 2010:277–278). Following this calculation approach, global biomass supply augmented from ~33 MW in 1,000,000 BC to ~1 GW in 10,000 BC. By the Nativity, it had reached the amount equalling ~270 GW and in year 1500 it accounted for ~670 GW. In 1800, global biomass supply reached 1.4 TW and increased to 2.8 TW in 1900, to 4.5 TW in 1950, to 10.1 TW in 2000 and by year 2010, the quantity had augmented to 11.5 TW. In comparison to the results yielded by Haberl's (2000:39–41; 2006:93) lump-sum calculation of global biomass supply, our values are lower as they take into account the share of people living in non-industrial standards that have less energy at their disposal.

The results for the past decade yielded by our societal composition approach are very similar to the result of a study on global terrestrial biomass harvest carried out by Krausmann et al. (2008:471). Examining socioeconomic biomass flows and thereby focusing – among other issues – on NPP_h, their study suggests a global

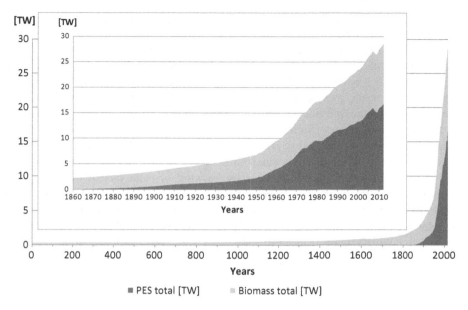

Fig. 15.3 Global primary energy supply
Source: Adapted from Dyckhoff et al. (2010:280)

NPP appropriation of 10.9 TW in above- and belowground terrestrial ecosystems in year 2000. Furthermore, our result of 10.1 TW in year 2000 compares to the approximations on NPP_h generated by HANPP, which amount to 9.6 TW (Haberl et al. 2007a) and 13.5 TW (Imhoff et al. 2004). In comparison to the absolute values on NPP appropriation generated by HANPP and other NPP appropriation definitions, the estimates produced on the basis of the energetic metabolism accounting concept are lower as they solely account for biomass that is embraced by the NPP_h component; they do neither include productivity changes compared to potential NPP (ΔNPP_{LC}) nor the NPP of human-dominated lands that actually remains in the ecosystem (NPP_t).

2.3.3   Completing Conventional Energy Statistics with Biomass Supply

The estimates on global biomass supply generated by the approach taking explicit account of societal composition are used to complement data on primary energy supply (PES) as stipulated by conventional energy statistics and balances. Figure 15.3 illustrates and clarifies humankind's actual energy supply from the Nativity until today, distinguishing between PES and biomass.

In correspondence with Fig. 15.1, conventional PES is reproduced below the supply of biomass to highlight the somewhat absurd deduction that becomes established when the data from conventional energy statistics is extrapolated into the past: it seems as if people living before 1860 hardly had any noticeable energy at their disposal. In contrast, Fig. 15.3 demonstrates the long-lasting exclusivity of biomass as energy carrier as well as its importance and magnitude: Humankind's real energy supply is almost twice as high as estimates on conventional PES suggest.

While biomass increased due to changes in societies' subsistence patterns (compare Sects. 2.3.1 and 2.3.2), conventional PES increased due to a shift from the traditional solar energy system to a fossil energy system. After coal constituted the first energy carrier to provide a relative energy surplus and to eliminate the centuries-long shortages of energy (Sieferle 1997:140), oil and gas supplemented the energy mix substantially in the course of the twentieth century. While supply enlarged through progress in the discovery and extraction of resources as well as accessible prices, it was also strongly influenced by growing demand incited by economic activities in the transportation, industrial and commercial sectors. Thereby, oil has played an important role as it provided the basis for favorable transport, which accelerated the international division of labor and economic globalization in the late twentieth century, again leading to further transport processes and ultimately oil consumption.

2.4 Juxtaposition of Results

Both the EF and HANPP aim at the quantification of humankind's draw on the earth's natural resources by taking a macro perspective approach. We conclude that when the carbon footprint is subtracted from the global EF and when this resulting EF is juxtaposed with the actual NPP of the ecosystem, humankind's draw on the ecosystem's biological material amounts to a minimum of 20%. According to HANPP, humankind appropriates 20–30% of the earth's actual or potential NPP. Taking account of further studies that quantify appropriated NPP of human-dominated lands (thus including NPP_h and NPP_t), the percentage range can be widened to the interval [20; 35]. Further approaches to estimate generation are either very narrowly conceived (yielding an appropriation of 3–4% when solely direct biomass consumption is considered) or very comprehensively defined (yielding an appropriation of 44% when all the NPP of human-dominated lands + ΔNPP_{LC} is regarded).

With the energetic metabolism accounting concept, a micro-perspective approach is taken. Haberl (2000) acquires data on the per capita biomass supply to members of each archetypical society that accounts for the amount directly consumed and for all losses occurred during energy procurement. He establishes global totals by arbitrarily allocating a constant biomass value of 70 GJ/yr. to each individual and multiplying this per capita supply by world population. His method yields an estimate on global biomass supply amounting to 14–15 TW in the past decade. Our approach (Dyckhoff et al. 2010), explicitly considering the share of people that live in non-industrial standards either in agricultural or in hunter-gatherer (both traditional and slum) societies and multiplying the amount of people of each society by the corresponding weighted per capita biomass estimate, yields a result for the past decade of 10–12 TW. These global totals established on the basis of the energetic metabolism accounting concept are from a content point of view comparable to the HANPP component NPP_h. Thereby, it is particularly our approach that yields an estimate that is close to the NPP_h approximations of 9.6 TW and 13.5 TW.

3   Conclusions

The first purpose of our work was the examination of two different approaches that quantify humankind's biomass demand and the earth's biomass supply. We analyzed the macro-perspective approaches EF and HANPP, and conclude that the two indicators provide valuable information for the quantitative ascertainment of global draw on the earth's biological material. After some adjustments undertaken on our part, the EF suggests that a minimum of 20% of the earth's NPP are appropriated by humankind. Considering the information provided by HANPP, humankind appropriates 20–30%. Alternative definitions of NPP appropriation suggest assuming a range of 20–35% or even of 3–44%. The latter range, however, is based on a very narrowly conceived and a very comprehensively regarded NPP appropriation definition.

The second purpose of our work was the complementation of conventional energy statistics with values on biomass so that the actual energy supply to humankind could be assessed. We examined the micro-perspective concept of energetic metabolism accounting and conclude that its conceptual framework is conducive to the approximation of biomass supply to individuals living in one of the archetypical societies. Using these per capita estimates and multiplying them either generally by world population as done by Haberl (2000, 2006) or generating weighted per capita biomass values and multiplying those by the population living in each defined society as proposed by Dyckhoff et al. (2010), annual approximations of global biomass supply can be yielded over long periods of time. Conventional energy statistics can thus be complemented with humankind's most vital energy carrier in order to provide a more comprehensive picture of actual energy supplied by natural resources to the global economic system. It is revealed that humankind's actual energy supply is about twice as high as conventional energy statistics essentially suggest, with biomass supply in the past decade amounting to either 14–15 TW – following Haberl's (2000, 2006) approach – or to 10–12 TW – following the approach conceived by Dyckhoff et al. (2010).

It becomes evident that global draw on the earth's NPP is considerable and both direct and indirect effects thereof are well-known: land transformation; alterations of biogeochemical cycles; enhanced greenhouse gas emissions in the atmosphere; loss of biological diversity; species extinction; and even loss of entire ecosystems, etc. (Vitousek et al. 1997:494–498). Given the increasing population size and (aspired) growing standards of living, biomass extraction will increase. If biomass is to substitute (partly) the supply of fossil energy carriers in the future, its removal from the earth's ecosystems will be even more reinforced. Most likely, the increased demand evoked by humankind's growing activities in sectors like agriculture, international commerce, transportation, and industrial production will not be met without substantial ecological detriments, which might eventually turn into political conflicts on a national and international level. Humankind will impinge on the ecosystem's natural limits and conflict interface is existent.

References

Ayres RU (1994) Industrial metabolism: theory and policy. In: Ayres RU, Simonis UE (eds) *Industrial metabolism*. UN University Press, Tokyo et al., pp 3–20

Beran L, Dyckhoff H (2012) Biomasse als industrieller Faktor einer nachhaltigen Weltwirtschaft? In: Corsten H, Roth S (eds) Nachhaltigkeit: Unternehmerisches Handeln in globaler Verantwortung. Wiesbaden, Gabler Verlag, pp 171–190

Boyden S (1992) Biohistory: The Interplay between Human Society and The Biosphere. *UNESCO Man and Biosphere Series*, vol 8. UNESCO and The Parthenon Publishing Group, Paris

BP (2011) BP statistical review of world energy June 2011. [WWW] <URL: http://www.bp.com/sectionbodycopy.do?categoryId=7500andcontentId=7068481> [Accessed 16 August 2011]

Braidwood RJ, Reed CA (1957) The achievement and early consequences of food-production: A consideration of the archeological and natural-historical evidence. Cold Spring Harb Symp Quant Biol 22:17–31

Campbell NA, Reece JB (2003) Biologie, Sixth edn. Spektrum Akademischer Verlag, Heidelberg/Bonn

Common M (1995) Sustainability and policy: Limits to economics. Cambridge University Press, Cambridge, NY/Melbourne

Cook E (1971) The flow of energy in an industrial society. Sci Am 224(3):135–144

Deevey ES Jr (1960) The human population. Scientific American, CCIII, pp 195–204

Dyckhoff H, Beran L, Renner T (2010) Primary energy supply and economic wealth. In: Brebbia CA, Jovanovic N, Tiezzi E (eds) Management of natural resources, sustainable development and ecological hazards II. WIT Press, Southampton, pp 271–282

Dyke JG, Gans F, Kleidon A (2011) Towards understanding how surface life can affect interior geological processes: A non-equilibrium thermodynamics approach. Earth Syst Dynam 2:139–160

Ewing B, Moore D, Goldfinger S, Oursler A, Reed A, Wackernagel M (2010) The ecological footprint atlas 2010. Global Footprint Network, Oakland

Fischer-Kowalski M, Haberl H (1997) Tons, joules, and money: modes of production and their sustainability problems. Soc Nat Resour 10(1):61–85

GFN (2012) Global Footprint Network. [WWW] <URL: http://www.footprintnetwork.org/en/index.php/GFN/> [Accessed 5 Sept 2012]

Grünbühel CM, Schandl H, Winiwarter V (1999) Agrarische Produktion als Interaktion von Natur und Gesellschaft: Fallstudie Sang Saeng. Interuniversitäres Institut für interdisziplinäre Forschung und Fortbildung, Wien

Haberl H (1997) Human appropriation of net primary production as an environmental indicator: implications for sustainable development. Ambio 26(3):143–146

Haberl H (2000) Energetischer Stoffwechsel und nachhaltige Entwicklung. Natur und Kultur 1(1):32–47

Haberl H (2001) The energetic metabolism of societies. Part I: accounting concepts. J Ind Ecol 5(1):11–33

Haberl H (2002) The energetic metabolism of societies. Part II: empirical examples. J Ind Ecol 5(2):71–88

Haberl H (2006) The global socioeconomic energetic metabolism as a sustainability problem. Energy 31:87–99

Haberl H, Wackernagel M, Krausmann F, Erb K-H, Monfreda C (2004) Ecological footprints and human appropriation of net primary production: a comparison. Land Use Policy 21:279–288

Haberl H, Erb K-H, Krausmann F, Gaube V, Bondeau A, Plutzar C, Gingrich S, Lucht W, Fischer-Kowalski M (2007a) Quantifying and mapping the human appropriation of net primary production in earth's terrestrial ecosystems. Proc Natl Acad Sci USA 104(31):12942–12947

Haberl H, Erb K-H, Krausmann F (2007b) Human appropriation of net primary production (HANPP). Internet Encyclopaedia of Ecological Economics. [WWW] <URL: www.ecoeco.org/pdf/2007_march_hanpp.pdf> [Accessed 13 Oct 2011]

Herendeen RA (2000) Ecological footprint is a vivid indicator of indirect effects. Ecol Econ 32:357–358

Imhoff ML, Bounoua L, Ricketts T, Loucks C, Harriss R, Lawrence WT (2004) Global patterns in human consumption of net primary production. Nature 429:870–873

Kapitza SP (2006) Global population blow-up and after. The demographic revolution and information society. Global Marshall Plan Initiative, Hamburg

Kitzes J, Peller A, Goldfinger S, Wackernagel M (2007) Current methods for calculating national ecological footprint accounts. Science for Environment and Sustainable Society 4(1):1–9

Kleidon A (2012) How does the earth system generate and maintain thermodynamic disequilibrium and what does it imply for the future of the planet? Philos Trans R Soc 370:1012–1040

Krausmann F, Haberl H (2000) From wood and rye to paper and beef: changes in the socioeconomic biomass and energy metabolism during 200 years of industrial modernisation in Austria. Third international conference of the European Society for Ecological Economics (ESEE) 3–6 May, Vienna

Krausmann F, Haberl H (2002) The process of industrialization from the perspective of energetic metabolism: Socioeconomic energy flows in Austria 1830–1995. Ecol Econ 41:177–201

Krausmann F, Erb K-H, Gingrich S, Lauk C, Haberl H (2008) Global patterns of socioeconomic biomass flows in the year 2000: a comprehensive assessment of supply, consumption and constraints. Ecol Econ 65(3):471–487

Kumar A, Ramakrishnan PS (1990) Energy flow through an Apatani village ecosystem of Arunachal Pradesh in Northeast India. Hum Ecol 18(3):315–336

Lee RB, Hitchcock RK (2001) African hunter-gatherers: Survival, history, and the politics of identity. Afr Stud Monogr Suppl 26:257–280

Maddison A (2010) Statistics on world population, GDP and per capita GDP, 1-2008 AD. [WWW] <URL: http://www.ggdc.net/MADDISON/oriindex.htm> [Accessed 16 Aug 2011]

Malanima P (2010) Energy in history. In: UNESCO (ed) Encyclopedia of Life Support Systems. EOLSS Publishers, UK

McEvedy C, Jones R (1978) Atlas of world population history. Penguin Books, Harmondsworth, Middlesex

Murdock GP (1975) The current status of the world's hunting and gathering peoples. In: Lee RB, DeVore I (eds) Man the Hunter, Fifth edn. Aldine, Chicago, pp 13–20

Netting RM (1981) Balancing on an Alp: Ecological change and continuity in a Swiss mountain community. Cambridge University Press, Cambridge

OECD (2011) Statistics from A to Z: Beta version. [WWW] <URL: http://www.oecd.org/document/0,3746,en_2649201185_46462759_1_1_1,00.html> [Accessed 15 Aug 2011]

Paeger J (2011) Eine kleine Geschichte des menschlichen Energieverbrauchs. [WWW] <URL: http://www.oekosystem-erde.de/html/energiegeschichte.html> [Accessed 28 Mar 2011]

Rojstaczer S, Sterling SM, Moore NJ (2001) Human appropriation of photosynthesis products. Science 294:2549–2552

Sieferle RP (1997) Rückblick auf die Natur: Eine Geschichte des Menschen und seiner Umwelt. Luchterhand, München

Smil V (1991) General energetics. John Wiley and Sons, New York

Survival International (2011) Die Yanomami. [WWW] <URL: http://www.survivalinternational.de/indigene/yanomami#main> [Accessed 7 Jan 2011]

UN (1952) World energy supplies in selected years, 1929–1950. UN Publications, New York

UN-HABITAT (2010) Slum population projection 1990–2020. [WWW]<URL: http://ww2.un-habitat.org/programmes/guo/documents/Table4.pdf> [Accessed 21 Aug 2010]

USBC (2011) International data base: Total mid-year population for the world: 1950–2050. [WWW] <URL: http://www.census.gov/population/international/data/idb/worldpoptotal.php> [Accessed 24 Nov 2011]

van den Bergh J, Verbruggen H (1999) Spatial sustainability, trade and indicators: An evaluation of the ecological footprint. *Ecol. Econ.* 29:61–72

Venetoulis J, Talberth J (2008) Refining the ecological footprint. Environ Dev Sustain 10(4):441–469

Vidal J (2007) World Bank accused of razing Congo forests. The Guardian, 4 October

Vitousek PM, Ehrlich RH, Ehrlich AH, Matson PA (1986) Human appropriation of the products of photosynthesis. Bioscience 36(6):368–373

Vitousek PM, Mooney HA, Lubchenco J, Melillo JM (1997) Human domination of earth's ecosystems. Science 277(5325):494–499

Whittaker R, Likens GE (1973) Primary production: the biosphere and man. Hum Ecol 1(4):25–36

Wikipedia (2010) Bushmen. http://en.wikipedia.org/wiki/Bushmen

Wright DH (1990) Human impacts on energy flow through natural ecosystems, and implications for species endangerment. Ambio 19(4):189–194

Chapter 16
Climate Change and Tonga Community Development: Thinking from the Periphery

Mark Matsa

Abstract The minority Tonga Community of the Great Zambezi River Basin (Basilwizi) in Binga District of north-western Zimbabwe suffers a double tragedy which threatens peoples' livelihoods. First, they were forcibly displaced from the resource-rich flood plain of the Zambezi River to the dry, marginal escarpments of the same river to facilitate the construction of the Kariba Dam in the late 1950s. The same community is now suffering from the impacts of climate change and variability which have rendered their environment even drier and less productive. In spite of this, and with little outside assistance, this subsistence, semi-pastoral community seems determined to prevail. This study assesses the relationship between climate and environmental change in the context of Tonga minority rural community development. Interviews and questionnaires were used to collect qualitative data from Tonga elders and other key informants in the district. Direct observations were used to identify in-situ environmental changes and coping strategies used to ameliorate the effects of climate change and variability. Results show that although the Tonga community is getting some assistance from NGOs and the central government, the assistance is not sustainable partly because it doesn't incorporate Tonga traditional knowledge systems which have been the bedrock of Tonga Community resilience for generations. This study posits that for meaningful climate-compatible development to take place in Binga, a community derived development 'basket of priorities' be used as a basis for sustainable community development. An identify-define-initiate-lead (IDIL) and a protect-empower-capacitate-facilitate (PECF) model is thus suggested to help the Tonga community cope with climate change impacts more sustainably.

Keywords Climate change · Basilwizi · Tonga community · Zambezi river · Sustainable development

M. Matsa (✉)
Department of Geography and Environmental Studies, Faculty of Social Sciences, Midlands State University, Gweru, Zimbabwe
e-mail: matsam@staff.msu.ac.zw

© Springer International Publishing AG, part of Springer Nature 2019
M. Behnassi et al. (eds.), *Human and Environmental Security in the Era of Global Risks*, https://doi.org/10.1007/978-3-319-92828-9_16

317

1   Introduction

While the international community discusses and predicts the potential impacts of climate change on a global scale, for indigenous and minority local communities, these impacts are already a sad reality (IPCC 2007; McLean 2010; Chagutah 2010; Ravindranath and Sathaye 2002). Mainly as a result of historical, social, political, and economic marginalisation and exclusion, many indigenous communities live in fragile ecosystems that are already suffering from the impacts of other stressors, and thus are more susceptible to climate change.

Much of the climate change literature reflects the approach of Western science which uses a knowledge base built on systematic observation to monitor changes in climate, provide forecasting services, and plan adaptation options. This is despite the fact that systematic observations and data availability are limited in many fragile ecosystems like islands, mountainous, coastal, and arid and semi-arid ecosystems in which most of the world's indigenous people live (Environmental Assessment and Saskatchewan's First Nations 2008).

In recent years, there has been an increasing realisation that the observations and assessments of indigenous groups are a valuable source of local level information, provide local verification of global models, and are currently providing basis for local community-driven adaptation strategies that are way past the planning stage and are already being implemented and tested (Bohle 2001; Baird 2008; McLean 2010). Indigenous people depend directly on diverse resources from ecosystems and biodiversity for many goods and services, and these communities have historical climate data that ranges from temperature, rainfall, and the frequency of climatic events as well as current fine-scale information that relates across all sectors, including water resources, agriculture and food security, human health and biodiversity.

Indigenous peoples' ability to link events in the natural world to a cycle that permits the prediction of seasonal events has been a key element of the survival of indigenous communities, and this has enabled them to develop localised seasonal calendars that are finely tuned to local conditions and natural events (Declaration of the Indigenous Peoples of the world to COP17 2012). In some countries, sophisticated indigenous knowledge-sharing programmes have been established for indigenous forecasting abilities, such as the Indigenous Weather Knowledge Website Project (IWKWP) in Australia, which displays the seasonal weather calendars of Aboriginals of central and Northern Australia and is mainly funded by the Australian Government (Australian Government Department of Meteorology 2013).

Preserving indigenous knowledge is proving essential to the effectiveness of community level responses to climate change (Boven and Morohashi 2002; McLean 2010). This is taking place in different ways; for example, some Russian communities have created simple dictionaries which capture the climate knowledge inherent in local languages. In Canada, the Canadian 'Voices from the Bay' project put ancestral knowledge of the Hudson Bay bio-region into writing so that it is appropriately transmitted and incorporated into environmental assessments and policies and communicated effectively to scientists, the interested public, and the youth of the participating communities (Voices from the Bay 1998).

In areas where there is relatively long history of scientific and indigenous collaboration like the Arctic, there are many case studies of indigenous participation in scientific data collection projects (McLean 2010). Modern forms of technology like Global Positioning Systems (GPS), for example, are incorporated into traditional activities by their Inuit to capture information from hunters, which are then combined with scientific measurements to create maps for use by the community (Thorpe et al. 2002). In Papua New Guinea, recording Hewa indigenous knowledge of birds, that would not tolerate habitat alteration or shortened fallow cycles in a way that is useful for conservation purposes, was used to predict the impact of human activities on biodiversity (UNESCO 2012).

Temperatures in south-western Zimbabwe are severely hot, ranging between 32 °C and 38 °C in summer. Daily minimum temperatures have risen by 2.6 °C over the last century while daily temperatures have risen by 2 °C during the same period (Chenje et al. 1998; Zhakata 2008). In addition to being the warmest in instrument record, the period 2000 to 2009 has also been the driest for Zimbabwe (GoZ-UNDP/ GEF 2011). Furthermore, from 2000–2010, the length and frequency of dry spells during the rainfall season has been increasing while the frequency of rain days has been reducing. A 10-year cycle of drought has since been noticed in Zimbabwe (Feresu 2010; Oxfam International 2011; Change in Southern Africa 2012).The very regular and predictable rainfall regimes of the 1970s which farmers remember vividly have since disappeared (Matsa and Matsa 2013).

Binga is one community in Zimbabwe which has survived droughts, floods, locusts, army worm and quelia birds outbreaks with very little, and sometimes no assistance from outside; this means that they have some unique symbiosis with their environment. Community members have some special traditional environmental knowledge enabling them to cope with such environmental disasters which are, to a large extent, manifestations of climate change. This study argues that solutions to problems of climate change should, therefore, not be from the 'outside' to the 'inside' but should be from inside to outside. Since it is the communities who know best their environment from time immemorial, their indigenous environmental knowledge and coping strategies should be respected. Community indigenous knowledge should not only be the starting point, but should lead in the formulation of any coping strategies to climate change and variability. Adaptation strategies or policies to climate change should not be taken from outside rather they should be tailor-made based on national circustances. Immitated policies are bound to fail.

2   The Study Area

Binga District is located north of Matabeleland north province on the southern edge of Lake Kariba (Fig. 16.1). Most of the district is characterised by high annual average temperature of about 30 °C, and low but erratic rainfall of less than 600 mm

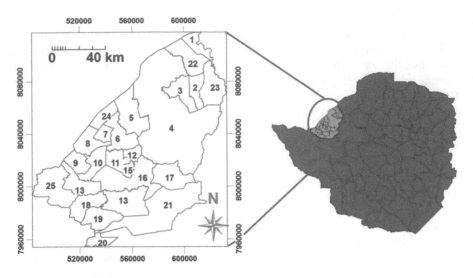

Fig. 16.1 Map of Binga district
Source: Department of geography and environmental studies, MSU

per annum. Much of the terrain is hilly. Except for valleys, soils are generally shallow and poor. A former tsetse fly infested territory, Binga district is sparsely populated, with a population of 138,074 of which 63,512 are males and 74,562 females (ZimStats 2012). Average household size is 4.4 and 90% of the population is chronically poor (Basilwizi Trust 2010; Zimbabwe Poverty Assessment Study Survey 1996). Before the displacement of the mid-1950s to make way for what became Lake Kariba, the Tonga enjoyed relative 'subsistence affluence' based on the environmental wealth of the Zambezi Valley flood plain, which supported cultivation throughout the year. The riverine ecosystem also provided Tonga communities with fish and game.

The construction of Kariba Dam led to the Zambezi River flood plain and its environs being declared protected areas. To prevent soil erosion and siltation, both settlement and cultivation were prohibited in the vicinity of the lake. The Tonga were thus resettled on the southern Zambezi escarpment, too dry for reliable dryland cultivation. Flood plain cultivation is no longer possible. The Tonga have never recovered from the devastating effects of the relocation (Tremmel et al. 1994). Main sources of livelihood are farming, fishing, weaving and braiding for tourists. Wild animals frequently destroy crops and attack people, the law prohibits animal killing.

Binga district ranks highest in terms of percentage of households considered living in dire poverty (Parliament of Zimbabwe 2012). Infrastructure is poor. Most parts of the district are inaccessible especially during the wet season. Schools are few, so are functional boreholes. Water is a serious problem for the whole district. Growing, consumption and smoking of cannabis is an important part of traditional Tonga culture.

3   Methodology

A combination of qualitative ethnographic research design and quantitative research design was used in this study. Ethnography allowed the researcher to unearth knowledge internalised in participants minds, for example, through focus group discussions, which allowed studying participants as a cultural group as they interacted in their day-to-day living. Quantitative design was used to establish relationships and differences in variables like temperature and rainfall experienced in the three districts over generations.

An analysis of relevant documents in text format was done to gain preliminary knowledge of Tonga minority communities in Binga district. The chief executive officer for Binga district was interviewed to get insight into government's involvement in climate change and climate variability issues in the region. A randomly selected traditional leader (local chief) interviewed in order to get the traditional perception of climate change issues. Since communities in the study area are agropastoralists, relevant government departments like Agricultural Research Technical and Extension Services (AGRITEX) and Livestock Development Programme (LDP) also provided data through interviews. Developmental support organisations, like Communal Areas Management Programme for Indigenous Resources (CAMPFIRE), were also important sources of data since their programme is tantamount to strategies used by communities to cope with climate change and variability.

Focused group discussions of between 8 and 12 discussants were used to give independent voice mainly to women, who, because of patriarchal domination, do not usually narrate their experiences as they experience them, but instead answer questions as expected of them by society. Discussions on climate change issues and how communities are coping with them were initiated at social gatherings like food-for-work programmes, boreholes, school development gatherings, and similar gatherings to get communities' views. Structured observation was used to verify communities coping strategies in-situ. It was used to 'ground truth' findings from other research techniques like interviews, questionnaires, focus group discussions, and ethnography. Questionnaires were distributed to randomly selected households per ward to confirm or refute information obtained from in-depth interviews and other methods. For Binga's 24 wards, 100 questionnaires were administered.

4   Results

4.1   Evidence of Climate Change in Binga District

Respondents in Binga district revealed that there is overwhelming environmental evidence indicative of a changing climate in their respective communities. These include, among others, drying of perennial rivers and springs, late onset of the rains,

Table 16.1 Environmental evidence of climate change in Binga district

Environmental evidence	Frequency Binga N = 100			
	Evident	%	Not evident	%
Drying of perennial rivers and springs	100	100	0	0
Late onset of the rains	100	100	0	0
Early cessation of the rains	100	100	0	0
Change in wind patterns	80	80	20	20
Diminishing pastures	100	100	0	0
Cold season warmer	68	68	32	32
Hot season hotter	96	96	4	4

Source: Field data

early cessation of the rain season, change in wind patterns, diminishing pastures, warmer winters and hotter summers (Table 16.1).

Tonga communities in Binga district have experienced a number of events pointing to a changing climate in the form of environmental evidence, like the drying of perennial rivers and springs, late onset and early ceasation of the rains, all of which were confirmed by all respondents (Table 16.1). All respondents (100%) also attributed diminishing pastures to climate change, with 80% saying that winds which used to bring rainfall are now less reliable. While a high percentage of 96 believed that the hot season is becoming hotter, a comparatively smaller percentage of 68 agreed that the cold season is now warmer than in the past decades.

4.2 Local Leaders Perceptions and Experiences on Climate Change

In Binga district, an interview with Sianzyundu village head, Mr. Mweembe, who is also deputy head of Sianzyundu secondary school, revealed that all the district's big rivers – like Sibungwe, Mativi and Mlibizi – used to have sacred natural pools, but these have since disappeared due to heavy siltation. He observed that rains have completely taken a different calendar since the 1980s, and this can be emphasised by analysing the Tonga nomenclature of months of the year. The month of February is called *Milonga* in Tonga because all rivers would normally be flooded during that period. It is also called *Mulumi* (man) because it used to be the wettest month, hence 'husband' in terms of rainfall amount. January is called *Mukazimaziba* [(wife of the rainy month *Milonga* (February)] because rains were lower than in February. Both January and February are now dry months because of the changing climate. March is called *Miyoba*, meaning a continuous period of an overcast sky with sporadic rains. In recent decades March hardly receives any rainfall.

Binga is generally not cold, but used to witness a cold spell characterised by some blowing wind, in July. The month of July was named *Kunkumuna masamu,* referring to the shedding of leaves by trees when the wind blows. Some of the trees would start shooting leaves in August, hence the month is called *Itwi* in Tonga, referring to the shooting of new leaves which continue until September when all trees would have had new leaves. Shooting of leaves made people start eating fruits like *inji* (red ivory), *mbubu (umviyo)*, and *mabiyu* (baobab). Currently, the shooting of new leaves begins in early November, and some hard woods stretch up to December without new leaves. Fruits like *inji* and *mabiyu* are now ripening in February and March.

In Tonga, October is called *Kavumbi kaniini* referring to small clouds which do not usually yield a lot of rain. November is called *Ivumbipati*, meaning a big cloud. The month used to be characterised by thick cumulonimbus clouds which would provide the first heavy rains that would give relief to people, livestock and wildlife. These early rains were characterised by lightning and thunder. Farmers used these rains to plant various crops. December is called *Nalupale*, meaning the month of weeding. The month used to provide a dry spell which allowed farmers to weed. In recent years, however, December has been offering the best rains, and farmers who miss December rains would have lost a whole season.

According to headman Mweembe, the Tonga used to do *shadduf* irrigation on some of the large perennial pools along large rivers like Sibungwe, Mativi and Mlibizi. Families grew tomatoes and vegetables like *bbobola* (pumpkin leaves for relish) throughout the year. They used to harvest twice annually because of abundant water from these perennial pools, which also provided a lot of fish that used to swim upstream from Lake Kariba. Manjolo springs wetland, which used to be very large, is now only about an eighth of its 1980 size.

Other evidence pointing to a changing climate according to Munkuli of Tinde ward is that yields are no longer enough to last households up to the next harvest. Tinde used to be a fairly wet ward, where farmers grew maize as the staple crop but they have since abandoned it for small grains. In Lunga ward (Chuunga village), farmers used to grow traditional Tonga tobacco, *kalukotwe*, sweet potatoes, green vegetables, all of which require a lot of water but in recent decades yields have been disappointing due to moisture stress. Although some women still do basketry using sisal and reeds, these are no longer as readily available as before because of changing soil moisture availability resulting from environmental change and climate variability.

4.3   Voices of Tonga Women on Climate Change

Focus group discussions at Lunga, Sinansengwe, Lubu, Tinde and Lubimbi revealed that the Tonga has always been a fishing community, having been removed from the Zambezi River valley by the colonial government in the 1960s to make way for the construction of Kariba Dam. They continued to fish from small dams and perennial pools, all of which have now completely silted. Wild fruits – like *inji (nyii), mbuubu (umviyo), mbuunya (umtshwankela), manego (matohwe)* – which they used to sell are now very few and not even enough for the community itself. From *mbuubu* and *mabiyu* (baobab) they used to prepare porridge for the whole family. Herdboys would milk cows while herding cattle, mix *mabiyu* with fresh milk which immediately turned sour. They would then drink the thick milk. Families would also roast *inteme (mazhumwi)* seeds, add salt or sugar for more taste.The whole family would have that for a meal. Currently, baobab fruits are very few as most trees no longer bear fruit. Cattle for milk are now also few. All these environmental cushions of household welfare and food security have since been eroded by the changing climate.

Women used to gather berries *(bwidi, babbonga, ntinde, masabayu)*, edible roots as dietary supplements, but most of these are now extinct. Thatch grass (hyparrhenia), used to abound in the district and women used to cut this easily for thatching their huts. Currently, however, it takes men to go up the mountain open spaces to harvest thatch grass. Mountains are infested with wild animals and are, therefore, unsafe for women. In the absence of thatch grass, most families are now banking on millet stalks for thatching. Gathering fruits and drying them for selling also used to be a very important economic activity. Currently, fruits have become scarce as trees hardly reproduce.

In Binga district, women complained during focus group discussions that the *kakunka* tree *(murara)*, which they used to get from river banks for their basketry craft work, has since depleted. This limits their survival options.

4.4   Household Strategies for Coping with Climate Change

In the extreme western part of Binga district, Tonga communities rely quite significantly on traditional household coping strategies to cushion themselves against climate change (Table 16.2).

These strategies include: planting early maturing varieties of the staple maize crop; planting drought resistant crops like sorghum, millet and rapoko; practising probability planting where different types of crop seeds are strewn onto a piece of land at the same time with the hope that even if some fail to reach maturity, others

Table 16.2 Indigenous household strategies for coping with climate change

Strategy	Frequency Binga N = 100			
	Helpful	%	Not helpful	%
Planting early maturing varieties of staple maize	100	100	0	0
Planting drought resistant crops	96	96	4	4
Collecting/drying wild fruits for future use	80	80	20	20
Drying some crops for future use	96	96	4	4
Probability planting	100	100	0	0
Eating wild fruits as household meals	86	86	14	14

will. Eating wild fruits as household meals and collecting and drying wild fruits for future use both have relatively lower percentages because most fruit trees only bear fruit when it rains. It is only during wet years that fruits are collected for immediate household use or dried for future use.

Another coping strategy involves planting long-term varieties like sorghum, millet, round peas, and groundnuts with short-term varieties like maize, water melons and green and yellow melons at the same time. This strategy helps farmers get some harvest from short-term varieties in case of a short rainfall season and, at least, some harvest from long-term varieties if the wet season experiences rainfall evenly distributed throughout the season.

Early planting is yet another coping strategy used by Tonga communities. This allows for replanting in case of poor germination due to a deceptive rainfall pattern. Farmers have also adopted multiple planting dates as an important coping strategy to climate change and variability. This is meant to spread the risk in case of climate-induced phenomena like drought, floods, quelia birds, locusts, and rats.

Polyculture is also an important coping strategy among the Tonga. This is whereby multiple crops like drought resistant millet, sorghum and water melons are intercropped with legumes like ground nuts and round nuts which require higher amounts of moisture. This is meant to guard against total loss in case of extreme climatic events like drought or excessive rainfall.

4.5 Experiences of Community Elders and Local Leaders

Interviews with farmers revealed that although Tonga communities still have faith in their traditional slash-burn-dig and sow farming technique, they have since improved it to proper conservation farming (zero tillage) where they are now digging deeper spaced basins capable of holding water for a prolonged period of time. Apart from serving water, this technique gives the crop more time to utilise soil moisture for maximised growth.

On average, women in Binga travel between 7–12 km to fetch water, and to cope with water problems, some families now have two homes, one closer to the Zambezi River for water and fish and another in areas with better crop yields like Tinde ward. This is a change in that homes closer to the Zambezi are illegal. Farmers, however, have them as a response to shrinking livelihood options as wetlands, rivers and streams in Tinde ward dried as they succumbed to climate change and climate variability.

As another way of coping with difficulties associated with climate change, gender roles have shifted, with both men and women now doing fishing for money to fend for their families. Women, however, now do have chores than men who usually spend time drinking beer and smoking traditional marijuana. Women thus fend for the family more than men. They fetch firewood, water and work in the fields. They dig fields for cropping using hoes in their hands. Cattle are usually not for farming but for prestige in the home. Besides, many families do not have cattle. Some women seasonally abandon their less productive fields in places like Kariyangwe, Lunga, Siachilaba, and Sinampande and go to the comparatively more productive neighbouring district of Gokwe where they work for food for their families. Some even seek employment in distant towns like Hwange and Bulawayo, something which only used to be done by men.

4.6 Observed Climate Change–Related Household Coping Strategies

Observation in Binga district revealed a lot of craftwork selling by both men and women along the Binga-Kamativi highway and at most service centres and growth points in the district. Women make a lot of weave and basketry items from sisal and reeds that they collect from river banks and from the banks of Kariba Dam. Men make different traditional curios from local hardwood trees like pterocarpous angelenses (mukwa), baikiaea plurjuga (teak) and afzelia quanzes (mukamba) found mainly on the Situmba escarpment (Fig. 16.2). While such work has been used in Tonga homes as household furniture, commercialising it is a new phenomenon which is necessitated by climate change challenges in the community. These pieces are sold very cheaply as the carpenters only need to raise whatever amount to take back home.

Fig. 16.2 Traditional carpentry pieces along Binga Highway
Source: Field data (Consent granted on photograph use)

Table 16.3 Indigenous strategies for livestock sustenance under climate change

Strategy	Frequency			
	Binga N = 100			
	Helpful	%	Not helpful	%
Sending cattle to far-off places for better pastures	68	68	32	32
Preserving crop residue after harvest	50	50	50	50
Sourcing tree leaves/twigs/fruits for livestock	50	50	50	50
Reducing the size of livestock herds during drought periods	74	74	26	26

Source: Field data

4.7 Community Pastoral Strategies under Climate Change

To cope with these changes, 74% (Table 16.3) of Tonga respondents in Binga district reduce the size of their livestock herds during drought periods by either selling to abattoirs or through barter trade with other communities.

This is a strategy to avoid stock loses during the drought period. Sixty-eight percent of the respondents send their cattle to far-off places with better pastures during dry seasons. Other attempts to save livestock include preserving crop residue like maize and sorghum stalks after the harvest and sourcing tree leaves, twigs and fruits for them.

4.8 Community Leaders, Livestock and the Climate Change Threat

According to headman Mukuli (85) of Binga district, in traditional Tonga culture, livestock were meant for social status and pride in the community. However, because of continued livestock loses to incessant droughts, wild animals and increasing incidences of disease (Hargreaves et al. 2004; Davies 2007), some households are now trading their livestock to towns like Hwange and Bulawayo for money. This has caused a shift in attitude and traditional-customary values. Some farmers now practice barter trading, exchanging livestock for food, either within the district or with merchants from other districts. For example, a goat can be exchanged for a 5 kg or 10 kg bucket of maize depending on size, and a cow for 100 kg of maize.

Mweende (87) added that potato-like traditional wild roots *(makuli),* which they never knew could be food during their youth, are now being eaten in the whole district during drought periods. During such times, families would drastically cut down on meals, eating only one meal per day at night.

Table 16.4 Constraints encountered by Tonga farmers in coping with climate change

Constraint	Frequency			
	Binga N = 100			
	Agree	%	Disagree	%
Planting time no longer predictable	100	100	0	0
Harvesting no longer certain	100	100	0	0
Water sources now too far	60	60	40	40
No access to weather forecasts	80	80	20	20
Weather forecasts unreliable	20	20	80	80

Source: Field data

4.9 Constraints of Tonga Farmers in Coping with Climate Change and Variability

In Binga district, the major disruptions brought by climate change to the Tonga community in their subsistence farming life include planting time, which all respondents (100%) qualified as no longer predictable, and harvesting time, which they (100%) also described as no longer certain (Table 16.4).

Although they are the traditional owners of the mighty perennial Zambezi River, which boasts of the mighty Victoria Falls (Mosi-oa-Tunya), and Lake Kariba, which covers 517,998 km^2 of dammed water, the Tonga themselves are a thirsty community. Water is a perennial problem in the district. 60% of respondents complained that water sources for both their livestock and domestic use are too far. Radio and television services are only barely accessible in very few areas (BRDCSP 2011), and respondents, qualified weather forecasts from these services unreliable and, hence, not useful for their farming activities. The rest of the respondents (80%) in the district have no access to weather forecasts and are solely dependent on intuition in their farming activities.

4.10 Seed as a Climate Change–Related Constraint for Tonga Farmer Communities

The Tonga are a highly conservative community, and this is reflected even in their farming practices. They are one of the very few communities in Zimbabwe which still practice shifting cultivation (chitemene), although they confine this within their district. Their major staple crops are sorghum and millet, whose seeds they do not normally buy, but are traditionally preserved in granaries from previous harvests for replanting (Table 16.5).

Maize, the country's major staple crop, needs seasonal rainfall amounts of between 600 mm and 800 mm, and yet due to climate change Binga's average seasonal rainfall has dropped to between 400 mm–600 mm (Zhakata 2008). The district, therefore, does not harvest much maize but if it does, 82% of the Tonga prefer

Table 16.5 Frequencies of household sources of seed in Binga district

| Crop | Source of seed Binga N = 100 | | | |
	Buy	Govt	NGOs	Granary
Maize	78	30	0	82
Sorghum	26	36	0	100
Millet	18	12	0	86
Rapoko	8	2	2	10
Groundnuts	26	18	2	36

Source: Field data

planting maize seeds from their granary than buying from seed houses. In the absence of traditionally treated maize seeds, farmers buy 'high yielding varieties' from agricultural inputs supplies.

4.11   Community Leaders' Perceptions on Climate Change Constraints

Headman Mweembe observed that as a peripheral minority district, Binga has less access to modern facilities like information communication technology (ICT) which could expose its people, especially the youth, to the innovative dynamic outside world. It is a population of conservatives where the basis of household wealth and livestock, is owned by the very elderly who usually interpret the outbreak of disease as punishment from the ancestors. Instead of treating their livestock, they hold rituals to appease the spirits.

Mweembe reasoned that since the district is remote and 'backward' socio-economically, many non-governmental organisations (NGOs) carry out their experiments there. The district thus becomes a guinea pig and a testing ground for many projects and programmes, most of which have unproven results and are, therefore, unsustainable.

On food security, one interviewee noted with concern that donors bring in food handouts instead of tapping water from Lake Kariba which would bring a permanent solution to the district's incessant water problems. The interviewee argued that NGOs should capacitate people rather than make them dependents. He suggested that they should emulate the government's food-for-work programmes which organises villagers into permanent work groups (*gombe* or *nhimbe* style). Such work groups would then be assigned specific sustainable developmental tasks which benefit them as communities.

Headman Mweembe, however, notes that Binga community lack entrepreneurial skills, observing, for example, that most fishermen operating in the district are not locals. He adds that although subjects like Geography are taught at schools and help pupils appreciate the importance of the environment, the curriculum needs to be revisited to suit the current climatic trend. The village head condemned the country's curriculum as being too academic, suggesting that pupils should be taught life-oriented instead of exam-oriented skills. He lamented that their children pass building and agriculture but cannot build; neither can they match their parents in farming skills.

The AGRITEX officer in Binga revealed that local communities are realising very little from the CAMPFIRE programme and this is the main reason why they are invading national parks. For example, Chizarira National Park has been invaded by Sinasengwe, Nsenga and Nabusenga communities while people from Tyunga ward have also settled in Chete safari area.

4.12   Climate Change Constraints and Tonga Women

In Binga district, discussants complained that the entire district needs water, and yet it is on the edge of the country's largest surface water body, Lake Kariba, which can easily be tapped to water and green the whole district. Women from Sikalenge fetch water, bucket-on-head, from Lake Kariba some 10–15 km away because they do not have scotch-carts. Earth dams dry around August and people start fetching water from the Lake. During this period, commuter omnibus operators experience brisk business, using their trailers to fetch as much water as they can and selling it to the thirsty community at US$1 per 20 litre bucket.

In Simatelele ward, which is closest to the Zambezi River, farmers who wish to venture into fishing cannot do so easily because they cannot afford fishing licences which cost US$35 on the spot. Fishing without a licence attracts a penalty of US$50 on the spot. In addition, the offenders' catch for the day is confiscated and they are forced to buy a fishing licence. Some fishermen complained that they attract very few customers because they fish in remote areas of the Zambezi for fear of the police and Campfire authorities. Commercial kapenta fishing using rigs (fishing boats) is even unimaginable for most Tongas because one rig cost US$14000 according to Fletcher of Katuya Co-operative. Average kapenta fish catch is 53 bags per month which they sell at US$150 each. Each co-operative has an average of 10 members. It is because of such high costs that most commercial fishermen along the Lake Kariba in Binga come from large towns and cities like Harare and Bulawayo.

Farmers in all districts bemoaned the prevalence of cattle rustlers who steal cattle, goats and sheep at night. They target mainly female-headed households where they know women would not go out for fear of possible harm or even death.

Fig. 16.3 Water containers children use to carry water back home after school
Source: Field data (Consent granted on photograph use)

4.13   Climate Change Impacts on Children's Education

At Sianzyundu secondary school in Binga, where pupils from distant homes were given bicycles by World Bicycle Services and Catholic Relief Services in 2010 and cycle 8–12 km, the researcher observed that most pupils carried with them 20 or 25 litre buckets to school so that they carry water back home from school (Fig. 16.3). This means, pupils lose time queuing to fetch water from the school borehole. Cycling back home with loads of water on their bicycles make them get home tired and not in the best disposition to do schoolwork. The school has a borehole whose water table is at 93 metres according to the school head.

4.14   Government Initiatives to Help Farmers Cope with Climate Change

According to the District AGRITEX Officer, in Binga district, AGRITEX trains farmers on conservation farming. It conducts field days, establishes demonstration plots through its extension workers in villages and wards. It also carries out workshops to promote small grains which are more suitable for the district than the national staple maize crop. Conceding that the issue of climate change is confusing farmers, because of the sharp rainfall variability from one season to the next, the AGRITEX Officer said some farmers still want to grow maize ahead of other crops because small grains do not have ready market. Farmers also argue that there are too many and straining stages involved in processing small grains, and besides, *sadza* from small grains is not as palatable as that from maize. Tonga communities originally came from the wet Zambezi Valley where they grew maize and other crops on the fertile alluvial soils of the Zambezi River before being forced to the current poorer dry escarpment during the construction of the Kariba Dam in the 1950s.The farmers also argue that small grains like sorghum and pearl millet are consumed by quelia birds resulting in low yields.

The Livestock Production Department (LPD) promotes the rearing of different types of livestock which include cattle, donkeys, goats and sheep. The District AGRITEX Officer revealed that from a livestock rearing perspective, Binga district can be divided into two different areas: southern part and northern part. Southern Binga on one hand, is cattle area because it has more pastures than acacia. It also has fewer cattle diseases. On the other hand, northern Binga is goats' area because of more acacia and other thorn tree species preferred by goats. Cattle do not usually do well in the northern part because of shortage of pastures, the mountainous terrain, more diseases and the prevalence of wildlife, including crocodiles from the Kariba Dam.

Lungwalala Dam, whose construction was facilitated by government in 1992, is one of the largest dams in Matabeleland north Province. The dam supports a vibrant irrigation scheme in Kariyangwe ward. Nabusenga Dam is another dam with a functional irrigation scheme in the district. During the time of the study (2012–2014), Government was initiating an irrigation scheme at Bulawayo kraal which, once complete, would be the largest in Matabeleland Province, according to Kohli, an Irrigation Department officer. The scheme, supervised by Vice President Mujuru's office, is expected to grow crops like maize, sorghum, and potatoes among others. Bulawayo kraal has fertile soils and is sparsely populated, and thus it is expected to accommodate a number of farmers from the Binga community. The scheme is expected to draw water from Kariba Dam.

Binga Rural District Council has on its books five CAMPFIRE projects, which include Siamuloba Campsite in Simatelele ward, Musiinji Cultural Village in Pashu ward, Sinakoma Elephant Dung Paper Making Project in Sinakoma ward, Buvubi Leather Project in Nakusenga ward and Banyama Mbibesu General Dealer shop in Sinansenga ward. The viability of these projects, however, dropped drastically when white partners withdrew in protest of the fast-track land reform and resettlement programme (FTLRRP) of the decade 2000–2009.

4.15   Initiatives by NGOs to Help Farmers Cope with the Impacts of Climate Change

NGOs have been quite active in Binga district, most significantly since the 1990s to 2013. Projects like community gardens, borehole drilling, chicken projects and cattle fattening have been initiated by various organisations. Chicken, goats and cattle pass-on projects have also been carried out in various wards throughout the district. Save the Children was also very active during the difficult decade 2000–2009, which was characterised by drought, poor and, sometimes, no harvests, famine and serious food shortages even in shops. This was mainly as a result of the so-called targeted sanctions by Western countries in response to the government's fast-track land reform programme which forcefully dispossessed white farmers of 'their' trademark large tracts of fertile land, and distributed it among majority, but landless, black farmers.

Siachilaba women fish market is another project worth noting, where some women were organised and funded by the Catholic Commission for Justice and Peace (CCJP) to engage in fish marketing since 1999. The women act as a conduit between fishing cooperatives on Lake Kariba and customers in transit to and from Binga Centre. They also have customers as far a field as Hwange, Bulawayo, Gweru and Masvingo.

4.16 Perspectives of Tonga Communities on Sustainability of Interventions

Although the Tonga are conservatives culturally, climate-change-inflicted hardships seem to be forcing them to appreciate foreign assistance. Table 16.6 shows that a sum of 88% of the respondents rate NGOs interventions between sustainable and very sustainable, which confirms their gratitude to assistance provided by these organisations.

Government assistance has a comparatively low combined rating of sustainability (sustainable plus very sustainable) of 50%. Most interviewees and focus group discussants said government assistance in the district is minimal. Tonga indigenous knowledge systems have a high combined rating (sustainable plus very sustainable) of 72% which means even under threat of climate change, many Tongas still seek solutions in their indigenous knowledge systems.

Binga Chief Executive Officer, Mr. Muzamba, revealed that more than 11 NGOs operate in the district among the Save the Children, Christian Care, ADRA, LEAD Trust, CARITAS, Dabani Trust, UNDP, Masilizwi (People of the Great River), KMTC (Kulima mbobumi Training Centre), CCJP and ZUBO. These NGOs are involved in various activities and projects which include food distribution, chicken,

Table 16.6 Interventions sustainability rating

Intervention	Binga					
	N = 100					
	Not sustainable	%	Sustainable	%	Very sustainable	%
Indigenous knowledge	28	28	40	40	32	32
Government interventions	50	50	44	44	6	6
NGO interventions	12	12	41	41	47	47

goat, pig and community garden projects. Some distribute agricultural inputs and, in conjunction with AGRITEX, educate farmers on conservation farming techniques.

CCJP built a cold room for women fish marketing entrepreneurs at Siachilaba Service Centre while at Binga centre, Basilizwi Trust in collaboration with ZUBO built a similar but larger cold room (Kujatana co-operative cold room) for use by all fishing cooperatives operating near Binga centre. The District Agritex Officer, Mombe, said LEAD Trust is involved in the promotion of small grains; pearl millet, sorghum and finger millet, and emphasises the quality of seeds that farmers should grow. He also revealed that ICRISAT used to have a project in Ward 9 where they were doing seed multiplication. The project has since stopped because of lack of funds. KMTC has built two earth dams in Ward 10, one in Ward 9 and another two in Ward 13, mainly in an attempt to save livestock. Livestock were depending on Lake Kariba which is too far from homesteads and, besides, farmers were losing a lot of their livestock to both crocodile attacks and stock thieves at the Lake. KMTC also gave goats to some households to breed. Save the Children is mainly involved in paying fees for orphans and vulnerable children (OVC). It also constructs toilets at schools and is involved in nutrition gardens. In Kariyangwe ward, it offered sewing machines to the community so that they could sew school uniforms for their children.

4.17 Climate Change and Binga's Development Basket of Priorities

The Tonga realise that they have no capacity to mitigate climate change impacts. They also realise that the ecological environment, which has been their trusted source of livelihood is under threat from the changing climate. A viable approach to helping these communities cope with climate change impacts would be to examine what these communities identify as their development basket of priorities. The first step would include identification of their immediate needs at household level which is where climate change impacts are usually experienced and felt hardest. At this level, Tonga communities in western Zimbabwe need food for themselves and pastures for their cattle. They also need their domestic pillars of wealth like cattle, goats, sheep and chickens. They want to preserve their indigenous varieties of drought tolerant crops like millet and sorghum. Water is a key driver of household economies in the sub-region, yet it is a seriously inadequate resource. These communities have traditionally settled along rivers, streams and close to wetlands which provide water for their small household gardens. An identification of household needs would thus help inform both government and NGOs about which developmental infrastructure each community requires to support household needs and aspirations.

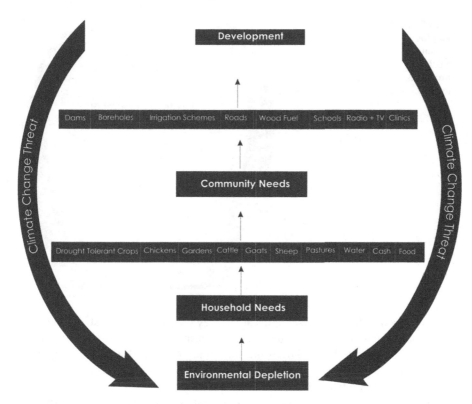

Fig. 16.4 Tonga communities' development basket of priorities
Source: Designed from field data

Since water is the lifeline of Binga district, dams, boreholes and irrigation schemes should take precedence. Communities would, however, also require good roads, schools and clinics as support infrastructure for their household needs. For informed farming activities, radio and/or television weather bulletins would be important. Since the natural ecosystem provides resources for most of minority communities' indigenous knowledge about food, medicine and preservatives for their crops, it would also be important to plant and conserve indigenous trees and their biodiversity. Besides, trees are the basic source of fuel in these communities. Figure 16.4 shows the community development basket of priorities as informed by findings from this research. It tries to suggest and emphasize that since climate change hit hardest marginal communities like the Tonga in western Zimbabwe, policies and solutions to such climate change problems should start from these marginal areas. In other words, we should think from the periphery and with the periphery towards the centre, and not vice versa, as is the case at present.

An identification of community development needs in itself may, however, not be conclusive as an approach for minority farmer communities to cope with climate change. Roles of communities themselves and the government must be clarified.

Fig. 16.5 The IDIL-PECF
model for Tonga
community development in
Zimbabwe
Source: Designed from
field data

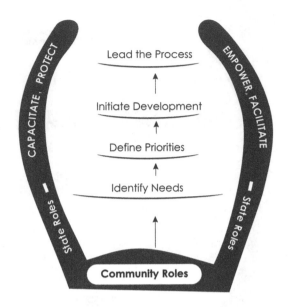

Communities usually cooperate in initiatives they themselves suggest. On the basis of the conduct, interaction with study participants and results, this study suggests an identify-define-initiate-lead (IDIL) role for communities and a protect-empower-capacitate-facilitate (PECF) role for the state to help communities in western Zimbabwe cope with the dire climate change impacts.

In this model, communities should identify their needs, define their development trajectory, initiate development activities, and lead the development process itself. However, minority communities are by their nature vulnerable to outside influence and manipulation. In line with the developmental state approach, which recognises the state as being nationalistic in nature, and approach to development, the state should of necessity protect minority farmer community initiatives through supporting policies and programmes; capacitate communities financially, infrastructurally and materially, and empower communities by making them owners of resources in their immediate environs. For example, water and fish resources in Lake Kariba should benefit the Tonga first and foremost. Above all, the state should facilitate development through providing expertise in various fields, including livestock management to these communities as they battle for survival against the climate change threat. The IDIL-PECF model is illustrated in Fig. 16.5. It is a cupful solution to community development in the face of climate change and variability.

Using this approach, Tonga communities would develop from within, but under the protection of the state. Other outside players and benefactors may only come to aid the process, which is inward-looking and fundamentally uses local indigenous resources, and respects community indigenous knowledge systems.

5 Conclusion and Recommendations

Minority Tonga communities have known climate change not from books, seminars or workshops, but have experienced it in various ways, which include mainly recurrent droughts, increasingly hot temperatures and periodic cyclonic episodes which cause serious floods. Consequently, the farmers have seen these climatic extremes triggering environmental change, which includes the drying of wetlands, rivers and streams from excessive evapotranspiration. The changing climate has also contributed to the depletion of rangelands and the ecosystem in general, on which these communities depend for various services like food, medicine and pastures for their livelihoods. This has negatively impacted the general livelihoods of minority Tonga communities who have depended almost exclusively on the natural environment. Climatic changes and variability have resulted in limited household economic options for minority farmer communities. Livestock, especially cattle, which traditionally are the bedrock of community economic affluence, is hardest hit by rapidly degrading pastures. Multiple deaths of cattle during recurrent droughts have heavily depleted household herds. Cattle have always been safety nets for the Tonga. Their depletion means hunger, starvation, school dropouts and general socio-economic insecurity which push the economically active population out of these communities in search of better fortune elsewhere.

With the help of the Forestry Commission, Binga Rural District Council should have a deliberate policy of tree planting and caring for both indigenous and exotic species in all wards of the district. The programme should involve all schools and the community at large. This will increase transpiration and possibly rainfall amounts. The Tonga type of *chitemene*, very common in the district, where most households have large areas of land, where they practice this slash-burn-dig-and-sow type of agriculture rotationally, must be discouraged by AGRITEX as it is responsible for the serious deforestation in the district.

Given Binga's diverse physical environment, solutions to the district's problems are likely to be variable through interventions by Government, NGOs and other well-wishers: Wards closest to Lake Kariba (Simatelele, Tyunga, Sianzyundu, and Saba) should be capacitated in fisheries by being provided with fishing rigs through local fishing cooperatives. Government should subsidise charges for fishing by local communities. Wards like Lusulu, Tinde and Siabuwa may do well if Government and NGOs support them with crop (millet and sorghum) inputs. Sianzyundu and Simatelele have potential for goat-keeping. Manjolo community can be sustainable with a serious investment in market gardening, given its wetland and natural springs.

In light of the fact that most Campfire projects in Binga district are currently not functional and yet they can be important safety nets against climate change and variability, there is dire need for Government to inject a substantial amount of money into these projects in order to resuscitate them. Binga is a water stressed district, and to boost its water supply both for domestic and wildlife use, Government could identify a stakeholder who would sink boreholes in the district. The district has very few dams, so several small dams could be constructed to hold water up to the next

rain season. This will cushion communities, livestock and wildlife from the vicis-
situdes of climatic changes and variability.

References

Australian Government Department of Meteorology-Indigenous Weather website (2013) [Online]
    http://www.bon.gov.au/iwk/. Accessed 15 Jan 2013
Baird R (2008) The impact of climate change on minority and indigenous peoples minority rights
    group international, April 2008. London. [on-line] Available from http.//www.minorityrights.
    org. Accessed 20 Mar 2012
Basilwizi Trust (2010) Strategic Plan 2010–2015. Basilwizi Trust, Bulawayo
Bohle HG (2001) Vulnerability and Criticality. Perspectives from Social Geography. In:
    *IHDP Update, Newsletter of the International Human Dimention Programme on Global
    Environmental Change (IHDP)*, vol 2, pp 1–4
Boven K, Morohashi J (eds) (2002) Best Practices Using Indigenous Knowledge. Nuffic, The
    Hague
BRDCSP (2011) Binga Rural District Council Strategic Plan (2011–2015). Binga Rural District
    Council, Binga
Chagutah TC (2010) Climate change vulnerability and adaptation preparedness in Southern Africa.
    Zimbabwe Country Report. Cape Town: Heinrich Boll Stiftung Southern Africa
Change in Southern Africa (2012) [Online] Available from http://www.oxfam.org/grow. Accessed
    30 May 2012
Chenge M, Sola L, Paleczny D (1998) The State Of Zimbabwe's Environment. Harare: Ministry of
    Mines, Environment and Tourism. Government of Zimbabwe
Davies J (2007) Farmers current coping strategies. In: *Building Adaptive to Cope With Increasing
    Vulnerability Due to Climate Change*. ICRISAT, Bulawayo
Declaration of the Indigenous Peoples of the World to COP 17 (2012) [online] Available from
    http://climateandcapitalism.com. Accessed 27 May 2012
Environmental Assessment and Saskatchewan's First Nations (2008) A Resource Handbook
    [Online] http://www.iisd.org/pdf/2008environmental-assessment_sask.pdf. Accessed 7 Jan
    2013
Feresu SB (ed) (2010) Zimbabwe Environmental Change: Our Environment, Everybody's
    Responsibility. Harare: Government of Zimbabwe's Third State of Environment Report
GoZ-UNDP/GEF (2011) Coping with drought and climate change. In: *Environmental Management
    Agency*. Sable Press (Pvt) Ltd, Harare
Hargreaves SK, Bruce D, Beffe ML (2004) Disaster Mitigation Options for Livestock Production
    in Communal Farming Systems in Zimbabwe. ICRISAT, Bulawayo
Intergovernmental Panel for Climate Change (IPCC) (2007) Forth Assessment Report (AR4)
    Climate Change 2007: Impacts, Adaptation and Vulnerability. Contribution of Working
    Group II to the Fourth Assessment Report of the Intergovernmental Panel on Climate Change.
    Cambridge University Press, Cambridge
Matsa M, Matsa W (2013) Traditional Adaptation Mechanisms to cc and Variability among
    Women in South west zimbabwe. In: Mulinge MM, Getu M (eds) Impacts of cc and Variability
    on Pastoral Women in Sub-saharan africa. Fountain Publishers, Kampala, pp 297–242
McLean KG (2010) Advance Guard: Climate Change Impacts, Adaptation, Mitigation and
    Indigenous Peoples: A Compendium of Case Studies
Oxfam International (2011) Overcoming the Barriers. How to Ensure Food Production Under
    Climate
Parliament of Zimbabwe (2012) Binga North Constituency Profile. Parliament Research
    Department, Harare

Ravindranath NH, Sathaye JA (2002) Climate change and developing countries. Kluwer academic publishers, London

Thorpe T, Eyegetok N, Hakongak, N, Kitimeot Elders (2002) Nowadays it is not the same: Inuit Qaumjimajatuqangit, Climate an Caribou in the Kitimeot Region of Nunavut, Canada [Online] http://inuitcircumpolar.com/files/icc-files/FINALPetitionICC.pdf. Accessed 10 Jan 2013

Tremmel M, Valley Tonga People (1994) The Poople of the Great River. Mambo Press in association with Silveira House, Harare

UNESCO (2012) Best practices on indigenous knowledge [Online] http://www.unesco.org/most/bpik21-2.htm. Accessed 10 Jan 2013

Voices from the Bay (1998) Traditional ecological knowledge of Inuit and Cree in the Hudson Bay Bioregion [Online] http://www.carc.org/voices_from_the_bay.php. Accessed 13 Jan 2013

Zhakata W (2008) Climate change in Zimbabwe in ICRISAT Building adaptive capacity to cope with increasing vulnerability due to climate change. ICRISAT, Andhra Pradesh

Zimbabwe Poverty Assessment Study Survey (1996) 1995 poverty assessment survey: preliminary report: Harare: Ministry of Labour, Public Service and Social Welfare

ZimStat (2012) Census 2012 Preliminary Report. UNDP, Harare

Chapter 17
Environmental Amenities and Regional Economic Development - The Case of Todgha Oasis

Mhamed Ahrabous, Fatima Arib, and Aziz Fadlaoui

Abstract Historically, natural resources have provided location-specific advantages for communities at various stages of their development. Agriculture and rural communities are regarded as producers of commodities and environmental and social services. This paper aims at analyzing the role of environmental amenities in rural economic development. In a first part, we present a survey of existing literature on the link between amenities and rural development. In a second part, we give an example from the Oasis of Todgha (Southeastern Morocco), and we analyze how environmental amenities can contribute to the economic development of the region.

Keywords Environmental amenities · Economic development · Natural resources · Agriculture · Rural area · Oasis of Todgha

1 Introduction

In the 90s, the concept of 'amenity' was at the center of the debate on the dynamics of rural development, because of its potential in terms of employment and income that it was supposed to generate in the development of agriculture and other activities (Taylor et al. 2000; Fleming et al. 2009). The term 'amenity', with an ancient origin and forgotten for a long time, was reinvented in the 90s by the Organization for Economic Cooperation and Development (OECD) with the context of discussions on the reform of the Common Agricultural Policy for rural areas. It had then a great success in connection with the concept of 'multifunctionality' because it

M. Ahrabous (✉)
Hassan II Institute of Agronomy and Veterinary Medicine, Rabat, Morocco

F. Arib
Faculty of Law, Economics and Social Sciences, Cadi Ayyad University of Marrakech /
Researcher at the Center for Research on Environment, Human Security and Governance (CERES), Marrakesh, Morocco

A. Fadlaoui
Environmental Economist, National Institute of Agronomic Research (INRA),
Meknes, Morocco

© Springer International Publishing AG, part of Springer Nature 2019
M. Behnassi et al. (eds.), *Human and Environmental Security in the Era of Global Risks*, https://doi.org/10.1007/978-3-319-92828-9_17

legitimized the 'second pillar' of rural development, which would better take in consideration the strengths and constraints of the environment in rural spaces thanks to public policy support (OCDE 1999).

Natural amenities have a positive effect on economic growth and an equalizing effect on income distribution (Kwang-Koo et al. 2005). Natural resources continue to play an important role in defining the structure and viability of rural communities. Historically, natural resources have provided location-specific advantages for communities at various stages of their development (Marcuiller and Clendenning 2005). In early stage, extractive industries used natural resources as raw materials for processed goods, thus creating plentiful job opportunities (Waltert and Schläpfer 2007; Power 2005; Goodstein 2004). In the framework of a spatial equilibrium model, Roback (1982) shows that exogenous local amenities that are valued by employees are capitalized into rents and wages. In an extension of this model, Moretti (2004) distinguishes low-human-capital employees from high-human-capital employees, finding that only the latter value the local amenity. This model suggests a spatial equilibrium, with a larger share of high-human-capital employees in the high amenity location.

It seems reasonable to assume that the amenities affecting property prices are partly the same as those promoting amenity-based development processes (Waltert and Schläpfer 2007). High quality natural and environmental amenities have also attracted increased housing density change and commercial development, which have resulted in the conversion of high natural amenities and land resources to development (Taylor and Smith 2000; Klein and Reganold 1997; Nickerson and Hellerstein 2003; Daniels 1991). As a result, a number of countries have initiated some forms of natural resource and land conservation initiatives to manage the increasing pressure on natural and environmental resources (Green 2001; Kwang-Koo 2005). These initiatives illustrate the importance of utilizing natural and environmental services and proper conservation and protection requirements.

Several recent empirical studies (such as Gottlieb 1995; McKean et al. 2005) support the general conclusion that amenities tend to be weakly related to business location or employment growth. Many authors (such as Green 2001; McGranahan 1999) found that a natural amenity index was related to employment change over the past 30 years. Thus, high amenity counties had an average of three times as many jobs during this period than those that scored low on the amenity index (Green 2001). However, employment change was much more variable during this period than population change, especially for high amenity counties.

What is the role of environmental amenities in rural development? In order to analyze this question, we will first present a theoretical framework of environmental amenities and rural development (2), and then we will discuss the role of amenities and natural resources to develop the Oasis of Todgha (Southeastern Morocco) (3).

2   A Theoretical Framework

2.1   Multifunctionality of Agriculture

The concept 'multifunctional agriculture' appeared in 1992 during the Rio Summit alongside that of 'sustainable development'. It has been used to describe the many roles of agriculture in addition to food or fiber production (Pezzini et al. 2001; Handerson et al. 2000). A normative vision of the multifunctionality of agriculture was thus put forward as a means of implementing sustainable economic and social development in human societies (Laurent 2002). In this perspective, the normative approach of multifunctionality is based on the idea that agriculture does not only supply food products and raw materials, but plays two other major functions as well: social and environmental.

- Through the social function, agriculture contributes to the viability of rural areas and to local development by creating jobs in the primary production sector as well as in the processing and supply circuits. Therefore, agriculture plays a key role in maintaining, recreating or reinventing rural social bonds.
- The environmental function of agriculture concerns the production of rural amenities whose existence confers an ecological and patrimonial value to rural areas (Allali 2002; Allali 2006). For example, the production of milk and butter of goats may also require maintaining upland pastures which increase the aesthetic value of scenic vistas mountain areas. In this example, we can consider animal waste as another positive externality of the primary agricultural activity of goat dairying.

According to the economic theory, efficiency is found if these externalities can be internalized so that primary production decisions are based on net social well-being (Pezzini et al. 2001; Allali 2006; Dupras et al. 2013; Estrada–Carmona et al. 2014). Given the improbability of this case, norms and regulations could be used to reduce social, social segregation, and poverty. The crux of the policy debate over 'Multifunctional agriculture' is to know if domestic subsidies are an appropriate way to positive domestic externalities (Marcuiller and Clendenning 2005; Candau et al. 2005; Travnikar and Juvančič 2013).

Increasingly, there is a recognition of the multifunctionality of agriculture, moving beyond providing agricultural commodity to a recognition of the landscape itself as a service provider to rural areas and the broader public in terms of quality of life determinants such as open space, landscape that enhance clean air and water and pleasing environments and scenery for general health and human well-being (Pezzini et al. 2001). For many years, the multifunctional character of agriculture has been very evident in public policies as a recognition of the existence of a social demand for non-market services with environmental nature (Romstad 2004; Randall 2002).

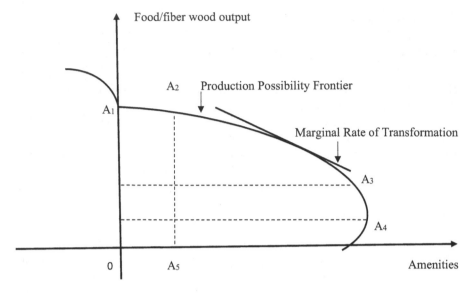

Fig. 17.1 Production possibility frontier between extractive and amenity uses
Source: Bonnieux and Rainelli (2000)

Figure 17.1 shows that the links between agricultural production (such as food, fiber, and wood) and amenities are relatively complex because they are complementary or substitutes according to the situation:

- A_1 marks a significant level of commodity output with no production of externalities.
- To the left of A_1, increasing production is done via intensification of the use of land; using more inputs creates negative externalities related to excess use of chemical products (fertilizers or pesticides), hence nonpoint pollution.
- From A_1 to A_2, there are several combinations of labor and capital for the same level of output. If capital is cheap relative to labor, the production system features large fields in order to work as fast as possible, where hedges and slopes are removed; positive externalities exist due to the maintenance of open space, but they are low. If labor is cheap relative to capital, then conserving hedges and slopes is justified, and as one moves closer to A_2, the production of amenities becomes significant with a small reduction in agricultural output. The relationship between the production of commodities and amenities is weakly complementary.
- To the right of A_2, there is a reduction in commodity production while environmental quality increases. The production system changes: first, 'reasoned' agriculture; then organic farming. Thus, from A_2 to A_3, the relationship between agricultural good production and environmental good production changes from complementarity to substitution.

- A$_3$ to A$_4$ commodity production varies with an almost constant level of amenities. This case corresponds to relatively extensive animal breeding with permanent pasture and low animal density. In this situation, there is, again, weak complementarity between the two categories of produced goods.
- From A$_4$ to A$_5$, the production of agricultural goods decreases, and so does the production of amenities. There is negative complementarity between the goods. This happens with very low animal density or extensive fallow. In this case, commodity output reduction is associated with the production of wasteland and the loss of biodiversity and landscape quality.

2.2 Amenities and Regional Economic Growth

In general, amenities can be defined as local attributes that provide comfort, enhance the attractiveness, or add value to a locality. As a regional growth factors, the amenities contribute to the quality of life of an area, along with other environmental, social and economic factors (Dissart and Deller 2000). In the literature, the relationship between amenities and regional development was discussed from the perspectives of human migration, the location of firms, and regional economic growth.

Amenities may play three roles in the economic development of rural areas (Deller et al. 2001). The first body of literature examined the effects of amenities as an economic development tool and was principally concerned with the extent to which a business location is affected by the quality of life.[1] The second part of literature analyzed the role of amenities on the redistribution of population from urban to rural areas, especially among retirees. Finally, recent research examined the effects of amenities on several attributes of the local economy, such as income inequality, fiscal health, and consumer spending (Dissart 2007).

While many rural areas[2] continue to have depopulation and economic decline, others are confronted with rapid immigration and offer jobs and income to the population. Much of this growth[3] is due to the presence and use of environmental amenities, broadly defined as qualities of a region that make it an attractive place for living and working. Instead of extracting and selling natural commodities in external markets, these communities have started building economies based on improving the quality of the environment (Allali 2006). During the last forty years, the United States had two distinct demographic changes: a regional trend (migration to the south and west of the country); and an evolution from urban to rural areas (1970 and

---

[1] we can define amenities as qualities which facilitate living and working in a region. So, those qualities attract firms to locate in the place.

[2] Rural areas with low amenities.

[3] This growth come from offring jobs, income to the population, establishment of new firms like hotels, restaurant, etc.

1990). This latter has been linked to several factors, including quality of life factors such as the change of lifestyle, a better quality of the environment, a lower cost of living, and opportunities for recreation (Beesley and Bowles 1991; Greenwood 1985).

The rise of daily, residential and occasional mobilities gradually disconnected the areas of wealth creation, from other spaces spending that wealth (Talandier 2014). Thus, production and consumption are increasingly determined by geographical laws. This spatial and temporal disconnection of supply and demand has given rise to an invisible flow of wealth (Davezies 2008), which largely determines the economic springs territories. So, to consider the creation of wealth of a territory in the sole prism of activities producing goods and services exported 'elsewhere' is not any more enough for understanding the processes and stakes of regional development. At this income earned from the production, is added unproductive income which still too rarely considered in economic development strategies. Indeed, the spending of tourists, the salaries of 'commuters', the pensions, salaries of civil servants, the various allowances, welfare benefits and services are all revenues that fuel local economies regardless of their ability to produce and export goods and services. Thus, the development of a region depends on its ability to capture these revenue streams, as much as to create wealth (Dissart, 2005; Allali 2006; Talandier 2014).

Many authors have written about the role of amenities to achieve rural development (Kusmin 1994; Gottlieb 1994; Gottlieb 1995; Granaham 1999; Deller and Tsai 1999; English et al. 2000; Green et al. 2005; Dissart and Vollet 2011). In his work *'Amenities as an economic Development Tool: Is there enough Evidence?'* Gottlieb (1994) concludes that amenity factors do not have a strong influence on firm location decisions. There is a general assumption that amenities should have a much stronger influence on the location decisions of high technology firms than employers in other industries, but Gottlieb suggests there is very little empirical evidence to support this argument. He argues that an amenity strategy for economic development makes the most sense at a regional scale, primarily because of the commuting behavior of workers.

In a review of factors associated with the growth of local and regional economies, Kusmin (1994) concludes that most studies have found that climate influences business activity. Warmer temperatures are generally more attractive to business. She warns, however, that some climate effects may be better captured by regional control variables.

However, several recent empirical studies tend to support this general conclusion that amenities tend to be weakly related to business location or employment growth (Power 1988, 1996; Power et Barrett 2001; OECD 1999; Mc Granahan 1999; Deller and Tsai 1999; Isserman 2000; Dissart and Deller 2000; Green 2001; Nzaku and Bukeny 2005; Marcouiller et al. 2005; Green et al. 2005; Kwang-Koo 2005; Deller and Lledo 2007). Unfortunately, much of the current thinking is based on empirical evidence with little, if any, theoretical foundations (Deller and Lledo 2007).

Fig. 17.2 Wages, rent, unemployment and amenities
Source: Deller (2009)

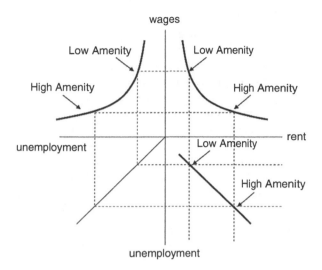

Using neoclassical growth theory, Marcouiller (1998) and Marcouiller and Clendenning (2005) suggest that natural amenities and quality of life factors act as non-market latent inputs into regional economic growth and development. No longer do traditional extractive industries (i.e. agriculture, forestry, and mining) or manufacturing form the backbone of the rural economy (Deller and Lledo 2007). Today, capital is no longer viewed as simply the machinery or public infrastructure used in production, but rather the more relevant form of capital takes a latent non-market attribute. On the other hand, the importance of landscape and forest to a regional economy is no longer just timber production but rather the natural amenities and recreational attributes of the landscape (White and Hanink 2004).

Figure 17.2 shows the link between the existence of amenities and wages, rent and unemployment in the region. First, regions have different equilibrium levels of wages and unemployment levels, and high amenity regions have lower wages. This can be explained by the fact that if a worker places high values on consuming amenities, such as climate or nature-based recreational activities, the decision to remain in a high amenity area and become unemployed is perfectly rational (Blanchflower and Oswald 1994, 1995; Deller 2009). On the other hand, there is a negatively sloped function linking wages and unemployment (Fig. 17.1). Roback (1982, 1988) argues that unemployment is an option for the worker.

From the works of Roback, we can find the result that high amenity regions will tend to pay lower wages. Inversely, low amenity areas will have to pay higher wages to compensate workers for living in a low amenity region. If migration is costly, workers in high amenity areas may look at a trade-off between working for low wages and remaining in the area with unemployment, but continuing to consume the amenity (Deller 2009). For low amenity regions, the gap between high wages and minimal income from support programs should be sufficiently large to discourage unemployment as an option.

The idea that amenities and quality of life in general play an important role in regional economic development is not necessarily new within the regional growth literature. Graves' studies take this direction. He was the first to make popular the argument that rising income and wealth leads to an increased demand for location-specific amenities (Graves 1979, 1980, 1983). In his works, Graves argued that the historically poor performance of migration prediction studies was directly attributable to their failure to account for amenity factors. On the other hand, theoretical works of Roback (1982, 1988) and Blanchflower and Oswald (1996) suggest that amenities and quality of life factors are capitalized into wages, rents, and unemployment in a manner that could hinder broader economic growth policies. In general, people are willing to accept lower levels of income, pay higher rents, and risk higher levels of unemployment to live in a high amenity and quality of life region.

3 Role of Natural Amenities in Regional Economic Development: The Case of Todgha Oasis

3.1 Economic Dynamic and Social Interactions

Todgha Oasis is located in the south of the High Atlas and north of the Little Atlas in southeastern Morocco (Fig. 17.3). The city is Tinghir (Fig. 17.4). Its name originally referred to the foothills of the Atlas Mountains, but its area has expanded to encompass surrounding villages and refers to the entire oasis.

The city of Tinghir has a population of 40,000 and the area has 90,000 inhabitants, according to the most recent census (2014). The predominant ethnic group is Amazighs, and the city is at the center of one of the most attractive oases in southern Morocco. Lush palm trees cover about 30 Kilometers on 500 to 1500-meter wide tracts along the Todgha valley. After the Todgha Gorge, Wadi Todgha has a difficult passage along the southern slopes of the Atlas Mountains (Tizgui); it then flows across the plain, meandering slightly over 20 kilometers to Ferkla (region of Tafilalet). The palm oasis, dense and widespread, is irrigated by a network of pipes and irrigation canals (Haddache 2012). Occasional heavy rains are absorbed in a few days.

The economy of the region is based on agriculture, trade and tourism (Haddache 2009). In addition, many families rely for their livelihoods on migrants' money transfers, especially from Europe. Social and cultural activities are growing; education projects for young children as well as literacy projects aimed at adults (particularly women) are increasing in many villages. These projects are supported by local authorities and non-governmental organizations.

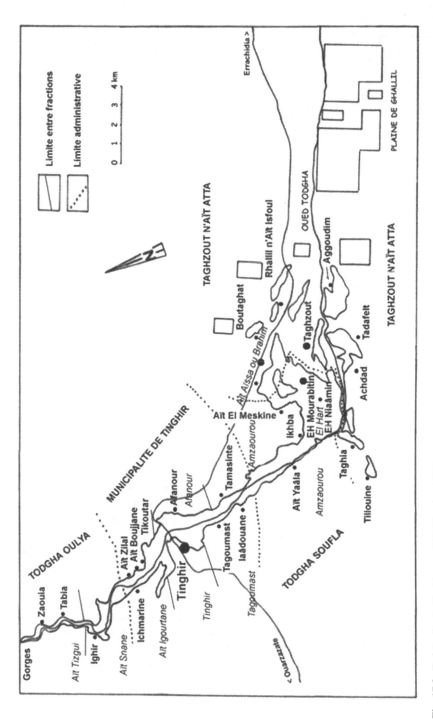

Fig. 17.3 Todgha Oasis (De Haas and El Ghanjou 2000)

Fig. 17.4 Geographical situation of Oasis of Todgha
Source: Googlemaps 2014

The economy of Tinghir is based on trade activity, taking advantage of its geographic location which places the region at the crossroad of surrounding towns and villages (Ait Atta in the southeast, Aït Merghade in the east, Aït Hdidou in the north and Aït Yaflmane in the west). Tinghir has become a large shopping center where the trade of various products (especially first-consumption) is performed daily and during major parties and *moussems*. On a second level, the economy of Tinghir relies on tourist activities. Indeed, Tinghir enjoys a lush composed of a vast oasis of palm and olive trees and Todgha Gorge (large mountains of the High Atlas), and also historic buildings including old *kasbahs* and *ksour*. On a third level, Moroccans from Tinghir living abroad (mainly France, Belgium, Spain, and the Netherlands) periodically inject capital into the local economy.

The role of the agricultural sector is increasingly minimized and loses its place as the first activity in the past because of its archaism, limited cultivated lands, and the inability of the activity to meet peoples' needs. Regarding industry, it is almost lacking, except for some craft activities (i.e. metalwork, carpentry, and sheet metal). Tinghir also hosts a mining activity in the Silver Mine of 'Imiter', but provides no real contribution to the local economy as a reliable factor for development since the activity is limited to the extraction and export of ore by a company[4]. Moreover, the massive use of chemicals and highly carcinogenic products within this mine has resulted in dangerous contamination of the groundwater, thus generating a public

---

[4]A private company.

Fig. 17.5 Aerial view of Todgha in July 1949
Source: Archives of Vincennes-Paris

health problem through the appearance of numerous pathologies among the local population and around the mine since the beginning of operations.

3.2   The Pressure of Urban Development on Natural Resources

After the era of 'Siba'[5] and restoring security and stability in the region by French authorities ('pacification' according to the colonial terminology), people have deserted their *ighreman* (*ksours*) and built new houses; a dynamic which helped extend the city over the urbanistic plan westward (Figs. 17.5 and 17.6).

Since then, the hydraulic works which are on the west of Tinghir were ravaged by the urbanization. Among these 'buried' works, we can quote the *khettara* of Tagoumast. Tagoumast is part of the municipality of Tinghir (Fig. 17.5). It has a *Khettara* which we cannot determine the exact date of construction. Nevertheless, the length of this *Khettara* is 2300 meters. According to the people of the locality, this *khettara* is abandoned for a long time. Before 1956, a trickle of water flows into

[5] During its history, Morocco has experienced periods of sociopolitical instability. Some regions of Morocco did not know the influence of the State, especially in the Middle and the High Atlas, in the Rif and in the Chaouia. Moreover, there were internal struggles within the 'Makhzen' family, the Moroccan state apparatus. The term 'Bled El Makhzen', which indicates the space where the State exercises its authority, is opposed to 'Bled Siba', a space not subjected to the central authority of the country.

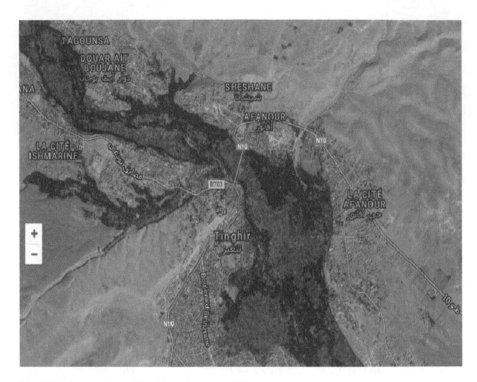

Fig. 17.6 Aerial view of Tinghir in December 2016
Source: Google Earth

the *khettara* but insufficient. After the independence of the country, the extension of urban planning has sounded the death knell of this *khettara*. By coincidence, the source of this *khettara* was where it built the center of agricultural development of Tinghir.

Thus, urban planning has completely degraded natural resources in Tinghir (Haddache 2009). The extreme scarcity of land will change the landscape of the oasis. In addition to the agricultural density which is going to penalize the farm, the population density is very high especially in the urban area of Tinghir (Haddache 2012). People are attracted to this center for many reasons, this is why we can find a high density, almost 888 inhabitants/Km² (Haddache 2009). By comparing the Figs. 17.5 and 17.6, we can observe the level of pressure on the land around the city center, which may exacerbate the pressure on farmland and natural resources. Moreover, the pressure on land can be combined with pressure on water resources. The drinking water supply of a city that is growing rapidly will heighten competition between the various uses of the water in favour of the domestic consumption, especially during periods of drought. This situation may challenge the sustainability of agriculture in the oasis.

4 Conclusion

In this work, we attempted to explain the role of natural amenities in regional economic growth. Amenities can be broadly defined as qualities of a region that make it an attractive place to live and work (Power 1988). In the search for forward-looking policy strategies, building on environmental amenities is emerging as an important area of policy action; thus supplementing traditional, agriculture-oriented rural policies and placing rural policy in the broader field of regional development (OCDE 1999). Because many natural and cultural features are public goods, with limited markets and hazy property rights, public policies are needed to strike the delicate balance between supply and demand, between use and conservation (OCDE 1999). This is not easy because of the many, multidimensional issues that natural amenities and regional development raise in terms of both policy and research.

For most part, amenities represent goods and assets that are not effectively regulated by markets (Power 1988). There are a number of problems in establishing the commodity character of amenities. The supply of them cannot be easily increased, while the demand grows significantly with development (Green et al. 2005). In many cases, amenities are public goods and it is difficult to make users pay to benefit from these resources (i.e. it is difficult to charge people who derive benefits from a rural landscape). Many of the beneficiaries from the promotion of amenities may live in urban areas, while most of the costs associated with this development are borne by residents in rural areas.

References

Allali K (2002) Revue sur les externalités environnementales de l'agriculture au Maroc, disponible sur ftp://ftp.fao.org/es/ESA/roa/pdf/2_Environment/Environment_MoroccoNA.pdf (Visité le 04/11/2015)

Allali K (2006) Agricultural landscape externalities, agro-tourism and rural poverty reduction in Morocco. Available at: ftp://193.43.36.92/es/esa/roa/pdf/oct05_env_morocco.pdf. Accessed 04 Nov 2015

Beesley KB, Bowles RT (1991) Change in the county-side: the turnaround, the community, and the quality of life. The Rural Sociologist 11(4):37–46

Blanchflower D, Oswald A (1994) The Wage Curve. MIT Press, Cambridge, MA

Blanchflower DG, Oswald AJ (1995) An Introduction to the Wage Curve. J Econ Perspect 9(3):153–167 Summer 1995

Blanchflower DG, Oswald AJ (1996) Efficiency wages and the German wage curve. Mitteilungen aus der Arbeitsmarkt-und Berufsforschung 29(3):460–466

Bonnieux F, Rainelli P (2000) Aménités agricoles et tourisme rural. Revue d'économie régionale et urbaine, vol 5. pp 803–820

Candau J, Deuffic P, Ferrari S, Lewis N, Rambonilaza M (2005) Equity within institutional arrangements for the supply of rural amenities, In Green GP, Deller SC, Marcouiller DW, Amenities and Rural Development - Theory, Methods and Publci Policy , Edward Elgar, Cheltenham, UK – Northampton, MA, USA, pp. 48–62

Daniels TL (1991) The purchase of development rights: preserving agricultural land and open space. J Am Plann Assoc 57(4):421–431

Davezies L (2008) La République et ses territoires. La circulation invisible des richesses. Edition Seuil, col. La République des Idées

De Haas H and El Ghanjou H (2000) General introduction to the Todgha Valley : population, migration, agricultural development, IMAROM, Working Paper Series no. 5, January 2000

Deller S (2009) Wages, rent, unemployment and amenities. J Anal Policy 39(2):141–154

Deller S, Lledo V (2007) Amenities and rural Appalachia economic growth, agricultural and resource economics review 36/1 (April 2007) pp 107–132

Deller SC, Tsai TS (1999) An examination of the wage curve. A research note. J Reg Analysis policy 28(2):3–12

Deller SC, Tsai T, Marcouiller DW, English DBK (2001) The role of amenities and quality of life in rural economic growth. Am J Agric Econ 83(2):352–365

Dissart JC (2005) Installations récréatives extérieures et développement économique régional : le cas des zones rurales isolées aux États-Unis. Revue d'Économie Régionale et Urbaine 2:217–248

Dissart JC (2007) Landscapes and regional development: What are the links?. Cahiers d'Economie et de Sociologie Rurales, INRA, 2007, 84–85, pp 61–91. <hal-01201159>

Dissart JC, Deller SC (2000) Quality of life in the planning literature. J Plan Lit 15(1):135–161

Dissart JC, Vollet D (2011) Landscapes and territory-specific economic bases. Land Use Policy 28(3):563–573

Dupras J, Revéret J-P, He J (2013) L'évaluation économique des biens et services écosystémiques dans un contexte de changements climatiques, Ouranous, Février 2013

English DBK, Marcouiller DW, Cordell HK (2000) Linking local amenities with rural tourism incidence: estimates and effects. Soc Nat Resour 13(1):185–120

Estrada-Carmona N, Hart AK, DeClerck FAJ, Harvey CA, Milder JC (2014) Integrated landscape management for agriculture, rural livelihoods, and ecosystem conservation: An assessment of experience from Latin America and the Caribbean. Landsc Urban Plan 129(2014):1–11 https://doi.org/10.1016/j.landurbplan.2014.05.001

Fleming DA, Granahan DA, Goetz SJ (2009) Natural amenities and rural development: the role of land-based policies, Rural Development Paper N°45, The Northeast Regional Center for Rural Development, Pennsylvania State University

Goodstein ES (2004) Economics and the environment, 4th edn. Hoboken, Wiley

Googlemaps (2014) https://maps.google.com/. Accessed 11 Apr 2014

Gottlieb PD (1994) Amenities as an economic development tool: is there enough evidence? Econ Dev Q 8:270–285

Gottlieb PD (1995) Residential Amenities, Firm Location and Economic Development. Urban Studies 32(9):1413–1436. https://doi.org/10.1080/00420989550012320

Granaham DA (1999) Natural amenities drive rural population change. Food and rural economic division, economic research service, department of agriculture. Agricultural Economic Report n° 781

Graves P (1979) A life-cycle empirical analysis of migration and climate by race. J Urban Econ 6:135–147

Graves P (1980) Migration and climate. J Reg Sci 20:227–237

Graves P (1983) Migration with a composite amenity. J Reg Sci 23:541–546

Green GP (2001) Amenities and Community Economic Development: strategies for Sustainability. J Anal Policy 31(2)

Green PG, Deller SC, Marcouiller DW (2005) Amenities and rural development theory, methods and public policy, ISBN: 1 84542 126 4, Published by Edward Elgar Publishing Limited Glensanda House Montpellier Parade Cheltenham Glos GL50 1UA UK

Greenwood MJ (1985) Human migration: theory, models, and empirical studies. J Reg Sci 25(4):521–544

Haddache M (2009) Gestion sociale de l'eau et mutations socio-économiques dans l'oasis de Todgha, Mémoire du Diplôme des Etudes Supérieures Approfondies, Université Cadi Ayyad, Faculté des Sciences Juridiques Economiques et Sociales

Haddache M (2012) Savoirs hydrauliques et mutations socioéconomiques dans l'oasis de Todgha (Sud-Est, Maroc). Asinag 7(2012):111–122

Henderson, Jason R, McDaniel K (2000) The impact of scenic amenities on rural employment growth, Selected Paper, 2000 AAEA Meetings, Tampa, Florida, August

Isserman AM (2000) The competitive advantages of rural America in the next century. Int Reg Sci Rev 24:35–58

Klein LR, Reganold JP (1997) Agricultural changes and farmland protection in western Washington. J Soil Water Conserv 52(1):6–12

Kusmin LD (1994) Factors associated with the growth of local and regional economies: a review of selected empirical literature. In: Economic Research Service, United States Department of Agriculture – AGES 9405. USDA, Washington, DC

Kwang-koo K, Marcouiller DW, Deller SC (2005) *Natural Amenities and Rural Development: Understanding Spatial and Distributional Attributes*, Growth and Change Vol. 36 No. 2 (Spring 2005), pp. 273–297. Available at: http://urpl.wisc.edu/sites/urpl.wisc.edu/files/people/marcouiller/publications/growthchangefinal.pdf (Accessed 2015-11-04)

Laurent C (2002) Multifonctionalité et éligibilité aux aides PAC dans l'UE. Economie Rurale 268–269:144–158

Marcouiller DW (1998) Environmental resources as latent primary factors of production in tourism: the case of forest-based commercial recreation. Tour Econ 4:131–145

Marcouiller DW, Clendenning G (2005) The supply of natural amenities: moving from empirical anecdotes to a theoretical basis. Link: http://citeseerx.ist.psu.edu/viewdoc/download;jsessionid=48D1EB1A2E9E1CDCD3B7E4B953B711C6?doi=10.1.1.195.6225andrep=rep1andtype=pdf (Accessed 2015-11-03)

Marcouiller DW, Deller SC, Green GP (2005) In: Green GP, Deller SC, Marcouiller DW (eds) Amenities and rural development: policy implications and directions for the future. Edward Elgar, Northampton, pp 329–336

McKean JR, Johnson DM, Johnson RL, Taylor RG (2005) Can superior natural amenities create high-quality employment opportunities? The case of non consumptive river recreation in Central Idaho. Soc Nat Res Int J 18(8):749–758. https://doi.org/10.1080/08941920591005304

Moretti E (2004) Estimating the social return to higher education: evidence from longitudinal and repeated cross-sectional data. J Econ 121(1–2):175–212

Nickerson CJ, Hellerstein D (2003) Protecting rural amenities through farmland preservation programs. Agric Resour Econ Rev 32(1):129–144

Nzaku K, Bukenya JO (2005) Examining the relationship between quality of life amenities and economic development in the Southeast USA. Review of Urban and Regional Development Studies 17(2):89–103

Organisation de Coopération et de Développement Economique (OCDE) 1999 Cultiver les aménités rurales : une perspective de développement économique, Paris, p 122

Pezzini M and Wojan TR (2001) Leveraging Amenities for Rural Development: Direction, Dialogue, and Negotiation, Exploring Policy Options for a New Rural America, a conference sponsored by the Center for the Study of Rural America, Federal Reserve Bank of Kansas City

Power TM (1988) The Economic Pursuit of Quality. M.E. Sharpe, Armonk

Power TM (1996) Lost Landscapes and Failed Economies: The Search for a Value of Place. Island Press, Washington DC

Power TM (2005) The supply and demand for natural amenities: an overview of theory and concept, in *Amenities and Rural Development,* Published by Edward Elgar Publishing Limited, Glensanda House, Montpellier Parade, Cheltenham, Glos GL50 1UA, UK

Power TM, Barrett RN (2001) Post-Cowboy Economics: Pay and Prosperity in the New American West. Island Press, Washington, DC

Randall A (2002) Valuing the outputs of multifunctional agriculture. Eur Rev Agric Econ 29(3):289–307

Roback J (1982) Wages, rents, and the quality of life. J Polit Econ 90:1257–1278

Roback J (1988) Wages, rents and amenities: differences among workers and regions. Econ Inq 26(1):23–41

Romstad E (2004) Policies for promoting public goods in agriculture. In: Brouwer F (ed) Sustaining Agriculture and the Rural Environment: Governance, Policy and Multifunctionality. Edward Elgar, Cheltenham, pp 56–77

Talandier M (2014) Retombées socio-économiques des aménités culturelles et naturelles dans les territoires de France métropolitaine, L'Observatoire de l'Economie et des Institutions Locales, Rapport remis au PUCA en juin 2014

Taylor L, Smith K (2000) Environmental amenities as a source of market power. Land Econ 76(4):550–568

Travnikar T, Juvančič L (2013) Application of spatial econometric approach in the evaluation of rural development policy: the case of measure modernisation of agricultural holdings. Stud Agric Econ 115(2013):98–103

Waltert F, Schläpfer F (2007) The role of landscape amenities in regional development: a survey of migration, regional economic and hedonic pricing studies, Working Paper No. 0710, Socioeconomic Institute, University of Zurich

White KD, Hanink DM (2004) Moderate environ-mental amenities and economic change: the nonmetro-politan northern Forest of the northeast U.S., 1970–2000. Growth Chang 35(1):42–60

Chapter 18
Economic Growth and Environment: An Empirical Analysis Applied to Morocco, Algeria, Tunisia, and Egypt

Aïcha EL Alaoui and Hassane Nekrache

Abstract The main objective for many developing countries in the coming years is to improve the economic growth, which is perceived as necessary to meet the increasing demand of their populations, to improve their well-being, and to help manage existing environmental challenges. This work attempts to investigate the links between economic growth and environment in four countries from the MENA region (Morocco, Algeria, Tunisia, and Egypt, hereafter 'MATE'). To do so, two steps are followed: in the first one, a basic Environmental Kuznets Curve (EKC) equation for each country over the period 1970–2010 is tested to measure the effect of economic growth on environmental quality, and to determine the possibility of the existence of an EKC; in the second one, a few variables are introduced in the basic EKC equation (model tested in the first step) such as economic openness indicator and enrollment and urbanization rates. The purpose is to measure the possible influence of these variables (including economic growth) on environmental quality, and also to determine the possibility of the existence of an EKC. The results of both models show that the linkages between economic growth and environment are still uncertain, complex and ambiguous. It is not possible to find a unique form of this linkage and each variable introduced in the model can give some explanation where the application of EKC is unclear and uncertain. Therefore, these countries through policymaking, and the involvement of private actors (such as corporations and NGOs), must apply preventive and precautionary measures to reduce environmental damage. These measures must be adapted to specific economic and environmental conditions benefiting from the experiences and good practices developed in other regions and avoiding others' past mistakes related to pollution, regional development, and natural resource management.

Keywords Growth · Environmental degradation · EKC · OLS · CO_2 emissions

A. EL Alaoui (✉)
Sultan My Slimane University, Chief of Research group of "the social economy and social justice", Beni Mellal, Morocco

H. Nekrache
Statistician and Demographer Engineer, The High Commission for Planning, Rabat, Morocco

© Springer International Publishing AG, part of Springer Nature 2019
M. Behnassi et al. (eds.), *Human and Environmental Security in the Era of Global Risks*, https://doi.org/10.1007/978-3-319-92828-9_18

1   Introduction

Economic growth remains important for all countries, whether developing, less developed or developed. It may affect people's well-being, i.e. health, education, employment, quality of life, etc. It may also affect government's stability, from social and nutritional security to political stability, and population's welfare. The recent example is the 'Jasmine' revolution started in Tunisia. The principal reasons behind this revolution are, inter alia, high rate of unemployment, high index of corruption, poor living conditions, lack of democracy (free elections), and deficiency of freedoms (freedoms of expression and of the press).

Economic growth requires the combination of different types of capitals in order to produce goods and services (World Bank 2006). These include:

- *Produced capital*, which means machinery, buildings, roads and rail network.
- *Human capital*, which refers to education, health, knowledge and skills. In the early 60s, economists have accorded a large importance to this concept, especially, with the works of Becker (1962), Schultz (1961, 1962), Mincer (1958, 1962), Kiker (1966), and Blaug (1976).
- *Institutional and social capital*, which involves the quality of political institutions represented by the extent of their connections to the society and their respect to the norms, values and human rights. This concept was popularized, namely, by Bourdieu (1985), Coleman (1988a, b), Putnam (1993), and Portes (1998);
- *Natural capital*, which is related to the natural resources such as air, water, minerals, the extracted raw materials (such as gas, phosphate, and petroleum), and animals (such as fish, cow, and pig). This capital is vital for securing both sustainable economic growth and development, not only for the present but also for the future generations. Natural capital is defined by the Global Development Research Center[1] as "the environment stock or resources of Earth that provide goods, flows and ecological services required to support life". This concept is used in many studies, especially in the work of Costanza and Daly (1992).

The links between economic growth and the four capitals mentioned above is complex and strong. This study focuses only on the links between economic growth and the environment/the natural capital[2]. Indeed, the environment plays important direct and indirect roles in supporting and sustaining economic activities (agriculture, fisheries, tourism, manufacturing, and services): directly by providing raw materials and minerals required as inputs for production processes; and indirectly by providing the required ecosystems services (such as rivers, ocean, air, etc.).

However, the intensive and irrational economic growth has caused many changes to the environment, especially, since the industrial revolution. In its report, the

---

[1] http://www.gdrc.org

[2] This study uses the concept of the environment because it is general and includes different aspects of life and resources in the Earth.

IPCC's Fifth Assessment (AR5) showed that "since the beginning of the industrial era, oceanic uptake of CO2 has resulted in acidification of the ocean; the PH of ocean surface water has decreased by 0.1 (high confidence), corresponding to 26% increase in acidity, measured as hydrogen ion concentration" (IPCC 2014:4). The key environmental changes can be summarized in three aspects: the ozone layer; the temperature change; and the biodiversity loss.

- The first aspect of environmental damage is the ozone layer, which is a thin layer of stratospheric gas that protects life on Earth by absorbing the solar UV radiations and preventing them from reaching the Earth's surface (Daniel 1999:10). During the last years, the ozone layer became extremely fragile because of its low concentration of ozone (O3). However, the pollution causes destruction of this layer notably via the reactions that take place between O3 compounds and pollutants. It consequently exposes humans to sunlight, and therefore causes many health problems such as skin cancer.
- The second aspect of environmental damage is the change in the earth's temperature: the atmosphere and the oceans have warmed, the amounts of snow and ice have diminished, and the sea-level has risen. The IPCC's Fifth Assessment Report (AR5) documented that "the number of cold days and nights has decreased and the number of warm days and nights has increased on the global scale" (IPCC 2014:7). Moreover, this report confirms that "each of the last three decades has been successively warmer at the Earth's surface than any preceding decade since 1850", (IPCC 2014:2). Thus, the global average land and ocean surface temperature warming combined is estimated of 0.85 [0.65 to 1.06] °C2 over the period 1880 to 2012 (IPCC 2014:2). In addition, the glacier areas have continued to shrink almost worldwide in response to the increased surface temperature and the changing snow cover since the early 1980s.
- The third aspect of environmental damage is the biodiversity loss which refers to a substantial decrease of non-human species worldwide. Indeed, the anthropogenic activities and their impacts on the environment are now the main driver behind the extinction and scarcity of many species, whether insects, animals, or plants. The extinction's rate has currently reached a higher level of 100 to 1000 times the natural rate (Chivian and Bernstein 2010:5).

These three aspects of environmental change have caused direct and/or indirect problems such as increasing the risk of famine, contagious maladies (malaria, Ebola...), flooding, and water shortage (Khagram et al. 2003; Bass 2006; Martino and Zommers 2007). "The harmful effects of the degradation of ecosystem services are being borne disproportionately by the poor, are contributing to the growing inequities and disparities across groups of people, and are sometimes the principal factor causing poverty and social conflict" (Bass 2006:2). More precisely, the environmental damage will be experienced by developing countries and the poorest people, especially in Sub-Saharan Africa, South Asia, Southeast Asia, and Latin America regions. In urban area, the risks for peoples, assets, economies, and ecosystems have increased due to air pollution, drought, and water scarcity

(IPCC 2014:15). In rural area, the major impacts are on water availability and supply, food security, infrastructure, and agricultural incomes (IPCC, 2014:16).

There is a clear conscience about environmental challenges, from averting dangerous climate changes to halting biodiversity losses and protecting our ecosystems. However, developed economies have partially reduced the environmental damage through various measures (political, legal, economic, technological, educational, etc.), including the relocation of some production processes to developing countries, thus exporting their pollution to other regions. It is true that these foreign investments are important and vital for host countries since they contribute to their economic growth and help reduce poverty, migration and unemployment. However, these dynamics should be accompanied by the necessary measures (such as financial and tencnology transfer) to help reducing environmental impacts, especially an era of global change where the environment is increasingly perceived as a common concern for all countries.

The main objective for many developing countries in the coming years is to improve the economic growth, which is perceived as necessary to meet the increasing demand of their populations, to improve their well-being, and to help manage existing environmental challenges. Against this background, this paper attempts to investigate the links between economic growth and environmental damage in four countries from the MENA region (Morocco, Algeria, Tunisia, and Egypt, hereafter 'MATE'). The work is organized as follows: the second section reviews a sample of theoretical and empirical studies that focus on the linkages between economic growth and the environment; the third section presents the economic and environmental situation in Morocco, Algeria, Tunisia, and Egypt; the fourth section is dedicated to the presentation of both methodology and main results; finally the last section serves to sketch the main components of a strategy to induce environmental improvement in MATE with other relevant conclusions.

2   Theoretical and Empirical Discussions about the Links between Economic Growth and the Environment

Environmental issues received growing attention throughout the 60s via the publication of Rachel Carson's Silent Spring in 1962, which examined the impact of man's indiscriminate use of chemicals in the form of pesticides and insecticides, mentioned by Cole (1999). In the early 70s, Ehrlich and Holdren (1971, 1972), Commoner et al. (1971), and Commoner (1972a, b) identified three factors that created environmental impact (I): increasing human population (P); increasing economic growth or per capita affluence (A); and the application of resource depleting

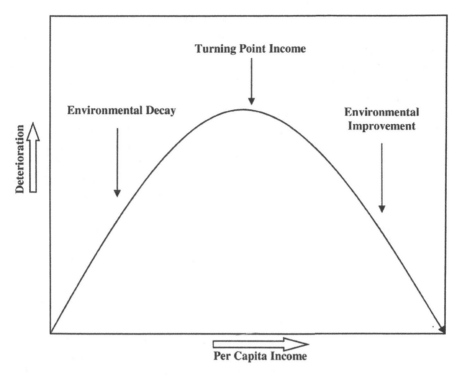

Fig. 18.1 Environmental Kuznets curve
Source: Yandle et al. (2002:3)

and polluting technology (T). These factors were considered as the worst for the planet and are linked by the following equation named IPAT[3]:

$$Impact = Population x Affluence\ x Technology$$

According to IPAT equation, the attention was growing to examine the links between economic growth and environmental quality. This relationship is represented by the Environmental Kuznets Curve (EKC), which refers to the hypothesis of an inverted U-shaped relationship between various indicators of environmental degradation and per capita income. In the early stages of economic growth, degradation and pollution increase, but beyond a certain level of per capita income, which will vary for different indicators, the trend reverses, so that a high income level of economic growth leads to environmental improvement. This implies that the environmental impact indicator is an inverted U-shaped function of per capita income. Typically, the logarithm of the indicator is modeled as a quadratic function of the logarithm of income. An example of an estimated EKC is shown in Fig. 18.1. The EKC takes the

---

[3] For more explication see Chertow (2001). The author tried to track the various forms the IPAT equation to examine which variables was worst for the planet.

name of Simon Kuznets (1955)[4] who hypothesized that income inequality first rises and then falls as the economic development proceeds from a certain threshold's economic growth.

The idea of this model is that population enrichment was accompanied by the demand for a cleaner environment. At the lowest income's level, the main preoccupations for a poor person are to afford the basic necessities for himself and his family such as food, shelter, water, and clothing, leaving a little place for other concerns as environmental issues. At the highest income's level, a rich person is more sensitive to environmental issues. What is true at the individual attitude is also valid at the national level. When an individual or a country becomes rich, it is easier to sacrify à part of the income to protect the environment. Many researchers have focused on the relationship between economic growth and environment such as Grossman and Krueger (1991, 1995); Beckerman (1992); Shafik and Bandyopadhyay (1992); Panayotou (1993; 1997; 2003); Shafik (1994); Selden and Song (1994); and Cropper and Griffiths (1994)[5]. Moreover, the empirical studies related to this subject have grown rapidly during last decades, especially in developed countries. This analysis represents a sample of these studies.

The first estimation of the EKC was established by Grossman and Krueger (1991), which analyzed the environmental impact of the North American Free Trade Agreement (NAFTA)[6]. The authors distinguished three separate mechanisms that can affect the level of pollution and the rate of depletion of scare environmental resources. These effects are the scale, the composition and the technique effects[7]. The authors used a cubic function to estimate the concentration of pollutants in the air (sulfur oxides (SO2), suspended particles, and dark matter (thin smoke)) in urban areas using the Global Environmental Monitoring System (GEMS) dataset as part of a study on the potential environmental impacts of NAFTA. The authors suggested that trade liberalization generates some benefits, such as increased income growth which tends to alleviate pollution problems, and increased specialization in sectors that cause less than average amounts of environmental damage. They suggested, also, that "the environmental impacts of trade liberalization in any country will depend not only upon the effect of policy change on the overall scale of the economic activity, but also upon the induced changes in the intersectoral composition of economic activity and in the technologies that are used to produce goods and services", (Grossman and Krueger 1991:36). Similar findings had been reported by Shafik (1994), he concluded that "some environmental indicators improve with

---

[4] Simon Kuznets (1901–1985) was an American economist, demographer and statistician of Ukrainian origin. He won the Nobel Prize in 1971.

[5] For a chronological presentation of the EKC, see Stern (2004). This author confirmed that the EKC concept was popularized through World Bank Development Report (1992).

[6] The NAFTA came into effect on January 1, 1994, creating the largest free trade region in the world. It is an agreement signed by Canada, Mexico, and the United States, creating a trilateral trade bloc in North America. For more detail see www.international.gc.ca/trade-agreements-accords-commerciaux (Global Affairs Canada).

[7] For more details, see Grossman and Krueger 1991, pp.3–4.

rising incomes (like water and sanitation), others worsen and then improve (particulates and Sulfur oxides), and others worsen steadily (dissolved oxygen in rivers, municipal solid wastes, and Carbon emissions)" (Ibid:769–770).

"Has past economic growth been associated with the accumulation of natural capital or the drawing down of natural resource stocks? Is the accumulation of physical and human capital from complement to or a substitute for the accumulation of natural capital? How do these relationships vary across different environmental resources? And how have macroeconomic policies affected the evolution of environmental quality?" (Shafik and Bandyopadhyay 1992:1). The authors tried to respond to these questions exploring the links between economic growth and environmental quality by analyzing the patterns of the environmental transformation of several countries at different income levels. The authors tested three models (log-linear, log-quadratic and log-cubic) to explore the shape of the links between income and each environmental indicator[8], which was used as the dependent variable in a panel regression using data from up to 149 countries over the period 1960–1990. Excluding deforestation and dissolved oxygen, they found that income has the most consistently significant effect on eight of environmental indicators than that of policy variables, i.e. the variables related to trade policy and political and civil liberties. Lack of clean water and urban sanitation decline uniformly over time with increasing income. River's quality tended to worsen with increasing income. The two indicators of air pollutants – Suspended Particulate Matter (SPM) and SO2 – confirmed the EKC hypothesis. Both per capita municipal waste and carbon dioxide emissions increased with rising income: "access to clean water and sanitation have elasticities of -0.48 and -0.57 respectively, implying that a 1 percent increase in income results in about 0.5 percent more people in the population are served by improved facilities" (Shafik and Bandyopadhyay 1992:22).

In another background paper in World Development Report, Beckerman tried to analyze the link between economic growth and environmental quality, namely local air quality and access to drinkable water and sanitation. The author has clearly described this link arguing that "there is a clear evidence that, although the economic growth usually leads to environmental deterioration in the early stages of the process, in the end the best way to attain a decent environment in most countries is to become rich" (Beckerman 1992:482). The author found that there is a strong positive relationship between income level and environmental quality. Although the environment in developing countries may get worse, he confirmed that "in the longer run they will be able to reverse the trends in more common forms of air pollution, and attain levels of water supply and sanitation essential to an acceptable, decent and healthy standard of living" (Ibid:21).

Examining the effect of population pressures on forest ecosystems in 64 developing countries over the period 1961–1988, Cropper and Griffiths documented that if

---

[8] They estimated for 10 environmental indicators which are "the lack of clean water, lack of urban sanitation, ambient levels of SPM, ambient SO2, change in forest area between 1961–1986, the annual rate of deforestation, dissolved oxygen in rivers, fecal coliforms in rivers, municipal waste per capita, and carbone missions per capita", (Shafik and Bandyopadhyay 1992, p.5).

there are "two countries with rapid population growth and significant forest resources but with different levels of per capita income, the country with the highest income is likely to be deforesting less rapidly. As income grows, people will switch to energy sources other than firewood and will use modern agricultural techniques that reduce the demand for agricultural land", (Cropper and Griffiths 1994:250). The authors showed that the Kuznets curve for deforestation was verified. Thus, an increase of the growth rate of per capita income by eight percentage points reduces the rate of deforestation by one-tenth of a percentage point.

Several studies have focused on the links between international trade and environmental quality, and have confirmed that the international trade can improve the environmental quality. Accordingly, the international trade would accelerate income; so it can allow a quick passage to the ascending part of the curve. Grossman and Krueger (1991:21) showed that trade liberalization generates an increase in income levels and then it can strengthen the incentives for 'environmental dumping'. Hence, they proposed that free trade can protect the environment. Lopez (1994:163) showed that "economic growth and trade liberalization decrease the degradation of natural resources if and only if producers internalize their stock feedback effects on production". He concluded that the effect of trade liberalization depends on three assumptions: (i) the manufacturing sector is protected vis-à-vis to the primary sector; (ii) the productive stock effects of the resource occur entirely in the primary sector; and (iii) the productive sector is characterized by constant returns to scale technology (Ibid:183). Antweiler et al. (2001) investigated how the openness to trading opportunities affects pollution concentrations by developing a theoretical model to divide trade's impact on pollution into scale, technique, and composition effects. The authors concluded that "free trade is good for the environment" (Ibid: 878).

The turning points[9] come somewhere between $4000 and $5000 per capita GDP, measured in 1985 US dollar (Grossman and Krueger 1991:5). 'Similar' results are found by Cropper and Griffiths (1994) which the turning points are $4760 per capita income for Africa and $5420 per capita income for Latin America. However, these points vary substantially across environmental indicators[10]. Shafik and Bandyopadhyay (1992) found that the turning points are $3280, $1375, and $1375 (per capita income in 1985 US dollar) for sulfur dioxides, SPM, and fecal coliform, respectively.

Other studies[11] have estimated the turning point to be generally higher. The turning points vary for the different pollutants[12], but almost in every case they occurred at an income of less than $8000 U.S dollars in 1985, (Grossman and Krueger 1995,

---

[9] Stern (2004:1425) presented in Table 1 a summary of turning points for sulfur emissions and concentrations assigned at the several studies. See also Table 1 of Cole (1999:92).

[10] For more details, see Shafik (1994).

[11] See for example Selden and Song (1994), Grossman and Krueger (1995), and Cole et al. (1997).

[12] They focused on four types of indicators: concentrations of urban air pollution; measures of the state of the oxygen regime in river basins; concentrations of fecal contaminants in river basins; and concentrations of heavy metals in river basins.

p.369). Selden and Song's (1994) estimates are under $10,000 per-head (1985 US dollar). These authors tested four indicators of air pollution (SPM, SO_2, NOx, and CO_2) in their model using the GEMS aggregate emissions data obtained from the World Resources Institute[13]. But, Cole et al. (1997) used carbon dioxide, carbonated fluorocarbons (CFC) and halons, methane, nitrogen dioxide, sulfur dioxide, suspended particulates, carbon monoxide, nitrates, municipal waste, energy consumption and traffic volumes to examine the EKC. They have estimated the turning points for different pollutants (from a low $5700 to a high $34,700 in 1985 US dollar).

The EKC has been the subject of growing criticism (Arrow et al. 1995; Ekins 1997; Torras and Boyce 1998; Perman and Stern 1999; Stern and Common 2001); Cole and Neumayer 2005). Some authors have confirmed that the EKC is just a utopia because the solution of environmental degradation is not related only to economic growth and higher income, but there are several other factors that can play an important role in improving the state of ecological systems such as education, quality of institution, and civil society[14]. Some critics have argued that the EKC suffers from severe methodological problems that cast doubt on the reliability of its results (Cole and Neumayer 2005:298). The authors documented that the rich countries have become clean up, at least partly, by exporting the dirty production processes to poorer countries. This fact may therefore explain the reductions in local air pollution experienced in most developed countries found in many studies.

Arrow et al. (1995) highlighted that the inverted-U relation is evident in some cases but not evident in all cases implying that economic growth is not sufficient to induce environmental improvement in general. They concluded that "economic growth is not a panacea for environmental quality" (Ibid:521).

Stern and Common (2001) and Perman and Stern (1999) declared that the studies which used only OECD data will have to estimate an optimistic turning points with variables that are likely to be no-stationary. Consequently, the standard estimation will probably generate spurious results. Ekins (1997) argued, also, that estimated turning points are highly dependent on the choice of functional form, the data set, and the estimation method. The EKC literature is overly optimistic in suggesting the existence of a systematic inverted-U relationship between income and pollution (Ibid:805).

---

[13] http://www.wri.org

[14] For example, Panayotou (1993:2) proposed that "the state of natural resources and the environment in a country depends on five main factors" ignoring/neglecting other factors that impact economic growth. These factors are: "a) the level of economic activity or size of the economy; b) the sectoral structure of the economy; c) the vintage of technology; d) the demand for environmental amenities; and e) the conservation and environmental expenditures and their effectiveness".

3 Description of Economic and Environmental Situation in MATE

In MATE, economic growth differs significantly from a country to another and within the same country. The best growth rates of real GDP and of real GDP per capita were recorded during the period 1970–1989, and the highest rates were recorded by Egypt. However, Morocco grew speedily by 3.9% during the period 2010–2013 against 3.1%, 2.8% and 2.6% respectively in Algeria, Egypt, and Tunisia. These rates are lower than those recorded in Africa (all countries combined), South Asia, Sub-Saharan Africa (SSA), East Asia and Pacific (EAP), and China. These growths were accompanied by a rapid urbanization in all regions of the World, but it is more important in developed countries than that in developing countries. Roughly 80% of China and OECD populations live in urban area against only 41.5% in Africa (all countries combined) and 36% in Sub-Saharan Africa. In MATE, the majority of Algerian, Moroccan, and Tunisian populations live in cities, while Egyptian populations live in rural area. Table 18.1 gives an idea about economic growth and rapid urbanization known in majority regions of the world.

Consequently, living in urban areas has an important impact on both citizens' life-style and economic activities such as boosting demand for transport, telecommunication technology, manufactured goods, drainage, sanitation, and other demands linked to the urban consumption culture. Thus, these changes in the population's behavior will increase the environmental damage, especially for air and water. Table 18.2 provides an idea about the evolution of environmental damage measured by CO_2 emissions in MATE and in other regions of the World.

Table 18.2 shows that: (i) Africa's emissions are lower compared to those of the World; (ii) the highest CO_2 emissions per GDP are recorded in China and EAP-developing countries; (iii) CO_2 emissions per capita are recorded in OECD members followed by South Africa; (iv) Egypt's emissions per GDP are more important than those recorded in Algeria, Morocco, Tunisia, and those recorded in MENA; (v) Algeria's emissions per capita are higher than those recorded in Egypt, Morocco, and Tunisia, but lower than those recorded in MENA; (vi) MATE's emissions per GDP are higher than those recorded in Africa and the World, but MATE's emissions per capita are lower than those recorded in the World, and more important than those recorded in Africa.

Figure 18.2 shows that there is a relationship between CO_2 emissions per capita and real GDP per capita, but this relationship has not a unique form.

In Sub-Saharan Africa (SSA), combustible renewable and waste constitute more than 50 percent of energy use during the period 2000–2009 (Fig. 18.3). In Tunisia, combustible renewable and waste is important than that recorded in China. The lowest rates are recorded in Algeria, Morocco, Egypt, and MENA.

The highest energy use per capita is recorded in OECD members followed by South Africa and MENA-all income levels (Fig. 18.4). Algeria, Tunisia and Egypt have an average of energy use per capita more important than that in Africa (all countries combined). The lowest energy use per capita is recorded in Morocco; it is just more than 400 kg of oil equivalent per capita.

Table 18.1 Real GDP (g)[a], Real GDP per capita (gy)[b], urban and rural population, during 1970–2013

Countries/ Region of the world	g (%) Average of period:			gy (%) Average of period:			Urban population[c], % Average of period:			Rural population[d], % Average of period:		
	70-89	90-09	2010–13	70-89	90-09	2010–13	70-89	90-09	2000–13	70-89	90-09	2010–13
Algeria	5.0	2.7	3.1	2.0	0.9	1.2	44.0	59.5	68.5	56.0	40.5	31.5
Egypt	6.1	4.6	2.8	3.8	2.9	1.1	43.4	43.0	43.0	56.6	57.0	57.0
Morocco	4.6	3.8	3.9	2.3	2.4	2.5	40.9	53.1	58.4	59.1	46.9	41.6
Tunisia	5.4	4.8	2.6	3.0	3.4	1.5	50.3	62.9	66.2	49.7	37.1	33.8
China	9.2	9.9	8.8	7.4	9.0	8.2	20.0	36.2	51.2	80.0	63.8	48.8
EAP- all income levels[e]	4.9	3.6	4.8	3.1	2.6	4.1	27.9	41.5	53.3	72.1	58.5	46.7
EAP- developing only	7.8	8.4	8.1	5.8	7.2	7.4	21.9	36.7	49.3	78.1	63.3	50.7
LAC-all income levels[f]	4.0	2.9	3.8	1.8	1.4	2.6	63.7	74.7	78.8	36.3	25.3	21.2
LAC- developing only	4.1	2.9	3.9	1.8	1.3	2.7	63.1	74.3	78.5	36.9	25.7	21.5
MENA-all income levels[g]	5.2	4.6	4.0	2.2	2.4	2.1	49.0	58.5	63.2	51.0	41.5	36.8
MENA- developing only	4.1	4.3	2.3	1.3	2.2	0.6	46.5	55.3	59.6	53.5	44.7	40.4
OECD members[h]	3.3	2.2	1.8	2.4	1.4	1.2	70.3	75.9	79.4	29.7	24.1	20.6
South Africa	2.7	2.5	2.8	0.4	0.6	1.4	48.8	56.8	63.0	51.2	43.2	37.0
South Asia	4.3	6.0	6.4	1.9	4.1	5.0	21.9	27.5	31.6	78.1	72.5	68.4
SSA-all income levels[i]	2.9	3.5	4.2	0.1	0.8	1.5	22.1	30.7	35.9	77.9	69.3	64.1
SSA- developing only	2.9	3.4	4.3	0.1	0.7	1.5	22.1	30.7	35.9	77.9	69.3	64.1
Africa	3.9	4.1	4.7	1.1	1.7	2.3	27.2	36.9	41.5	72.8	63.1	58.5
World	3.5	2.6	2.9	1.7	1.3	1.7	39.3	46.7	52.3	60.7	53.3	47.7

Source: Calculated using WDI (2015)
[a]g is growth rate of the real GDP (2005 US$); [b]gy is growth rate of the real GDP per capita [real GDP per capita = GDP (constant 2005 US$)/total population]; [c]Urban population (%) represents share of urban population in total population; [d]Rural population (%) represents share of the rural population in the total population; [e]EAP is the East Asia and Pacific; [f]LAC is Latin America and Caribbean; [g]MENA is the Middle East and North Africa; [h]OECD is the Organization for Economic Co-operation and Development; [i]SSA is Sub-Saharan Africa

Table 18.2 CO_2 emissions in MATE and other regions of the World (1970–2009)

Countries/Region of the world	G-CO_2 (1)				P-CO_2 (2)			
	70–79	80–89	90–99	2000–09	70–79	80–89	90–99	2000–09
Algeria	0.9	1.1	1.3	1.0	2.1	3.0	3.1	3.0
Egypt	1.6	1.7	1.6	1.8	0.8	1.3	1.6	2.3
Morocco	0.6	0.6	0.7	0.7	0.6	0.8	1.1	1.4
Tunisia	0.7	0.8	0.8	0.7	1.0	1.6	1.8	2.3
China	7.2	5.7	3.7	2.4	1.2	1.8	2.5	4.1
EAP- all income levels (3)	0.9	0.9	0.8	0.9	1.9	2.3	3.1	4.2
EAP- developing only	4.3	3.8	2.8	2.1	1.1	1.6	2.3	3.4
LAC-all income levels (4)	0.6	0.6	0.5	0.5	2.1	2.3	2.4	2.7
LAC -developing only	0.6	0.6	0.5	0.5	2.0	2.3	2.3	2.5
MENA-all income levels (5)	1.0	1.1	1.2	1.2	3.0	3.5	4.1	5.2
MENA-developing only	1.2	1.4	1.6	1.6	2.0	2.3	2.8	3.5
OECD members (6)	0.7	0.5	0.4	0.4	11.0	10.4	10.8	10.9
South Africa	1.5	1.9	2.0	1.7	7.5	9.8	9.1	8.7
South Asia	1.3	1.6	1.8	1.6	0.4	0.5	0.8	1.1
SSA-all income levels (7)	1.0	1.2	1.1	1.0	0.9	1.0	0.9	0.8
SSA-developing only	1.0	1.2	1.1	1.0	0.9	1.0	0.9	0.8
Africa	0.6	0.6	0.6	0.5	0.9	1.0	0.9	1.2
World	0.9	0.8	0.7	0.6	4.2	4.1	4.1	4.5

Source: Calculated using WDI (2015)

[a]G-CO_2 refers to CO_2 emissions (kg per 2005 US$ of GDP) = CO_2 emissions/ Real GDP (constant 2005 US$); [b]P-$CO_2$ is CO_2 emissions (metric tons per capita) = CO_2 emission/total population; [c]EAP is the East Asia and Pacific; [d]LAC is Latin America and Caribbean; [e]MENA is Middle East and North Africa; [f]OECD is the Organization for Economic Co-operation and Development; [g]SSA is Sub-Saharan Africa

4 Methodology and Results

Estimating and quantifying the effect of economic growth on environmental quality vary according to the conditions of each country such as the economic growth rate, the degree of openness, the population density, and education and public policies. For this reason, two steps have been followed to investigate the links between economic growth and environmental degradation using a basic EKC equation used in many studies.

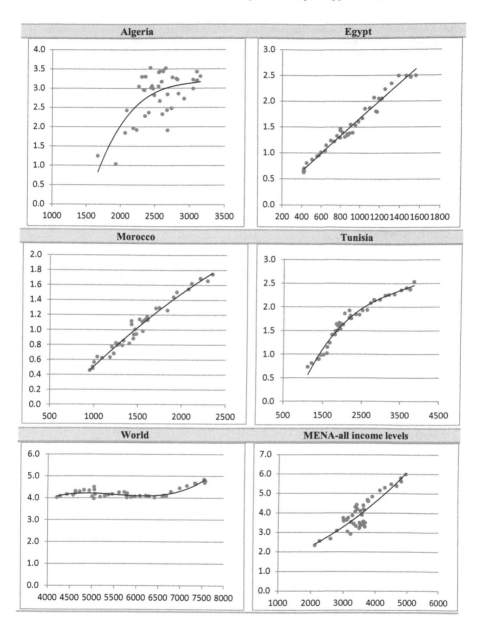

Fig. 18.2 Statistical relationships between CO_2 emissions (metric tons per capita) and real GDP per capita (2005 US$) of MATE, the world, and MENA regions (period 1970–2010)
Source: Elaborated using WDI (2015). E refers to CO_2 emissions per capita in level. Y refers to the real GDP per capita 2005 US dollars in level

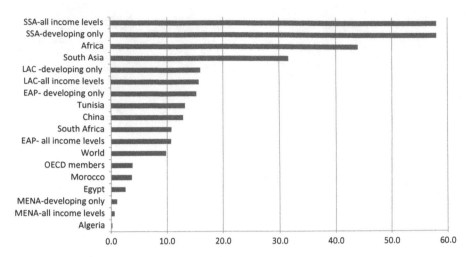

Fig. 18.3 Ranking of regions of the world by combustibles renewable and waste (% of total energy use) (period 2000–2009)
Source: Elaborated using WDI (2015)

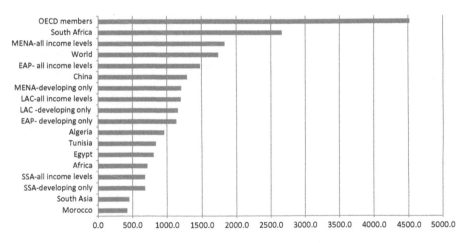

Fig. 18.4 Ranking of regions of the world according to energy use per capita (kg of oil equivalent per capita) (2000–2009)
Source: Elaborated using WDI (2015)

4.1 First Step

A basic EKC equation for each country over the period 1970–2010[15] is used to measure the effect of economic growth on environmental quality and to determinate the possibility of the existence of an EKC, i.e. the determination of the environmental curve in the form of an inverted U, which is estimated by the following form:

$$LE_{it} = a_0 + a_1 LY_{it} + a_2 \left(LY_{it} \right)^2 + \varepsilon_{it} \qquad\qquad \textbf{model 1}$$
$$For\ each\ i = Algeria, Egypt, Morocco\ or\ Tunisia.$$

Here, LE is the logarithm of the environmental degradation, LY is the logarithm of the per capita income, et refers to the error term, and t= '1970, 1981 ... 2010' year. The existence of an EKC implies that the coefficients a1 and a2 will be positive and negative, respectively, ($a_1 > 0$ and $a_2 < 0$). In that case, there is a level of real GDP per capita beyond which the environmental indicator begins to improve, the turning point (noted Ytp), therefore, is determined by: $\mathbf{Y_{tp} = -\dfrac{a_1}{2a_2}}$.

4.2 Cond Step

This step consists of introducing other variables[16] in the basic EKC model because this might have some impact on the level of environmental damage by decreasing or increasing it. These variables are:

- Urbanization: the increase of people living in urban areas often involves more waste production and energy consumption;
- Enrollment rates: because they have a direct and indirect impact on income and it may modify peoples' life style;
- Economic openness indicator measured by (X + M)/GDP, where X and M represent, respectively, exportation and importation.

$$LE_{it} = a_0 + a_1 LY_{it} + a_2 \left(LY_{it} \right)^2 + B.X_{it} + \varepsilon_{it} \qquad\qquad \textbf{model 2}$$
$$For\ each\ i = Algeria, Egypt, Morocco\ or\ Tunisia.$$

Where B is a parameter vector and X is an independent variables vector.

This work uses annual data taken from World Bank. Table 18.3 summarizes the descriptive statistics of all variables used in this work.

---

[15] The data of CO2 emission per capita is not available over the period 2011–2015.

[16] There are several factors that affect economic growth or environmental damage, but we cannot use all these variables, so we make some selection according to the availability of data regarding MATE and it importance.

Table 18.3 Statistic descriptive of the variables (sample: 1970–2010)

Variables	Notation: variables_code of country	Mean	St. Dev	Max	Min	Obs.
Real GDP per capita at 2005US$	Y_alg	2558.05	331.10	3143.63	1669.43	41
	Y_egy	886.72	320.90	1550.24	421.35	41
	Y_mor	1494.88	365.59	2348.59	953.93	41
	Y_tun	2263.14	724.89	3861.51	1119.71	41
Environment's Indicator: CO2 emissions per capita	E_alg	2.82	0.61	3.53	1.04	41
	E_egy	1.47	0.56	2.50	0.62	41
	E_mor	1.01	0.35	1.74	0.45	41
	E_tun	1.69	0.48	2.54	0.73	41
Enrollment rate measured by rate of primary completion	Pcr_alg	74.31	13.73	93.40	40.52	39
	Pcr_egy	77.81	20.29	105.91	34.64	39
	Pcr_mor	52.22	16.13	83.90	26.08	39
	Pcr_tun	79.18	13.98	101.72	55.02	39
Urbanization rate is the share of urban population in total population	u_alg	52.15	9.17	67.53	39.50	41
	u_egy	43.18	0.59	43.95	41.48	41
	u_mor	47.26	7.13	57.68	34.48	41
	u_tun	56.80	7.22	65.93	43.48	41
Economic openness indicator = (X + M)/GDP	open_alg	57.74	11.48	76.68	32.68	41
	open_egy	52.87	12.66	82.18	32.48	41
	open_mor	56.69	10.76	88.35	36.68	41
	open_tun	80.63	15.24	115.40	46.74	41

Source: Calculated using WDI (2015)
Code of country refers to alg = Algeria, egy = Egypt, mor = Morocco, and tun = Tunisia

Table 18.4 summarizes the regression results for each country based on the two models mentioned above (model 1 and model 2), differ with some specific additional independent variables (u, pcr and open).

Model 1 In MATE, real GDP per capita and its square are statistically significant and the coefficients attached to these variables are respectively, positive and negative. Therefore, these results prove the existence of an EKC and the levels of real GDP per capita beyond which the environmental indicator begins to improve, noted Ytp, are around $8000 per capita (2005 US dollar), except in case of Egypt whose turning point is higher (more than $26,000 per capita according to 2005 US dollar). This result can be partially explained by the feeble level of real GDP per capita in Egypt against those recorded in Algeria, Morocco, and Tunisia.

Model 2 In case of Egypt, real GDP per capita and its square have not expected signs. Therefore, the results cannot prove the existence of an EKC in this country. However, real GDP per capita and its square have expected signs in cases of Algeria, Morocco, and Tunisia; hence the results prove the existence of an EKC. But, the turning points of Morocco and Tunisia are estimated more than $8000 per capita (2005 US dollar); for the case of Tunisia, the turning point is estimated higher (more than $10,000 per capita according to 2005 US dollar).

Table 18.4 Results of models 1 and 2 from OLS estimation method, sample 1970:2010

		Algeria		Egypt		Morocco		Tunisia	
		Model 1	Model 2	Model 1	Mode 2	Model 1	Model 2	Model 1	Model 2
Constant	a0	-218.00	-2.38	-7.80	-4.28	-39.53	-10.29	-51.23	-47,63
	Std. dev	62.29	131.91	2.76	2.87	8.79	9.21	5.40	7,78
	t-stat	-3.50	-0.02	-2.83	-1.49	-4.50	-1.12	-9.49	-6,12
Independent variables LY	a1	54.87	0.64	1.38	-0.38	6.36	1.86	12.49	11,55
	Std. dev	15.99	33.47	0.83	1.03	2.41	2.48	1.40	1,99
	t-stat	3.43	0.02	1.67	-0.37	3.89	0.75	8.89	5,81
LY2	a2	-3.43	-0.03	-0.027	0.10	-0.54	-0.09	-0.75	-0,70
	Std. dev	1.06	2.12	0.06	0.08	0.16	0.17	0.09	0,12
	t-stat	-3.35	-0.01	-0.42	1.33	-3.28	-0.54	-8.19	-5,63
Pcr	b1		0.01		0.001		0.0003		-0,004
	Std. dev		0.01		0.002		0.0016		0,002
	t-stat		1.40		0.56		0.1777		-2,10
Open	b2		-0.01		0.0003		0.004		0,002
	Std. dev		0.01		0.001		0.002		0,001
	t-stat		-0.11		0.28		2.60		2,19
u	b3		-0.001		0.06		0.03		0,01
	Std. dev		0.01		0.03		0.00		0,01
	t-stat		-0.11		1.93		5.37		1,76
Turning point at 2005US$	Ytp	7987,28	10531.12	26254.02	—	8662.42	10461.87	8347.83	8305.85
	R^2	0.57	0.57	0.98	0.98	0.96	0.98	0.96	0.97
	F-stat-value	25.122	8.62	925.88	380.78	523.62	364.15	482.12	233.95
	Probability of F-stat	0.0033	0.0000	0.0000	0.0017	0.0000	0.0000	0.0000	0.0000

Source: Estimated using the available data

In Egypt, Morocco and Tunisia, the economic openness (open) is linked positively to CO_2 emissions per capita. These results mean that the openness increases the environmental damage. But, this variable is a negative sign in case of Algeria. This result can be explained by that Algeria is an exporter country of Oil and gas which represent more than 97% of total exports. However, urbanization rate (u) is linked positively to CO_2 emissions per capita in MATE. Rate of primary completion has no stable sign in model 2. This indicator is negative and significant in case of Tunisia and it is positive and not significant in other cases.

5 Environmental Strategies and Concluding Remarks

There are conflicts between economic growth and environment. Improving the quality of citizens' life cannot be realized, even if it is not sufficient, without boosting economic growth whether in developed or developing countries. But, this growth often generates negative environmental externalities which affect ecosystems' balance and reduce biodiversity, sometimes irreversibly. The links between these variables are still uncertain, complex, and ambiguous. Therefore, it is not possible to find a unique form of this links, and each variable introduced in model can give some explanation, as it is shown in this work, where the application of EKC is often unclear and uncertain.

These results mean that each country through policymaking and the involvement of other actors such corporations and non-governmental organizations must apply preventive and precautionary measures to reduce environmental damage. Such measures must be adapted with the specific economic and environmental conditions while benefiting from the experiences and good practices developed in other regions and avoiding others' past mistakes related to pollution, regional development, and natural resource management.

In parallel, it is necessary to establish a global political strategy to protect ecosystems and biodiversity in all countries because solidarity and participation of all people of the planet are important steps to reduce environmental damage. These steps mean that the present generations must not only think about future generations while using resources, but it must involve all people in improving and protecting the environment through solidarity actions, recreational activities and volunteering.

References

Antweiler W, Copeland BR, Taylor MS (2001) Is free trade good for the environment? Am Econ Rev 91:877–908
Arrow K, Bolin B, Costanza R, Dasgupta P, Folke C, Holling CS, Jansson BO, Levin S, Mäler K, Perrings C, Pimentel D (1995) Economic growth, carrying capacity and the environment. Science 268:520–521

Bass S (2006) Making poverty reduction irreversible: development implications of the millennium ecosystem assessment. IIED Environment for the MDGs' briefing paper. International institute on environment and development, London

Becker GS (1962) Investment in human capital: a theoretical analysis. J Polit Econ 70(5):9–19

Beckerman W (1992) "Economic development and the environment: conflict or complementarity", background paper for the world development report 1992. The World Bank, Washington, DC

Blaug M (1976) The empirical status of human capital theory: a slightly jaundiced survey. J Econ Lit 14(3):827–855

Bourdieu P (1985) The forms of capital. In: Richardson JG (ed) Handbook of theory and research for the sociology of education. Greenwood, NewYork, pp 241–258

Chertow MR (2001) The IPAT equation and its variants. Changing views of technology and environmental impact. J Ind Ecol 4(4):13–29

Chivian E, Bernstein A (2010) How our health depends on biodiversity. Center of Health and the Global Environment, Harvard Medical School, Boston

Cole MA (1999) Limits to growth, sustainable development and environmental Kuznets curves: an examination of the environmental impact of economic development. Sustain Dev 7:87–97

Cole MA, Neumayer E (2005) Environmental policy and the environmental kuznets curve: can developing countries escape the detrimental consequences of economic growth? In: Dauvergne P (ed) Handbook of global environmental politics. Elgar original reference. Edward Elgar, Cheltenham, pp 298–318 ISBN 9781843764663

Cole MA, Rayner AJ, Bates JM (1997) The environmental Kuznets curve: an empirical analysis. Environ Dev Econ 2:401–416

Coleman JS (1988a) Social capital in the creation of human capital. Am J Social 94:95–121

Coleman JS (1988b) The creation and destruction of social capital: implications for the law. Notre Dame J Law Ethics Publ Policy 3:375–404

Commoner B (1972a) A bulletin dialogue on 'The Closing Circle': response. Bull At Sci 28(5.): 17):42–56

Commoner B (1972b) The environmental cost of economic growth. In: Ridker RG (ed) Population, resources and the environment. Government Printing Office, Washington, pp 339–363

Commoner B, Corr M, Stamler PJ (1971) The causes of pollution. Environment 13:2–19

Costanza R, Daly HE (1992) Natural capital and sustainable development. Conserv Biol 6:37–46

Cropper M, Griffiths C (1994) The interaction of population growth and environmental quality. The American Economic Review. vol 84, No. 2, papers and proceedings of the hundred and sixth annual meeting of the American economic association (May, 1994), pp 250–254

Daniel JJ (1999) Introduction to atmospheric chemistry. Princeton University Press, Princeton

Ehrlich PR, Holdren JP (1971) Impact of population growth. Science 171:1212–1217

Ehrlich PR, Holdren JP (1972) A bulletin dialogue on 'The Closing Circle': critique. Bull At Sci 28(5.):16):18–27

Ekins P (1997) The Kuznets curve for the environment and economic growth: examining the evidence. Environ Plan A 29:805–830

Grossman GM, Krueger AB (1991) Environmental impacts of a North American free trade agreement. National bureau of economic research. Working paper No 3914, NBER, Cambridge, MA

Grossman GM, Krueger AB (1995) Economic growth and the environment. Q J Econ 112:353–378

Intergovernmental Panel on Climate Change (IPCC) (2014) Climate change 2014: synthesis report. Contribution of working groups I, II and III to the fifth assessment report of the intergovernmental panel on climate change [Core writing team, RK Pachauri and LA Meyer (eds)]. IPCC, Geneva, 151 pp

Khagram S, Clark WC, Raad DF (2003) From the environment and human security to sustainable security and development. J Hum Dev 4(2):289–313

Kiker BF (1966) The historical roots of the concept of human capital. J Polit Econ 74(5):481–499

Kuznets S (1955) Economic growth and income inequality. Am Econ Rev 49:1–28

Lopez R (1994) The environment as a Factor of production: the effects of economic growth and trade liberalization. J Environ Econ Manag 27(2):185–204

Martino D, Zommers Z (2007) Environment for development, Chapter 1 of report of Global environment outlook, GEO4, United Nations Environment Programme (UNEP)

Mincer J (1958) Investment in human capital and personal income distribution. J Polit Econ 66(4):281–302

Mincer J (1962) On-the-job training: Costs, returns, and some implications. J Polit Econ 70:50–79

Panayotou T (1993) Empirical tests and policy analysis of environmental degradation at different stages of economic development. Working paper WP238, technology and employment programme, International labour office, Geneva

Panayotou T (1997) Demystifying the environmental Kuznets curve. Turning a black box into a policy tool. Environ Dev Econ 2(04):465–484

Panayotou T (2003) Economic growth and the environment. Harvard University and Cyprus International Institute of Management, Harvard. https://www.unece.org/fileadmin/DAM/ead/sem/sem2003/papers/panayotou.pdf

Perman R, Stern DI (1999) The environmental Kuznets curve: implications of non-stationarity, working papers in ecological economics 9901, Centre for resource and environmental studies, Australian National University, Canberra

Portes A (1998) Social capital: its origins and applications in modern sociology. Annu Rev Sociol 24:1–24

Putnam RD (1993) The prosperous community: social capital and public life. Am Prospect 4(13):35–42

Schultz TW (1961) Investment in human capital. Am Econ Rev 51(1):1–17

Schultz TW (1962) Reflexions on investment in man. J Polit Econ 70(5):1–8

Selden TM, Song D (1994) Environmental quality and development: is there a Kuznets curve for air pollution emissions? J Environ Econ Manag 27:147–162

Shafik N (1994) Economic development and environmental quality: an econometric analysis. Oxf Econ Pap 46:757–777

Shafik N, Bandyopadhyay S (1992) Economic growth and environmental quality: time series and crosscountry evidence. Background paper for the world development report 1992, the World Bank, Washington, DC

Stern DI (2004) The rise and fall of the environmental Kuznets curve. World Dev 32(8):1419–1439

Stern DI, Common MS (2001) Is there an environmental Kuznets curve for sulfur? J Environ Econ Manag 41:162–178

Torras M, Boyce JK (1998) Income, inequality, and pollution: a reassessment of the environmental Kuznets curve. Ecol Econ 25(2):147–160

World Bank (2006) Where is the wealth of nations? measuring capital for the 21st century. The World Bank, Washington, DC

Yandle B, Vijayaraghavan M, Bhattarai M (2002) The environmental Kuznets curve: a primer. Research study 02–1. Montana, USA. Political Economy Research Center (PERC)

Chapter 19
The Role of Corporate Social Responsibility in Environmental Sustainability

Peter Karácsony

Abstract Traditionally, environmental protection has been considered to be 'in the public interest' and external to private life. Since the Brundtland Report was published in 1987 as a result of the work of the World Commission on Environment, business and management scholars have been grappling with the question of how and why corporations should incorporate environmental concerns into their own strategies. Today, many companies have accepted their responsibility to do no harm to the environment. In relation to corporate social responsibility (CSR), the business community uses frequently-used terms that are commonly considered to be related to corporate ethics. The goal of this paper is to present, to some extent, current practices and approaches to environmental aspects of corporate social responsibility in the case of some Hungarian organizations.

Keywords Corporate social responsibility · Environment · Hungary · Sustainable development

1 Introduction

The major industrial development that began in the nineteenth century was essentially based on exploiting resources more efficiently and maximizing profits. However, in the last third of the twentieth century, it became clearer that this path of development had lead to unreasonable exploitation of natural resources, and that sustainable social and economic development cannot be achieved (Porter and Van der Linde 1995; Sen et al. 2006). For some companies, investing in environmentally-conscious operations simply represents additional costs, while others have become more competitive due to this investment. According to Chen (2008), companies that invest in environmental issues are often able to improve their reputation and develop new market opportunities, as well as to increase their competitive advantages. The objectives of responsible corporate leaders are not only competitiveness and

P. Karácsony (✉)
Department of Economics, Faculty of Economics, J. Selye University, Komárno, Slovakia
e-mail: karacsonyp@ujs.sk

© Springer International Publishing AG, part of Springer Nature 2019   377
M. Behnassi et al. (eds.), *Human and Environmental Security in the Era of Global Risks*, https://doi.org/10.1007/978-3-319-92828-9_19

compliance with the law, but also long-term success, sustainability and value creation (Aluchna and Idowu 2017).

The most well-known definition of sustainable development is contained in the 1987 United Nations Environment and Development Commission's document "Our Common Future", also known as the Brundtland report, named after the Norwegian Prime Minister. 'Sustainable development' is defined in the report as "a development that meets the needs of the present without compromising the ability of future generations to meet their own needs" (Visser and Brundtland 1987). Many researchers believe that the Brundtland Report was the turning point in the emergence of the referential of corporate social responsibility (CSR) (Cohen and Winn 2007; Hueting 1990).

In this study, I will attempt to answer the following question: how important is the CSR in corporate environmental responsibility? One of the components of sustainable development is environmental damage caused by the companies during their production of products or services, while the other component is related to the consumption. The unreasonable use of resources in the company's production, the decline in biodiversity, poor water management and the increase in greenhouse gas emissions are all part of the environmental problem. Solving environmental problems and building a sustainable economy without strong corporate environmental responsibility is impossible.

2 History of CSR and Relationship with Environmental Protection

Conscious thinking about CSR was the first time in America in the 1950s. The first major study on corporate social responsibility Howard R. Browen's 1954 thesis was: Social responsibilities of the businessman. In the 1970s, the term 'Corporate Social Responsibility' spread, which was used both in professional and public relations. Most commonly, the following topics were concerned: consumer protection, workplace safety, fair pay, support for local communities etc.

In the 1970s, Milton Friedman's article, the Corporate Social Responsibility in Profit Growth explained that people are only ethically responsible for their actions, so companies can not have such a responsibility. He stated that social responsibility of companies is merely an increase in profits, and business leaders have no social obligation beyond compliance with basic rules. He felt that social responsibility was nothing but a charity for shareholders. He argues that solving social or environmental problems is within the sphere of government (Friedman 1970).

In the 1980s stakeholder theory points that the company has to work in a space where different stakeholders, that is, interest groups, all of which must be met. According to this, not only the opinion of consumers, but also employees, suppliers, capitalists, trade unions, civil organizations, local communities and everyone who can influence the achievement of the goals of the organization (Evan and Freeman

1983; Frederick 1986). For the 1990s, sustainability and environmental protection were even more focused, more and more people wrote environmental reports, and the theme of public discourse continued. Porter and Linde in their work *Green and Competitive*, explained that in the long term the company could be competitive if it uses environmentally friendly techniques in its production (Porter and Van der Linde 1995). In the 2000s, CSR became increasingly conscious and critical, and the pressure on the civil sector has been steadily increasing. In the interest of large corporations, which are committed to CSR, it became important that competitors who do not take social responsibility or simply perform a visual activity are at a competitive disadvantage. That is why they argued in the 2000s to create more corporate responsibility legal bases (Matten et al. 2003).

Daly (1994) has suggested that sustainable development can be operationalized in terms of the conservation of natural capital. CSR is an attempt by corporations to integrate social and environmental objectives on a voluntary basis into their businesses, and to shape their relationships with their stakeholders along these principles (Ozdora Aksak et al. 2016; Perrini 2006). In other terms, CSR means commitment in which the company pursues a voluntary, freely chosen business practice for the well-being of the community, supported by its resources (Trong Tuan 2012). The most common manifestations of CSR in a company's business practices are:

- facility design, including the development of environmental standards and recommendations (energy conservation, waste-free technologies);
- production process: waste reduction including hazardous waste, and a reduction in the use of chemicals;
- selecting suppliers with an emphasis on sustainable development of the most environmentally friendly raw materials and packaging materials;
- public or employee well-being programs;
- marketing activities; and
- charity programs, etc.

For future generations, the best we can do is sustainable growth, so we can grow on the economic and social levels without endangering the living conditions of future generations. What can a company manager do for this purpose? CSR provides a promising solution: making profit while paying attention to environment in the interest of future generations. The environmental policy formulated by a company also means that the organization also enhances environmentally conscious operations as its central goals (Khojastehpour and Johns 2014). CSR helps businesses achieve these environmental goals. If the organization is committed to CSR, it can have a number of positive effects, as the operation of an optimized system can reduce energy and raw material costs, authorities are more positive about the organization, the company's reputation may improve, and its social perception may be more positive. Leadership's commitment is decisive, as their roles greatly determine the success of the organization's environmental-conscious strategy (Khan and Lund-Thomsen 2011).

Environmentally-conscious corporate governance is a behavior which means that the management operating a company are aware that the products and services it produces may endanger human health and negatively impact the workplace, urban and natural environment. Therefore, companies should be committed to manage the environmental and social externalities of their activities, while taking into account global issues such as climate change (Behnassi et al. 2013). Within this perspective, the interest of the company is, among others, to improve the environmental image constantly, therefore, environmental policy must take into account the needs of consumers, the expectations of society, technological progress and the state of science as well. Products and services provided by the company should have no adverse environmental impact, with safe continuous use, and permit an effective energy and natural resource use (Radhakrishnan et al. 2014).

In an increased competitive environment, companies often ignore environmental factors in order to gain the greatest competitive advantage (Kolk 2016).

From an environmental point of view, companies can be divided into several groups (Fryzel and Seppala 2016). Companies with a relatively high level of environmental protection belong to the first group. These companies are pursuing an innovative environmental strategy as their environmental product and technology developments are designed to create a competitive advantage, and they are increasingly integrating environmental aspects into their innovation and communication policy.

Companies in the other group do not see market opportunities in their environmental performance, so they only try to meet the minimum for their further operation. They have not put their marketing or innovation strategy in the focus of their environmental performance, they have not yet discovered competitive advantages or market opportunities in the field of environmental protection (Kolk 2016).

Third-party companies do not show undertake an environmental activity: they do not have any environmental protection investments. Companies are not developing their environmentally-friendly technologies, because of the corporate costs are significantly increased by environmental protection (Altman 2001).

When assessing CSR, three aspects are commonly considered: social, corporate (ethical) and environmental aspects (Brønn 2011). From these three pillars, this chapter mainly focuses on the environmental dimension.

3 Materials and Methods

The principal objective of this study is to identify the level of CSR and its environmental aspects in Hungarian small and medium-sized industrial companies. To this end, a questionnaire was used for the data collecting method based on a quantitative research approach. This operation was conducted between 2014 and 2016 in person and electronically. In the course of the study, 328 questionnaires from 23 SMEs had been evaluated. SMEs are the most common form of enterprises in Hungary. They operate in all sectors of the Hungarian economy: general business, services, manufacturing, and agriculture. According to the European Commission, SMEs represent 99.9% of all businesses and create jobs for 73% of the working population.

What is an SME? As defined in the EU recommendation 2003/361, the main factors determining whether an enterprise is an SME are: staff headcount and either turnover or balance sheet total. Based on this, "the category of micro, small and medium-sized enterprises is made up of enterprises which employ fewer than 250 persons and which have an annual turnover not exceeding 50 million euro, and/or an annual balance sheet total not exceeding 43 million euro" (EU recommendation 2003/361).

As far as the environmental performance of companies operating in Hungary is concerned, there is an ever-increasing trend, though – in the case of small and medium-sized enterprises (SMEs) – it is far behind expectations. In my opinion, this lag is fundamentally attributable to two reasons: the lack of capital in the SME sector; and the low level or lack of environmental awareness.

4   Results and Discussion

CSR consists of several elements. In the first question, I examine the extent to which the particular elements are considered important by the surveyed companies. Among the five CSR elements, the commitment to employees (25 percent) is the most important, followed by quality (22 percent), responsibility for environmental protection (20 percent), commitment to the community (18 percent), and finally charity and support (15 percent) (Fig. 19.1).

When asked *How important is CSR for your business?*, more than half of respondents said that CSR is very important at their business, while almost a third of companies said it is important, and only 14 percent said CSR is of negligible importance for their business (Fig. 19.2). In addition, compliance with environmental standards is considered important by most respondents (89 percent).

The next question was: *Which area from CSR is the most important from the following environmental aspects?* For this question, the respondents could choose

Fig. 19.1 Importance of CSR elements, percentage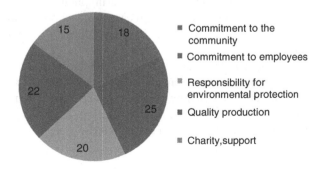

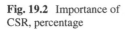

Fig. 19.2 Importance of
CSR, percentage

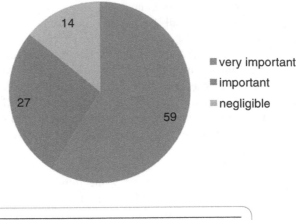

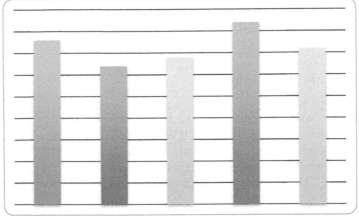

Fig. 19.3 The most important environmental aspects of CSR, Likert scale

more than one answer using the Likert scale (1 - very important, 5 - unimportant)
(Fig. 19.3).

Companies consider waste recycling (4.2) and the use of environmentally-
friendly materials (3.8) as the most important environmental protection intervention. Then follows low emissions of pollutants (3.6), environmentally friendly
production (3.4), and low energy consumption (3.2).

Each stage of production processes has different impacts on the environment.
The Fig. 19.4 shows how companies address the environmental impact of each
process.

It was stated that during the production process (4.2), the companies pay the
most attention to environmental protection, followed by the sales process (3.9)
(environmentally-friendly packaging) and acquisition processes (3.8) (acquiring
environmentally-friendly raw materials). Marketing (3.6) and logistics (3.2)
received the least amount of environmental care.

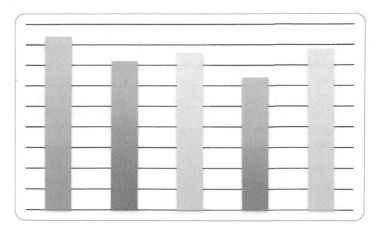

Fig. 19.4 The impact of each process on the environment, Likert scale

Fig. 19.5 Future plans for environmental protection, percentage

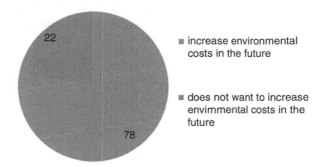

According to the interviewed managers, the environmental aspects of CSR should be promoted in the future, and more emphasis should be placed on them. Protecting the natural environment is important for everyone. About 78 percent of the interviewed leaders want to pay more attention to environmental protection in the future, while the remaining 22 percent seek to maintain the current state of affairs (Fig. 19.5).

5   Conclusion

The analysis shows that companies' efforts to develop sustainable development and to improve their environmental image can only be rewarded as part of a strategy. Indeed, a company that pays great attention to CSR must have a related strategy that can be followed in the short or medium term. Companies need to plan their environmental strategy not only in their own unique situation but also in their current social and political situation. In social responsibility, companies voluntarily apply social and environmental considerations in their own business and in their relationships with their stakeholders.

Efficient and relevant CSR programs create a 'win-win' situation as they bring advantages to both the company and the local communities or other stakeholders (enhancing corporate image, customer satisfaction, good conditions for workers, etc.). Therefore, CSR and environmentally conscious attitudes are indispensable to sustainable development. An increased number of companies are committed to this issue and clearly demonstrate that market-based interests and social-minded thinking are compatible.

References

Altman M (2001) When green isn't mean: economic theory and the heuristics of the impact of environmental regulations on competitiveness and opportunity cost. Ecol Econ 36:31–44

Aluchna M, Idowu SO (2017) Responsible corporate governance: an introduction. Responsible corporate governance. Springer International Publishing, pp 1–7

Behnassi M, Pollmann O, Kissinger G (eds) (2013) Sustainable food security in the era of local and global environmental change. Springer, Netherlands

Bowen HR (1954) Social responsibilities of the businessman. Am Cathol Sociol Rev 15(1):42

Brønn PS (2011) Marketing and corporate social responsibility. The handbook of communication and corporate social responsibility:110–127

Chen YS (2008) The driver of green innovation and green image - green core competence. J Bus Ethics 81(3):531–543

Cohen B, Winn MI (2007) Market imperfections, opportunity and sustainable entrepreneurship. J Bus Ventur 22(1):29–49

Daly HE (1994) Operationalizing sustainable development by investing in natural capital. In: Jansson AM et al (eds) Investing in natural capital: the ecological economics approach to sustainability. Island Press, Washington, DC

EU recommendation (2003) Commission recommendation of 6 May 2003 concerning the definition of micro, small and medium-sized enterprises (available on: http://eur-lex.europa.eu/legal-content/EN/TXT/?uri=CELEX:32003H0361)

Evan W, Freeman RE (1983) A stakeholder theory of the modern corporation: Kantian capitalism. In: Beauchamp T, Bowie N (eds) Ethical theory and business. Englewood Cliffs, New Jersey, pp 75–93

Frederick WC (1986) Toward CSR3: why ethical analysis is indispensable and unavoidable in corporate affairs. Calif Manag Rev 28:126–141

Friedman M (1970) The social responsibility of business is to increase its profits. Corporate ethics and corporate governance, pp 173–178

Fryzel B, Seppala N (2016) The effect of CSR evaluations on affective attachment to CSR in different identity orientation firms. Bus Ethics Eur Rev 25(3):310–326

Hueting R (1990) The Brundtland report. Ecol Econ 2(2):109–117

Khan FR, Lund-Thomsen P (2011) CSR as imperialism: towards a phenomenological approach to CSR in the developing world. J Chang Manag 11(1):73–90

Khojastehpour M, Johns R (2014) The effect of environmental CSR issues on corporate/brand reputation and corporate profitability. Eur Bus Rev 26(4):330–339

Kolk A (2016) The social responsibility of international business: from ethics and the environment to CSR and sustainable development. J World Bus 51(1):23–34

Matten D, Andrew C, Wendy C (2003) Behind the mask: revealing the true face of corporate citizenship. J Bus Ethics 45:109–120

Ozdora Aksak E, Ferguson MA, Atakan Duman S (2016) Corporate social responsibility and CSR fit as predictors of corporate reputation: a global perspective. Public Relat Rev 42(1):79–81

Perrini F (2006) Corporate social responsibility: doing the most good for your company and your cause. Acad Manag Perspect 20(2):90–93

Porter ME, Van der Linde C (1995) Green and competitive: ending the stalemate. Harv Bus Rev 73(5):120–133

Radhakrishnan MS, Chitrao P, Nagendra A (2014) Corporate social responsibility (CSR) in market driven environment. Proc Econ Financ 11:68–75

Sen S, Bhattacharya CB, Korschun D (2006) The role of corporate social responsibility in strengthening multiple stakeholder relationships: a field experiment. Acad Market Sci 34(2):158–166

Trong Tuan L (2012) Corporate social responsibility, ethics, and corporate governance. Soc Responsib J 8(4):547–560

Visser W, Brundtland GH (1987) Our common future ("The Brundtland Report"). World commission on environment and development. The top 50 sustainability books, pp 52–55

Postface

Human security inseparably links humans, social systems and ecosystems and strives to achieve freedom from fear, freedom from the impacts of natural hazards and freedom from want by reducing natural and social disruptions. Nevertheless, it is only at the first decades of the third millenium that the effects of global risks – such as those associated to environmental and climate change – are beginning to shape a new and more urgent need for the human security paradigm. Such a referential is still being debated and informed by many disciplines, studies, and global assessments. Such a shift in considering the human security paradigm is increasingly relevant both in terms of how to perceive global risks and how to address their security impacts.

Throughout this volume, it has been demonstrated that the complex and uncertain evolution of current risks is considerably altering our perception of security and shaping related decision-making processes at all levels. This change is triggered by the fact that emerging global risks are raising new and unavoidable questions with regard to human and environmental insecurity. Besides, perceiving and managing current risks from a human-environmental nexus connect these risks to a myriad of other issues.

More precisely, the volume's authors have investigated how global risks, especially environmental and climate change, may worsen threats to human and environmental security especially in developing countries. Through its theoretical essays and a range of country-level case studies and experiences, the volume aims to provide readers with access to valuable research material and insights about the interactions between global risks and human and environmental security from a range of perspectives. The majority of chapters demonstrate above all that global risks affecting human and environmental security arise through multiple and interconnected processes operating across diverse spatial and temporal scales. This complexity means that there is no single conceptual model or theory that can fully capture all interactions. Therefore, many risks to human and environmental security deserve additional investigation with the aim to provide decision-making processes with the needed referential and tools to manage the concerned linkages. This was apparent for instance regarding the links between environmental stress, resource scarcity and conflicts.

This volume is a contribution to the growing academic literature about human and environmental security and risks. It also provides policy agendas with insights and materials on how to manage current and future challenges with security implications. The framing of many global risks, especially those associated with environmental and climate change, as a human security concern provides an excellent opportunity to learn and exchange across different policy and research spheres. The Center for Research on Environment, Human Security and Governance (CERES), former NRCS, is engaged to continue investigating the linkages between human and environmental security and global risks with the aim to shape future research and policy agendas both locally and globally. Future projects will be conceived in such a way as to build on some of the many issues highlighted by volumes and conferences recently initiated by CERES. We hope that this process will open further points for discussion and offer readers some thought-provoking insights.

Biographical Notes of Contributors

Mhamed Ahrabous is a Ph.D. student in Environmental Economics at Hassan II Institute of Agronomy and Veterinary Medicine in Rabat. His works are focused on Environmental Services and Poverty in Rural Areas.

Amine Amar Dr. Amar is a statistician engineer from National Institute of Statistics and Applied Economics (INSEA, Rabat-Morocco). He holds a PhD from Mohamed V University of Rabat and has seven years experience in conducting surveys, statistical methodologies and economic analysis (High Commission of Planning, Morocco). Currently, he is senior official at the Moroccan Agency for Sustainable Energy (MASEN) and a Fellow Researcher at the Center for Research on Environment, Human Security and Governance (CERES, Morocco), the International Association for Research in Income and Wealth (IARIW, Canada), and the Economic Research Forum (ERF, Egypt). He has participated in many conferences and published many papers related to the elaboration and implemention of economic and statistical methodologies for some research fields, such as environment, finance, economy, and public policies.

Mohamed Aneflouss Dr. Aneflouss holds a PhD in Geography from Hassan II University of Mohammedia, Morocco. He is currently Professor at the Department of Geography of the faculty of Arts and Human Sciences of Mohammedia and Director of the Research Laboratory on Dynamics of Space and the Society.

Fatima Arib Doctor in Economics from the University of the Mediterranean – Aix Marseille II in France; Professor in economics of sustainable development at the Faculty of Law, Economics and Social Sciences, Cadi Ayyad University of Marrakech. She is responsible for the master of Environmental Economics in this faculty; expert in environmental economics and consultant for several national and international organizations. She is also responsible for sustainable development and great projects in the presidency of Cadi Ayyad University in Marrakech and Vice

© Springer International Publishing AG, part of Springer Nature 2019
M. Behnassi et al. (eds.), *Human and Environmental Security in the Era of Global Risks*, https://doi.org/10.1007/978-3-319-92828-9

President of the Moroccan Association for the Economy of the Environment AM-EconEnv. She animated several conferences and trainings in sustainable development, environmental economics, and climate change.

Arunesh Asis Chand Dr. Asis Chand is a PhD Scholar in Climate Change Governance at the University of the South Pacific, School of Geography, Earth Science and Environment, Suva, Fiji Islands and a recipient of the European Union Scholarship. He obtained his Bachelor of Science Degree in Marine Science and Master of Arts Degree in Governance from the University of the South Pacific in 2002 and 2010 respectively. He started his professional career as a Research Coordinator in 2003-2007 at the Ministry of Fisheries and Forestry under the Government of Fiji. He also served at Fiji National University from 2009-2011 as a Research Officer before joining the University of the South Pacific once more to complete his PhD in Climate Change Governance. His main areas of expertise, research and work include: climate governance and policies; disaster risk management; and environmental governance.

Lucia Beran Ms. Beran has been working as Research Assistant at the RWTH University's Chair of Business Theory, Sustainable Production and Industrial Management Control since 2008. Her research is focused on the significance of energy for the production of wealth. Thereby, she examines the historical development of world population, primary energy supply and economic performance by focalising on biomass as a neglected energy carrier in the economic system, the supply of which is critically evaluated against the background of sustainable development.

Kaderi Noagah Bukari Dr. Bukari holds a PhD from the University of Gottingen and the Zentrum für Entwicklungs Forschung (ZEF) / Center for Development Research, University of Bonn, Germany. He initially studied a Master of Philosophy (MPhil) in Peace and Development Studies at the University of Cape Coast, Ghana. His PhD thesis was on Fulani herder relations in Ghana, with a focus on how conflict, environmental change and cooperation interplay in these relations. His research interests cover resource conflicts, ethno-political conflicts, environmental/climate change, development studies and governance.

Seksak Chouichom Dr. Chouichom is Senior Business Development Officer at the Office of Technology and Innovation Management of the Thailand Institute of Scientific and Technological Research in Bangkok. He has conducted research and published widely on Thai jasmine rice and organic agriculture as part of his Ph.D. research. He holds degrees in Agricultural Economics with a focus on rural farming systems from Kasetsart University and Hiroshima University.

Elmar Csaplovics Dr. Csaplovics is Professor of Remote Sensing at the Institute of Photogrammetry and Remote Sensing, Department of Geosciences, TU Dresden. He holds a doctorate and a habilitation in remote sensing from TU Vienna (1982, 1992), was a post-doc research fellow at the National Institute of Agronomic Research (INRA) of Montpellier and at the Department of Geology, Geophysics and Geoinformatics, Free University Berlin from 1988-1992, and was appointed Professor of Remote Sensing at the TU Dresden in 1993. Dr. Csaplovics was also Visiting Professor at University College London (UCL) in 2007. He focuses in his research on remote sensing and applied geoinformation analysis for spatio-temporal monitoring and assessment of land use/land cover with emphasis on wetlands and semi-arid lands (desertification and degradation), as well as on landscape history and world conservation monitoring in transnational project cooperation.

Taisser H. H. Deafalla Taisser H. H. Deafalla, has obtained her B.Sc. (Hons) and M.Sc. degrees at the Faculty of Forestry, University of Khartoum (2004, 2011). Since 2013, she has been working in her PhD research at the Institute of Photogrammetry and Remote Sensing, Faculty of Environmental Sciences, University of Dresden, Germany. Mrs. Deafalla has professional experience in the fields of forestry and rural development where she worked with many local and international organizations. Currently, she is a researcher working with united nation volunteers. Moreover, Mrs. Deafalla participated in many conferences and international workshops and she has many scientific publications.

Harald Dyckhoff Harald Dyckhoff has been professor at the RWTH University's Chair of Business Theory, Sustainable Production and Industrial Management Control in Aachen since 1988. His main fields of work and research include production and decision theory, sustainable industrial creation of value as well as performance evaluation.

Aïcha EL Alaoui Ph.D, Associate Professor of Economics. She teaches macroeconomics, economics, national account, econometric, DSGE applied on GAMS, and statistics. Her research areas include the applications of the input-output analysis, the applications of modelling, environment and the economic growth and gender approach. University of Sultan My Slimane. Mghila, B.P. 592, CP: 23000, BeniMellal, Morocco. Email. aicha_elalaoui@yahoo. fr, Tel. +212663506211

Rachida El Morabet Rachida El Morabet holds a PhD in Biology (2004) from Mohammed V University of Rabat, an Inter-university diploma of Biotechnology from Paris (2000), and a Certificate from the University Pedagogy (teaching practices, learning theories, models assessment practices knowledge; educational resources and distance learning) from Hassan II University of Casablanca (2015). She is currently Assistant Professor at the Department of Geography and researcher at the Research Laboratory on Dynamics of Space and the Society (Research group: Dynamics of Natural Environments and their Impacts on Society and Territories), Faculty of Arts and Human Sciences, Hassan II University of Mohammedia,

Morocco. Dr. El Morabet is also a founding member of the Center for Research on Environment, Human Security and Governance (CERES) (2016). She is the owner of two Gold Medals with honors from the International Exhibition for Invention and Innovation, Casablanca (2001), issued for the theme: "For a Better Value of Coastal and Saharan regions in Morocco".

Mustafa M. El-Abbas Mustafa M. El-Abbas, is an assistance professor at the Faculty of Forestry department of forest management, University of Khartoum, Where heobtained his B.Sc. (Hons) and M.Sc. degrees (2001, 2006). Moreover, he has got a diploma of forest ecology and forest resource management, university of Helsinki, Finland in 2006. Dr. El-Abbas award his PhD in remote sensing and natural resource management from TU Dresden. Currently, he has been working as a researcher at the Institute of Photogrammetry and Remote Sensing, Faculty of Environmental Sciences, University of Dresden, Germany. He has professional experience in the fields of remote sensing and GIS in general and object-based approaches in particular with emphasis on natural resources management. Dr. El-Abbas participated in several international workshops and scientific forums. Additionally, he has many scientific publications.

Zhar Essaid Zhar Essaid: PhD student in the Laboratory "Dynamics of Space and the Society". Thesis topic: "Impact of hydro erosion on the surface in chiadma region, northern of Essaouira"

Aziz Fadlaoui Ph.D. in Agronomic Sciences and Biological Engineering (Catholic University of Louvain la Neuve, Belgium). He is an Agricultural Economist at National Institute of Agronomic Research. His works are focused on environmental economics, vertical integration in agriculture and contract farming, and climate change adaptation in agriculture.

Raj Kumar Gupta Raj Kumar Gupta is an anthropologist by education and journalist by profession. He did his Master of Science in Human Biology in 1977 and worked with three national dailies including National Herald, Indian Express in New Delhi and Ahmedabad. He was the chief of economic bureau with The Observer of Business and Politics in New Delhi. He personally knows the leaders of prominent environmental movements in India. He wrote extensively on international trade, food, environment, and social policies and their interlinkages.

Peter Karácsony Dr. Peter Karácsony was born in 1981 in Dunajská Streda, Slovakia. He started his university studies in 1999 at the University of West Hungary, and received his degree in Agricultural economics and European Union Management in 2005. In 2008, he finished his Ph.D research on international competitiveness. He has worked as a university professor since 2005. He currently works as an associate professor at the Széchenyi István University and J. Selye University. Besides his university works, he considers research and publication activities as a priority: he has published 35 reviewed scientific publications, co-authored five books, participated in 28 conference proceedings, and authored six other professional publications to date.

Lawrence M. Liao Lawrence M. Liao is associate professor of aquatic botany at the Graduate School of Biosphere Science, Hiroshima University where he conducts research on freshwater and marine algae, their taxonomy and economic utilization mainly focusing on species from around Southeast Asia. He obtained his degrees in marine biology from the University of San Carlos and the University of the Philippines Diliman, and Ph.D. in biology (major in phycology) from the University of North Carolina at Chapel Hill.

Mark Matsa Mark Matsa (PhD) is a Senior Academic at Midlands State University, Department of Geography and Environmental Studies, Faculty of Social Sciences. Dr Matsa earned a Masters in Environmental Policy and Planning from the University of Zimbabwe and a PhD in Geography and Environmental Science from the University of South Africa. Dr Matsa has written extensively on environmental change/climate change and sustainable development, food security, migration, gender and the environment, land reform in Zimbabwe among other geographical and environmental issues. He has published extensively in reputable journals and is a member of several Geographical and Environmental Associations in Zimbabwe (Geographical Association of Zimbabwe (GAZ)), East Africa (Organisation of Social Sciences Research in Eastern and Southern Africa (OSSREA)), the United States (American Association of International Researchers (AAIR) and Japan (International Society for Development and Sustainability (ISDS)). At Midlands State University he is the Chairperson of the Faculty of Social Sciences Research Seminar Series (SSRSS).

Said Mouak Said Mouak PhD student in the Laboratory "Dynamics of Space and the Society". Thesis topic: "Spatial and social disparities in the health status of kenitra"

Maizatun Mustafa Dr. Maizatun Mustafa is Associate Professor at Ahmad Ibrahim Kulliyyah of Laws, International Islamic University Malaysia. She acquired her Ph.D from the School of Oriental and African Studies, University of London, majoring in environmental law. She has researched and published on areas relating to environmental pollution and climate change. In 2016 she published her third edition of a book entitled ` Environmental Law in Malaysia' with the Kluwer Law International, the Netherlands.

Hassane Nekrache Statistician and demographer Engineer at High Commission for Planning, Morocco. His research areas include the applications of modelling and statistics. Email. n_hassane@yahoo.com, Tel. +212673342900

Mostafa Ouadrim Mostafa Ouadrim holds a PhD in Geography from the University Hassan II- Mohammedia. And currently, Vice Dean of pedagogy of the faculty of Arts and Human sciences - Mohammedia, professor in the Geography

Department and Member of the Laboratory: Dynamics of Space and the Society. Research Team: the Dynamics of Natural Environments and their Impacts on Society and Territories.

Szilárd Podruzsik Dr. Szilárd Podruzsik, Corvinus University of Budapest, Department of Agricultural Economics and Rural Development, Fővám tér 8, 1093 Budapest, Hungary, Tel. +36 (0) 1 482 55 76, E-Mail: szilard.podruzsik@unicorvinus.hu

Dr. Podruzsik is an economist. He holds a PhD in economics from the University of Economic Sciences and Public Administration in Budapest. He works for the Corvinus University of Budapest as a senior lecturer. His research fields are the agriculture and food industry. Currently his research focuses on the food consumer welfare, food logistics and its process optimisation. In his research he applies different models help to stimulate, estimate and evaluate the sector.

Leon van Rensburg Prof. Dr. Leon van Rensburg, School of Environmental Science and Development, North-West University, Potchefstroom Campus (PUK), Private Bag X6001, Potchefstroom 2520, South Africa, Tel. + 27 (0) 18 / 299 15 42, Fax + 27 (0) 18 / 299 15 44, E-Mail: Leon.vanRensburg@nwu.ac.za

Prof. van Rensburg started his academic career by obtaining his BSc, BSc Honns and MSc (all cum laude) at the PU for CHE (now NWU). Up to 1994 with the completion of his PhD he focused on understanding and quantifying plant responses to environmental stress, i.e. what plants experienced as being stressful environments and how they responded physiologically to be able to cope with the stressor/s. Up to his appointment as Director of the now Research Unit for Environmental Sciences and Sustainable Management he has conducted numerous research projects in various climatic regions (locally and internationally) and on a large number of different mine discard materials for various large.

Jürgen Scheffran Jürgen Scheffran is a Professor at the Institute of Geography and head of the Research Group Climate Change and Security (CLISEC) in the Excellence Initiative "Integrated Climate Systems Analysis and Prediction" (CliSAP), University of Hamburg. After attaining his PhD in Physics at Marburg University, he worked at Technical University of Darmstadt, the Potsdam Institute for Climate Impact Research, and as Visiting Professor at the University of Paris (Sorbonne). His research and teaching interests include: energy security, climate change and sustainable development; complex systems analysis and computer modeling; technology assessment, arms control and international security. He has published extensively in the field of climate change and conflict.

Hartmut Sommer Dr. Sommer Professor of Management and Trade, Technische Hochschule (University of Applied Sciences) Bingen, Department of Life Sciences and Engineering, main areas: conferences and seminars on business administration, international trade and finance, export management, international agricultural policy (especially for developing countries), information systems and training courses in commercial and simulation software, researcher and consultant in cooperation, international development policy and trade, rural development, decentralization and governance, education systems.

Papa Sow Papa Sow holds a PhD and MA in Human Geography from the Autonomous University of Barcelona (UAB). He is currently a senior research fellow at the Zentrum für Entwicklungs Forschung (ZEF)/ the Center for Development Research, University of Bonn under the West African Science Service Center for Climate Change and Adapted Land Use (WASCAL). He teaches and undertakes research in the fields of migration, climate change, land use and conflicts.

Azlinor Sufian Dr. Sufian is Assistant Professor at Ahmad Ibrahim Kulliyyah of Laws. Her areas of expertise are housing and construction laws, property laws and civil procedure. Her current research area includes green building and climate change.

Sharifah Zubaidah Syed Abdul Kader Dr. Sharifah Zubaidah Syed Abdul Kader is Associate Professor at Ahmad Ibrahim Kulliyyah of Laws, International Islamic University Malaysia. She obtained her doctorate from Bond University Australia. She has researched and published on areas relating to land development, river management and water resources management.

Aline Treillard As a PhD Student in environmental law and a teaching fellow in public law, her research concerns international biodiversity law, French law for the protection of nature, animal rights and human rights. Her thesis is about the concept of "ordinary nature" under the direction of the Professor Jessica Makowiak. She participates in the European Legal Observatory for Natura 2000.

Laurent Weyers Laurent Weyers holds a Master's degree in Law (Université Libre de Bruxelles, 2006) and an Advanced Master's degree in Public International Law (Université Libre de Bruxelles, 2013). From 2013 onwards, he has been a PhD researcher and member of the Centre de droit international et de sociologie appliquée au droit international. As the provisional working title of his PhD indicates – 'Quel rôle pour le droit international face au problème du changement du climat?' –, he has a strong interest for all climate related legal questions. Previous publications include: 'La chasse à la baleine dans l'Antarctique: Une application du principe de l'exercice raisonnable des compétences discrétionnaires de l'Etat' (R.B.D.I., 2013-2, pp. 618-642); 'La sécurisation du changement climatique vue par un juriste internationaliste. De l'art de sécuriser pour protéger le climat?' (Nicolas Clinchamps, Christel Cournil, Catherine Fabregoule, Geetha Ganapathy-Dore, Sécurité et Environnement, Bruylant, 2016, pp. 259-282).

Printed in the United States
By Bookmasters